W9-BNN-349

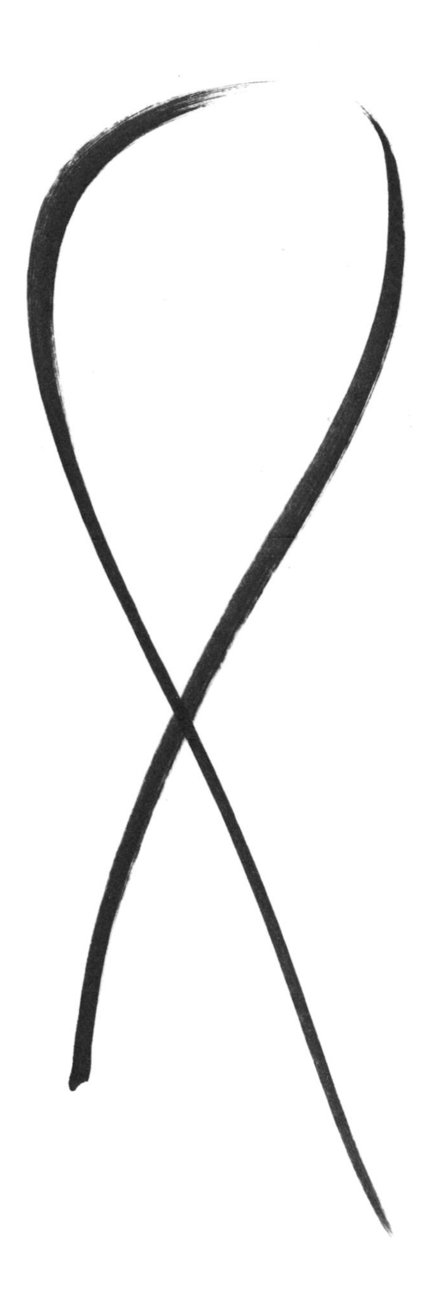

State of the World 2002

Other Norton/Worldwatch Books

State of the World 1984 through *2001* (an annual report on progress toward a sustainabie society)

Vital Signs 1992 through *2001* (an annual report on the trends that are shaping our future)

Saving the Planet
Lester R. Brown
Christopher Flavin
Sandra Postel

How Much is Enough?
Alan Thein Durning

Last Oasis
Sandra Postel

Full House
Lester R. Brown
Hal Kane

Power Surge
Christopher Flavin
Nicholas Lenssen

Who Will Feed China?
Lester R. Brown

Tough Choices
Lester R. Brown

Fighting for Survival
Michael Renner

The Natural Wealth of Nations
David Malin Roodman

Life Out of Bounds
Chris Bright

Beyond Malthus
Lester R. Brown
Gary Gardner
Brian Halweil

Pillar of Sand
Sandra Postel

Vanishing Borders
Hilary French

State of the World 2002

A Worldwatch Institute Report on Progress Toward a Sustainable Society

Foreword by Kofi A. Annan
Secretary-General, United Nations

Christopher Flavin
Hilary French
Gary Gardner

Seth Dunn
Robert Engelman
Brian Halweil
Lisa Mastny
Anne Platt McGinn
Danielle Nierenberg
Michael Renner

Linda Starke, Editor

W · W · NORTON & COMPANY
NEW YORK LONDON

The text of this book is composed in Galliard, with the display set in Franklin Gothic and Gill Sans. Book design by Elizabeth Doherty; composition by Worldwatch Institute; manufacturing by the Haddon Craftsmen, Inc.

First Edition

ISBN 0-393-05053-X
ISBN 0-393-32279-3 (pbk)

W. W. Norton & Company, Inc., 500 Fifth Avenue, New York, N.Y. 10110
www.wwnorton.com

W. W. Norton & Company Ltd., Castle House, 75/76 Wells Street, London W1T 3QT

1 2 3 4 5 6 7 8 9 0

♻ This book is printed on recycled paper.

Acknowledgments

This nineteenth edition of *State of the World* draws on the dedication and hard work of everyone on the Worldwatch staff. Backed by the generous support of funders and friends, the Institute's researchers, writers, editors, communications specialists, and administrative staff have our many thanks for working to complete this year's review of planetary health.

We begin by acknowledging the foundation community, whose faithful backing sustains and encourages the Institute's work. The John D. and Catherine T. MacArthur Foundation awarded funds specifically for *State of the World*. We also would like to acknowledge several other funders who generously support Worldwatch: the Geraldine R. Dodge Foundation, the Ford Foundation, the Richard & Rhoda Goldman Fund, the William and Flora Hewlett Foundation, the W. Alton Jones Foundation, the Charles Stewart Mott Foundation, the Curtis and Edith Munson Foundation, the David and Lucile Packard Foundation, The Summit Foundation, Surdna Foundation, Inc., Turner Foundation, Inc., the U.N. Environment Programme, the Wallace Genetic Foundation, the Wallace Global Fund, the Weeden Foundation, and the Winslow Foundation.

In addition, we are indebted to the Institute's individual donors, including the 1,300+ Friends of Worldwatch who, through their deep commitment to the Institute, are the best multipliers of our vision for a more sustainable world. We are indebted, as well, to the Worldwatch Council of Sponsors—Tom and Cathy Crain, James and Deanna Dehlsen, Roger and Vicki Sant, and Eckart Wintzen—who have consistently showed their confidence and support of our work with donations of $50,000 or more.

This year, we want to add our special thanks to John McBride and Kate McBride-Puckett and the McBride Foundation in appreciation for their work on population issues and their commitment to promoting environmental awareness and action. For their support of these issues and of the second annual State of the World Conference in Aspen in July 2001, we have dedicated the population chapter of this year's book to the McBrides.

Chapter authors were grateful for the enthusiasm and dedication of the 2001 team of interns, who cheerfully pursue obscure information leads and compile graphs and tables. Liza Rosen and Erik Assadourian tenaciously compiled informa-

ACKNOWLEDGMENTS

tion for Chapters 1 and 4; Marcella Athayde delayed returning to law school by a month to help complete the research for Chapters 3 and 6; Uta Saoshiro found information on both tourism and resource-based conflicts for Chapters 5 and 7; and Jessica Dodson provided invaluable research assistance for Chapter 8.

The immense job of tracking down and obtaining articles, journals, and books from all over the world fell to Research Librarian Lori Brown and office assistant Jonathan Guzman. As in past years, they controlled and organized the flow of information for researchers, keeping them up to date on the latest issues in their fields.

After the initial research and writing were completed, an internal review process by current staff members and Worldwatch alumni helped ensure that we would present our findings as clearly and accurately as possible. At this year's day-long review meeting, chapter authors were challenged, complimented, and critiqued by interns, magazine staff, and other researchers. Special thanks go to researchers Janet Abramovitz, David Roodman, Payal Sampat, and Molly O'Meara Sheehan and to former Worldwatcher John Young for their detailed reviews of chapters. The magazine staff of Ed Ayres, Chris Bright, and Curtis Runyan also lent their superb editing and writing advice to Chapters 3 and 6. This year, the Institute drew on the expertise and knowledge of Population Action International's Robert Engelman, who coauthored the chapter on "Rethinking Population, Improving Lives."

On the international front, we would like to thank the many Worldwatch supporters who provide advice and translation assistance from outside the United States. *State of the World* is published in 39 languages. Without the dedication of a host of publishers, nongovernmental organizations, and individuals who work to spread the Institute's message, we would not be able to live up to our name. Special thanks go to Eduardo Athayde in Brazil, Hamid Taravaty in Iran, Gianfranco Bologna and Anna Bruno in Italy, Soki Oda in Japan, Magnar Norderhaug in Norway, Jose Santamarta and Marie-Amelie Ponce in Spain, George Cheng in Taiwan, and Jonathan Sinclair Wilson in the United Kingdom.

Reviews from outside experts, who generously gave us their time, were also indispensable to this year's final product. We would like to thank the following individuals for the information they provided to authors or for their thoughtful comments and suggestions: Bina Agarwal, Bas Amelung, Stan Bernstein, Judith Bruce, Robyn Bushell, Steve Charnovitz, Nada Chaya, Richard P. Cincotta, Terry Collins, Frans de Man, Felix Dodds, Navroz Dubash, Megan Epler Wood, Taryn Fransen, David Gee, Ken Geiser, Adrienne Germain, Margaret E. Greene, Ronald Halweil, Carl Haub, David Hunter, Jodi Jacobson, Nadia Johnson, Rachel Kyte, Darryl Luscombe, Mia MacDonald, Bill Mansfield, Alan Miller, Sascha Mueller-Kraenner, Jim Paul, Anita Pleumarom, Sandra Postel, Jules Pretty, Jim Puckett, Kate Queeney, Maria Rapauano, James Rochow, Wolfgang Sachs, Richard Sigman, Axel Singhofen, Rosa Songel, J. Joseph Speidel, Joe Thornton, Joel Tickner, Norman Uphoff, Geoffrey Wall, Jack Weinberg, and Pam Wight.

Further refinement of each chapter took place under the careful eye of independent editor Linda Starke, whose gentle—and sometimes not so gentle—prodding ensured that we met our deadline with all our t's crossed and our i's dotted. After the rewrites—and many edits—were complete,

Art Director Elizabeth Doherty skillfully crafted the text, tables, and graphs of each chapter into the book you now hold. The page proofs were then ready for Ritch Pope, for the important task of preparing the index.

Writing was only the beginning of getting *State of the World* to readers. Credit also goes to our excellent communications department. Vice President for Communications Dick Bell and Public Affairs Specialist Leanne Mitchell worked closely with researchers to craft their messages for the press and the public. Niki Clark provided energetic and creative administrative support, aided by intern Susanne Martikke. And Sharon Lapier helped keep the department running, staffing the front desk and tracking the thousands of press clips we receive every year.

Sadly, Christine Stearn, our resident Web goddess, left for New York City in October after completing several major projects this year, including a new network operating system, a powerful search engine for the Web site, and research topic Web pages (with the assistance of summer intern Ryan Bowman). Although we will miss Christine, we are excited about the skills and sophisticated new network management experience that Patrick Settle has brought to Worldwatch as our new IT manager.

This year Elizabeth Nolan joined the Institute as Vice President for Business Development. She and Denise Warden coordinated all our activities with our publishers, and brought creativity and energy to our marketing efforts. Director of Finance and Administration Barbara Fallin kept us all in line by making sure the office runs smoothly. Joseph Gravely continued his reign as czar of Worldwatch's mail room. And Suzanne Clift ably assisted Worldwatch President Christopher Flavin while helping other researchers to arrange speaking engagements and make travel plans.

The Institute's foundation fundraising activities are now under the able leadership of Kevin Parker, our new Director of Foundation Relations, with assistance from Development Associate Mary Redfern. Both worked closely with current donors and funders, cultivating new relationships that will sustain the Institute's work for years to come. And at the end of the year, we were happy to welcome Adrianne Greenlees as our new Vice President for Development.

We would also like to express our gratitude to our long-time U.S. publisher, W. W. Norton & Company. Thanks to the dedication of their staff—especially Amy Cherry, Andrew Marasia, and Lucinda Bartley—Worldwatch publications are available from university campuses to small-town bookstores around the United States.

We are also grateful for the hard work and loyal support of the members of the Institute's Board of Directors, who have provided key input on strategic planning, organizational development, and fundraising over the last year.

In addition, we welcome with joy our newest edition to the Worldwatch family. Tyler Rene was born to Suzanne and Ronald Clift on July 12th, a poignant reminder and inspiration to us all of the need to build a healthier—and happier—world.

The year 2001 was thus one of many changes and new beginnings for Worldwatch. In May, the Institute's founder and first President, Lester Brown, left the staff to launch the Earth Policy Institute, a new kind of research organization that is focused on describing and encouraging the "eco-economy" needed in the new century. Lester was joined in this exciting endeavor by long-time Worldwatchers Reah Janise Kauffman (after 15 years at the Institute) and Millicent Johnson (after 11 years). Janet Larsen, who helped Lester with research during her year at Worldwatch, also joined Earth Policy as Staff Researcher. We are confident that Lester will continue to make important contributions to thinking on global environmental trends, and are pleased that he continues to work with Worldwatch as a member of our Board of Directors. Without Lester's vision and dedication, *State of the World* would not exist.

Hilary French
Project Director

Contents

CONTENTS

List of Boxes, Tables, and Figures

Boxes

8 Reshaping Global Governance

Tables

1 The Challenge for Johannesburg: Creating a More Secure World

2 Moving the Climate Change Agenda Forward

3 Farming in the Public Interest

4 Reducing Our Toxic Burden

5 Redirecting International Tourism

6 Rethinking Population, Improving Lives

7 Breaking the Link Between Resources and Repression

Figures

Foreword

Fifteen years have passed since the World Commission on Environment and Development presented its historic report, *Our Common Future*, to the United Nations General Assembly. The Commission's recommendations—presented unanimously, without reservations or footnotes—were courageous, visionary, and demanding. They called for a fundamental reordering of global priorities. They illustrated the inescapable links between environmental, economic, and social concerns. And they established sustainable development as the central organizing principle for societies around the world. At the Earth Summit in Rio de Janeiro in 1992, governments recognized the great wisdom of these findings; most important, they committed themselves to an unprecedented global effort to free our children and grandchildren from the danger of living on a planet whose ecosystems and resources can no longer provide for their needs.

The political and conceptual breakthrough achieved at Rio has not, however, proved decisive enough to break with business as usual. As the global community prepares for the World Summit on Sustainable Development in Johannesburg in September 2002, unsustainable approaches to economic progress remain pervasive. Indeed, it is too late for the Summit to avoid the conclusion that there is a gap between the goals and promises set out in Rio and the daily reality in rich and poor countries alike. But it is not too late to set the transformation more convincingly in motion.

The Johannesburg Summit can and must lead to a strengthened global recognition of the importance of achieving a sustainable balance between nature and the human economy. The responsibilities that flow from this recognition are not identical, since the nations of the world are at very different levels of development. Such differences notwithstanding, all of us should understand not only that we face common threats, but also that there are common opportunities to be seized if we respond to this challenge as a single human community.

If the World Summit in Johannesburg is to lead to effective strategies for sustainable development, we will also have to reinvigorate the fight against abject and dehumanizing poverty. We will have to assess the risks associated with globalization and the imperatives of global markets. We will need to breathe life into the treaty commitments and other agreements the international community has reached to save biodiversity, protect forests, guard against climate change, and stop the march of desertification. We will have to reinvent national and global governance. We will need new and additional financial resources. We will need strong partnerships among governments, nongovernmental organizations, the private sector, and others in a position to contribute, such as the academic and scientific communities. And we will need to do all of this while adhering to the principles of equity and solidarity found in the United Nations Charter and other guiding documents of international affairs.

This is no doubt an ambitious agenda, not least because the record of disappointment is already long, and the status quo remains deeply entrenched. *State of the World 2002* highlights both the obstacles and opportunities ahead. Readers may approve or reject the various assessments and proposals; I myself do not necessarily agree with all the ideas expressed here. But we can agree that the perilous state of our world is in an object of genuine, urgent concern. We have the human and material resources with which to achieve sustainable development. With leadership, creativity, and goodwill, at Johannesburg and beyond, a peaceful, prosperous common future can be ours.

Kofi A. Annan
Secretary-General, United Nations

Preface

The World Summit on Sustainable Development in Johannesburg in September 2002 will present a rare opportunity for national leaders from around the world to address some of the most fundamental issues facing the human race at the dawn of the new century: Will the global economy find a new balance with Earth's natural systems? And can we meet the basic needs of over a billion poor people today, as well as the additional 2–3 billion who will be added to the world's human population in the coming decades?

My Worldwatch colleagues and I decided in early 2001 that helping to define the agenda for the World Summit was the most important goal we could focus on in *State of the World 2002.* It has been 10 years since the historic Earth Summit in Rio de Janeiro—a good time to revisit the achievements since that gathering and to consider how to accelerate the pace of change in the decade ahead. The last 10 years have seen many disappointments as well as successes in the cause of creating a sustainable world, all of which, we find, offer important lessons.

The urgency of our effort jumped dramatically on September 11th. Early that morning, *State of the World* authors were just settling down to work in our Washington office when word began to filter in that first one, and then two, planes had hit the World Trade Center towers in New York, with a third plane hitting the Pentagon, just across the Potomac River.

As my Worldwatch colleagues and I recovered from the initial shock and confusion, we began to consider the deeper implications of September's tragedies. At their core, these disturbing events are powerful reminders that the ecological instability of today's world is matched by an instability in human affairs that must be urgently addressed. Meeting basic human needs, slowing the unprecedented growth in human numbers, and protecting vital natural resources such as fresh water, forests, and fisheries are all prerequisites to healthy, stable societies. Building a more sustainable and secure world—and one that is based on principles of universal human values and mutual support—could not be more urgent.

While the urgency of the task is new, the underlying themes are a direct extension of the Earth Summit agenda in 1992. At its heart was a global consensus that the world needed a new approach to development—one that ensures that human needs are met in a way that protects the natural environment without undermining the prospects of future generations. The Rio Summit led to some historic achievements: two landmark global treaties on climate change and biological diversity and a document called *Agenda 21*, a 40-chapter plan for achieving sustainable development.

These agreements reflected a significant shift in outlook and a broadening of horizons for the world community. But the intense public enthusiasm and media coverage that came with the largest ever gathering of world leaders gave a false sense of just how far the world had come in fundamentally reordering its priorities. *Agenda 21* itself was a relatively vague set of goals, lacking clear implementation plans or binding legal requirements.

As national governments prepare for the Johannesburg Summit—and reflect on the lessons of September 11th—two questions beg to be answered: Why has so little progress been made on the ambitious agenda that was laid down a decade ago? And what must be done to ensure that the next decade is one of sustainable social and environmental progress?

The answer to the first question is both simple and complex: governments and individuals around the world are still treating issues such as population growth, the loss of biological diversity, and the buildup of greenhouse gases in the atmosphere as if they were equivalent to local air or water pollution—problems that could be solved simply by ordering the addition of control devices. Humanity has not yet shown the ability to deal with fundamental global and long-term changes in the biosphere, particularly when they require a systemic response—the creation of fundamentally different technologies, the development of new business models, and the embracing of new lifestyles and values.

To date, our prodigious ability to expand our own numbers and levels of material consumption has greatly outpaced our ability to understand and respond to the scope of the problems we are creating for ourselves. Only recently have we been able to use satellite imagery to chart the destruction of vast areas of forest or to develop the computer models that allow us to project even roughly the kinds of changes in weather that will occur as we add more carbon dioxide to the atmosphere.

But the newly gained knowledge of scientists is hard to translate into the common language of average people or the specialized jargon of business executives or politicians. Stunning developments, such as the fact that half the world's wetlands have been destroyed—a goodly portion of them in the decade since Rio—are hard to grasp or to respond to. The fact that 12 percent of bird species are threatened with extinction is beyond our daily imagination. And the fact that 1.1 billion people lack access to adequate clean water—more than double the number who use computers—suggests a level of poverty that is inconsistent with our image of the twenty-first century.[1]

In his remarkable environmental history of the twentieth century, *Something New Under the Sun*, historian J. R. McNeill points to the unusual adaptability and cleverness of the human species—characteristics that allowed the extraordinary expansion of the human enterprise in the twentieth century. But this cleverness has not yet been turned away from its evolutionary focus—exploiting the rest of the natural world to meet human demands—and toward a new conception of an interconnected and mutually dependent world in which short-term exploitation will eventually cause injury to humanity.[2]

One of the major challenges that will be

faced by world leaders who gather in Johannesburg will be to develop a new concept of globalization—one that moves beyond the narrow focus on trade and finance that has distorted international discourse and that led to a large public backlash in developing and industrial countries alike. Forging a harmonious global community will only be possible if it is based on universal principles of respecting human rights, meeting basic human needs, and preserving the natural environment for future generations. In that endeavor, governments, international organizations, private companies, and citizens all have important roles to play.

The decision to follow the tradition of Rio by holding the World Summit in a southern country—and one with the unique history of South Africa—has sent a message of its own. While global environmental progress has languished in the last decade, South Africa has transformed itself from a divided country in which the majority was excluded from political power into a modern democracy that is moving to address a range of deep-seated social and environmental problems.

The stunning transformation of South Africa's political system after decades of downward spiral into ever more oppressive apartheid policies suggests that human beings are capable of dramatic and *rapid* change—when the conditions are right. In the case of South Africa, it required outside economic pressure, exerted by the world community. After years of claiming immunity from such pressure, the country's apartheid political structure suddenly cracked.

For all the promise of South Africa, world leaders traveling to Johannesburg will find strong reminders of many of the problems still plaguing much of the developing world: choking air pollution from the country's heavy dependence on dirty coal, 10 percent of its diverse indigenous flora threatened with extinction, some of the world's highest rates of infection with tuberculosis and HIV, and water shortages that plague a large share of the indigenous population. They will also see powerful reminders that only by bridging racial, ethnic, and economic gaps can these kinds of problems be overcome—as South Africa is beginning to do.[3]

Some of the outside pressures on the diplomats who will gather in Johannesburg will come from the biosphere itself. Global emissions of carbon have grown by an additional 400 million tons during the decade it has taken to agree to a modest climate protocol that grew out of a convention signed in Rio. And the proportion of the world's coral reefs that is threatened has grown from 10 percent to 27 percent, while the Convention on Biological Diversity signed in 1992 has languished. On the human front, a decade of unprecedented economic growth—adding over $10 trillion a year to the global economy—has left the number of people living in poverty nearly unchanged at more than 1 billion.[4]

Additional pressure for movement will come from the tragedies of September 11th and subsequent world events. It is now clear in a way that it never was before that the world of the early twenty-first century is far from stable. At a time when we are still adding a billion people to the human population every 15 years, many societies are struggling with the difficult transition from traditional rural societies to modern, urban, middle-class ones. In many of these societies, basic human needs for food, water, health care, and education are not being met, with over a billion people living on less than a dollar a day. Moreover, the lack of democratic political representation and the concentration of economic and political power in a few hands has created a fundamental instability in many nations—an instability that echoes around the world in the form of large-scale human migration, illegal drug exports, and, increasingly, terrorism.[5]

If the lofty social and ecological goals of the Rio Earth Summit had been achieved, it is possible that the crises of the last year would not have occurred. But these goals are monumental ones, and achieving them was bound to take time. In 2002, the challenge is even greater, but this very urgency may provide the kind of wake-up call that is needed if global priorities are to be reordered. In particular, meeting this challenge will require a common sense of mission that bridges rich and poor countries—overcoming a sort of global apartheid that was reflected in the divisions between rich and poor nations that deeply marked the Rio negotiations and that have continued all too strongly since then.

In the struggle to create a sustainable world, there are only allies, not adversaries. Johannesburg can be an important step in waking the world up to the scale of the challenge we face—and the commitments that will be required to address it. The eight chapters in *State of the World 2002* provide our vision of the transformation ahead, as well as our suggestions for concrete steps that can be taken at Johannesburg to start the world on a decade of social and environmental progress that is far more productive than the last one.

Christopher Flavin

Christopher Flavin
President, Worldwatch Institute
1776 Massachusetts Ave., N.W.
Washington DC 20036
worldwatch@worldwatch.org
www.worldwatch.org

November 2001

State of the World 2002

Chapter 1

The Challenge for Johannesburg: Creating a More Secure World

Gary Gardner

In the anxious days following the terrorist attacks on the World Trade Center and the Pentagon, world leaders described the global community as suddenly and irrevocably changed. On September 11, 2001, "night fell on a different world," in the words of President George W. Bush, largely because of a more broadly shared experience of vulnerability.[1]

"Americans have known wars," he observed, but rarely on their own soil. "Americans have known surprise attacks. But never before on thousands of civilians." The new experiences of that September morning produced a shift in national priorities, literally overnight.[2]

Those who would move the world rapidly toward sustainability must be amazed at the galvanizing power of the attacks. We are left to wonder: are tragedies of this magnitude needed to steer the world toward a new model of development, one built along the recommendations of the 1992 Earth Summit? If so, there is plenty to report. Imagine a prime minister or president at the World Summit on Sustainable Development in September 2002 reviewing events and findings of the past decade, in an echo of President Bush:

> *The human family has suffered sickness, but rare is the plague that can kill a third of a nation's adults—as AIDS may well do in Botswana over the next decade.... Our planet has regularly seen species die-offs, but only five times in 4 billion years has it experienced anything like today's mass extinction.... Nations have long grappled with inequality. But how often have the assets of just three individuals matched the combined national economies of the poorest 48 countries, as happened in 1997?*[3]

These trends are no doubt less riveting than the drama of a surprise attack. Yet they alert the world to a danger less visible than

Units of measure throughout this book are metric unless common usage dictates otherwise.

terrorism but over the long term more serious. These and other trends—from the loss of forests, wetlands, and coral reefs to social decay in the world's most advanced nations—warn us of creeping corrosion in the favored development model of the twentieth century. That model, used by developing as well as industrial nations, is materials-intensive, driven by fossil fuels, based on mass consumption and mass disposal, and oriented primarily toward economic growth—with insufficient regard for meeting people's needs. In 1992 the U.N. Conference on Environment and Development (the Earth Summit) challenged this model and offered a comprehensive alternative. It called the human family to a new experience—that of sustainable development.

Steps in the 1990s toward a more just and ecologically resilient world were too small, too slow, or too poorly rooted.

Ten years after the historic meeting in Rio de Janeiro, the world has begun to respond to this call—but only tentatively and unevenly. Steps in the 1990s toward a more just and ecologically resilient world were too small, too slow, or too poorly rooted. Wind and solar energy grew vigorously over the decade, for example, yet the world still gets 90 percent of its commercial energy from fossil fuels—whose carbon molecules play increasing havoc with our climate. Imaginative advances in the way goods and services are produced and consumed could generate manifold reductions in materials use and waste generation, yet most remain largely on the drawing board or are only at the pilot stage. And improvements in health and education, while laudable in many developing countries, were uneven—and by some measures may actually be unraveling in wealthy ones.[4]

Not surprisingly, then, global environmental problems, from climate change to species extinctions, deforestation, and water scarcity, have generally worsened since delegates met in Rio. Social trends have shown some improvement, yet gaping global disparities in wealth remain: one fifth of the world's people live on a dollar or less each day, even as the world's wealthy suffer from symptoms of excess, such as obesity. And a growing number of economies have a voracious appetite for materials. While recycling of glass, paper, and a few other household wastes is now common practice in many countries, most materials in industrial nations are used only once before being discarded. In sum, while awareness of the environmental and social issues central to sustainable development undoubtedly was raised in the 1990s, the new consciousness has yet to register improvements on the ground for most global environmental issues.[5]

Still, emerging awareness of the need for a sustainable path is an important start. More than ever, citizens, businesses, and government leaders understand that development is about more than economic growth—a key theme of the Earth Summit. *Agenda 21*, the action plan that emerged from the conference, addresses social issues, the structure of economies, conservation of resources, and problems of civil society. This broad panorama is consistent with the picture of development endorsed by the U.N. Development Programme (UNDP): expanding people's choices to lead the lives they value, especially choices that foster a long and healthy life, access to education, a decent standard of living, and participation in community life. Following the lead of the Earth Summit and UNDP, this chapter will assess development over the past decade with a broad lens, examining how well the world has advanced environmental protection, human health and education, and ecological economics since Rio.[6]

As nations gather in Johannesburg in September for the World Summit to recommit to a just and environmentally healthy world, delegates would do well to summon the singleness of purpose that characterizes the battle against terrorism. "We have found our mission and our moment," President Bush declared in response to the attacks in 2001. Imagine a global community with the same resolve—directed wholeheartedly to realizing the vision of development outlined at Rio. That is the potential and the hope for Johannesburg.[7]

The Toll on Nature

More than any previous international conference, the 1992 Earth Summit highlighted the central importance of the natural environment for a healthy economy. This idea found conceptual support in 1997 when environmental economist Robert Costanza and colleagues quantified the value of "nature's services"—things like the soil-holding capacity of tree roots and the flood protection offered by mangroves—at a minimum of $33 trillion annually, nearly twice the gross world product that year. Despite improved understanding of the importance of the natural environment for development, global response to environmental degradation was sluggish—even as nearly every global environmental indicator worsened.[8]

Leading the list of growing environmental problems is climate change, which gained prominence over the decade as scientists improved their understanding of the link between emissions of greenhouse gases, climbing global temperatures, rising sea levels, and the increased frequency and intensity of extreme weather events. (See Chapter 2.) Ice core readings suggest that current atmospheric carbon dioxide levels are at their highest level in 420,000 years; the global temperature record points to the 1990s as the warmest decade since measurements began in the nineteenth century; and scientists have documented a 10–20 centimeter rise in global average sea levels over the past century. Responding to these and other data, the Intergovernmental Panel on Climate Change, a group of more than 2,500 scientists from around the globe, warned in 1996 that a "discernible human influence" was evident in the changing world climate. By 2001, its Third Assessment Report was more definitive: "most of the warming of the past 50 years," it declared, "is attributable to human activities."[9]

Despite the growing evidence of a human-generated disruption of climate, global emissions of carbon—a key greenhouse gas—increased by more than 9 percent over the decade, although performance varied widely from nation to nation. Some countries, notably Germany, the United Kingdom, and former Eastern bloc nations mired in economic recession, reduced their emissions. Others, especially China, saw emissions increase with rapid economic expansion, but they also became more efficient, reducing the amount of carbon needed to build products or deliver services. Perhaps the most disappointing performance was that of the United States, which is responsible for nearly a quarter of global carbon output. Although armed with the wealth and technology to curb carbon emissions, and in spite of ample scope for cuts, U.S. emissions rose some 18 percent between 1990 and 2000. The capstone of American reluctance to address to climate change came in 2001, when the Bush administration abandoned the U.S.

commitment to the Kyoto Protocol, a key diplomatic initiative whose origins trace back to the Earth Summit.[10]

The connection between climate change and economic and human development became increasingly apparent as the 1990s unfolded. Insurance companies were among the first businesses to side with environmentalists in debates about climate change. Their epiphany came as damage claims from storms surged over the decade: claims from violent weather were greater in 1998, for example, than in the entire decade of the 1980s. Another group that emerged in the 1990s in response to the expected impact of climate change was the Alliance of Small Island States, a disparate group of island nations from all parts of the world with one thing in common: all face economic ruin—even physical extinction—from rising seas in a warming world. The group's 43 member states, representing about 5 percent of the world's population, were active in seeking commitments from other governments to reduce emissions of carbon. A host of other effects of climate change, from the impact on agriculture to the spread of disease and insect plagues in a warmer world, were cited as possible impediments to development over the decade.[11]

Another environmental issue that took on great importance over the decade was water scarcity. This leapt up the international issue agenda in the 1990s as projections of huge gaps between supply and demand and fears of conflict over water gained a high profile. (See Chapter 6.) A 1997 U.N. assessment of global fresh water found that about a third of the world lives in countries that find it difficult or impossible to meet all their water needs, a condition known as water stress. That share could double to two thirds by 2025, as population increase and economic growth combine to squeeze fixed supplies of water. Water stress typically shows up as a shortage of water for farms; agriculture is a prime target for water savings because this sector is often politically weak and can account for two thirds or more of a nation's water use.[12]

Water scarcity could also have serious developmental consequences through its impact on food supplies. Water-scarce countries increasingly turned to two coping strategies in the 1990s: tapping into groundwater reserves to maintain or expand agricultural production, and increasing food imports. But neither is likely to be a long-term solution for guaranteeing food supplies. Sandra Postel of the Global Water Policy Project estimated in 1999 that nearly 10 percent of the world's grain harvest is produced with water pumped from wells faster than it is replenished, notably in the agriculturally rich regions of India, China, and the Great Plains of the United States. Without a change in water use practices, and unless substitute sources of water are found, that share of the harvest will one day be unavailable—with disruptive consequences for nations that depend on those supplies.[13]

Importing food can mean major water savings for parched regions, since agriculture typically accounts for upwards of two thirds of a country's water use. Yet this option is open to only a limited number of nations. If many nations turn to foreign sources of food, and if global markets cannot meet the demand, malnutrition and civil unrest could result. Even if foreign markets are up to the challenge, increased food imports curtail a nation's opportunities for other imports or create greater foreign indebtedness.[14]

Poor people are especially hard hit in a water-stressed world. Of the world's people classified by the World Bank as having low incomes in 1995, more than a third lived in countries that faced medium-high to high water stress. Not surprisingly, more than a billion people in the world lack safe drinking water, and nearly 3 billion do not have access to adequate sanitation—two conditions with serious consequences for development. About half the people in the developing world suffer from diseases caused by contaminated water or food, and an estimated 14,000–30,000 people die each day from water-related diseases. That is the equivalent of several September 11th tragedies every day, year in and year out—but without the media attention.[15]

Response to the growing seriousness of water issues has been encouraging, but slow. Water analyst Peter Gleick sees the emergence of a "changing water paradigm" that reflects a more thoughtful perspective on water use. Policymakers and engineers, he says, increasingly prize water efficiency, include environmental values in water planning, emphasize meeting basic human needs for water, and show less enthusiasm for large dams.[16]

This new perspective is beginning to show itself on the ground. The United States, for example, withdrew 10 percent less water from rivers, lakes, aquifers, and other sources to support human activities in 1995 than it did in 1980, the peak year of use. Part of this drop is due to a restructuring of the economy away from water-intensive industries. But some of it may be due to the adoption of water efficiency standards, especially since 1992. In other countries, highly efficient drip irrigation technologies have been adopted for high-value crops, with dramatic water savings. And novel strategies such as water reclamation from sewage flows, water pricing that discourages waste, and increased interest in "dry" forms of sanitation are all aimed at increasing efficiency of use.[17]

Meanwhile, dams—for decades a widely accepted technology for supplying water reliably—have lost some of their luster. After a decade that saw protests over the environmental and social impacts of dams, in November 2000 the independent World Commission on Dams issued a critical assessment of a century of large dam building. The report acknowledged the contributions of dams to economic development, especially in providing irrigation water and electricity. But it criticized dam projects for their impacts on people and ecosystems: over the past century, for example, between 40 million and 80 million people were displaced by large dams, and 46 percent of the world's primary watersheds now have one or more large dams that disrupt river flows. The Commissioners called for including a broad range of perspectives—from those of displaced peoples to dam builders and environmentalists—in decisionmaking about dams. And it declared that decisionmaking should be informed by values of equity, sustainability, and accountability, providing a new framework for evaluating dams.[18]

About a third of the world lives in countries that find it difficult or impossible to meet all their water needs.

Concrete evidence of recent shifts in thinking about dams is found in the United States, where more than half of the removal or decommissioning of nearly 500 small dams since 1912 occurred in the 1990s. (See Figure 1–1.) Of those removed for environmental reasons, more than three quarters came down in the 1990s. Authorities even began to discuss partial removal of large dams in Idaho to restore salmon runs, and the Sierra Club, among many other environmental organizations, advocated removal of the giant Glen Canyon

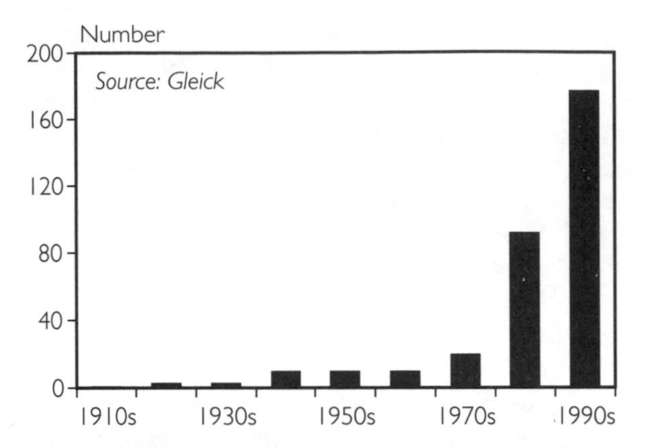

Figure 1–1. Small Dams Decommissioned or Removed in the United States, 1910–99

Dam in Arizona. Although the 500 removals and decommissionings represent just a tiny portion of the 80,000 dams and reservoirs built in the United States in the twentieth century, the rapid increase in the trend reflects a new caution about traditional strategies for supplying water.[19]

People endorsing this new attitude toward water increasingly see this resource as a security concern. While scarcity has seldom led to war in the past, areas prone to conflict over water appear to be on the rise. Analysis by Sandra Postel and Aaron Wolf shows that 17 river basins in 51 nations on five continents are at greatest risk of conflict because dams or other diversions are planned unilaterally by one or more nations, and because no mechanism for resolving disputes exists. Scarcity-induced conflict could limit the development potential of adversary nations, either through damage inflicted in violent conflict or through the diversion to the military of resources that might have gone to education, health, or other sectors important for development.[20]

Biodiversity loss, like water scarcity, received greater attention in the 1990s. Ongoing species extinctions demonstrate the urgent need to act. The World Conservation Union– IUCN documented at mid-decade the share of various wildlife groups threatened with extinction: vascular plants, 12.5 percent; birds, 11 percent; reptiles, 20 percent; mammals, 25 percent; amphibians, 25 percent; and fish, 34 percent. The greatest immediate cause of this assault on species is loss of habitat—a byproduct of human activities such as farming and livestock raising; mining, fishing, logging, and other extractive activities; and urban and industrial expansion.

In a 2000 update, IUCN found increases in the number of many species under threat, especially among mammals and birds. It also determined that 18 percent of the 11,000 threatened species are "critically endangered," the highest category of threat.[21]

One of the most important and threatened habitats is forests. The world continued to lose forested area in the 1990s, although the extent of loss is debated. The *Forest Resources Assessment 2000* put out by the U.N. Food and Agriculture Organization (FAO) cites a global loss of forested area of 2.2 percent over the decade. But that figure may be conservative. FAO includes plantation area in its forest totals, even though plantations lack the biological diversity of natural forests and cannot provide many of the same environmental services. And in an effort to standardize definitions globally, FAO has dropped the minimum tree coverage needed for an area to qualify as "forest" from 20 percent to 10 percent. This small definitional change nearly quadrupled Australia's forested area compared with the 1990 figure, leading the World Resources Institute (WRI) to note that "some parts of the Australian outback that are officially classified in Australia as desert . . . are now recorded by FAO as forest."[22]

WRI's own analysis of the FAO numbers indicates that excluding plantation forests from the calculations would more than double natural forest loss in tropical Asia and temperate Latin America. For the tropics as a whole, WRI estimates 17 percent more natural forest loss than FAO does. Although the data are confusing and less than reliable, it is clear that both groups report continued forest losses, a trend that threatens not only forest ecosystems, but also the more than 1.7 billion people in 40 nations with critically low levels of forest cover who rely on forests for fuelwood, timber, and other goods and services.[23]

As with water, the impact of deforestation is most devastating to the poor. Many of the world's rural poor who depend on wood for cooking and heating must walk great distances to find it, or must switch to dirtier fuels, such as animal dung. And forest-dwelling peoples, for whom the trees are a source of food, income, and cultural and spiritual wealth, can lose an entire way of life through deforestation. Of the 500 million people living in and around tropical forests, 150 million are members of indigenous groups that depend on forests and forest resources to sustain their way of life.[24]

Indirect effects of forest loss are also serious. Forests provide a host of environmental services: trees regulate the flow of water between soils and the atmosphere; their roots hold soils in place, preventing erosion; and their branches, bark, leaves, and soils provide habitat to the largest collection of biodiversity of any ecosystem on the planet. Deforestation means lost lives and livelihoods: in 1998 alone, forest clearing was blamed for contributing to a landslide in India that killed 238 people, and for worsening flooding in China that killed 3,000 and caused $20 billion in damage. Deforestation disrupts natural systems the way the attack on New York disrupted the urban system of telephone lines, transit routes, and commerce—and on a far larger scale, since serious deforestation occurs daily in dozens of countries.[25]

The damage from deforestation borne by developing countries is especially disturbing when linked to wasteful consumption habits. While 80 percent of the world's people do not have access to enough paper to meet minimum requirements for basic literacy and communications, wealthy countries consume paper at an astonishing rate. The average American, for example, uses 19 times more paper than the average person in a developing country, and most of it becomes trash: less than half of the paper used in the United States gets recycled.[26]

Deforestation in the 1990s, while tragic, was no surprise to scientists and policymakers, who had been tracking it for years. But another cause of biodiversity loss—the degradation of coral reefs—shocked the scientific community with the breadth and pace of its advance over the decade. Some 27 percent of the world's coral reefs are effectively lost, up from 10 percent in 1992, according to the Global Coral Reef Monitoring Network (GCRMN), a web of governments, nongovernmental organizations (NGOs), institutes, and individuals that tracks the health of reefs. Because coral reefs are second only to forests in biological wealth, such extensive losses inevitably take a great toll on many species as well.[27]

Degradation of coral reefs is closely linked to human and economic activities. Warming seas, likely the product of climate change, stress the corals to the point where they expel the algae that live within them, leaving the corals white, or "bleached." The bleaching event of 1998, one of the warmest years on record, damaged huge expanses of coral around the world and sharply increased the share of reefs damaged. Pollution from nutrients and sediment, mining of sand and rock, and use of explosives and cyanide for fishing also stress the world's reefs.[28]

At the same time, the loss of coral damages the prospects for a better life for coastal peoples. Nearly a half-billion people live within 100 kilometers of a coral reef, and many rely on reefs for food and jobs. About a quarter of the fish catch in developing countries comes from coral reef areas, which provide food for about a billion people in Asia alone. Reefs protect beaches from erosion, and help produce the fine sands that make many beaches attractive for tourism, a prime source of revenue in many tropical countries. In all, goods and services from reefs were valued in 1997 at $375 billion per year.[29]

Without "urgent management action" to stem the damage, GCRMN estimates that the share of reefs lost will climb to 40 percent by 2010. Some of these reefs have a "reasonable" chance of returning to health—but only if they are not stressed again soon, something difficult to assure in a warming world. But even in the most optimistic case, 11 percent of the world's reefs are now regarded as permanently lost.[30]

Despite the litany of discouraging trends, enough progress was made on at least one global environmental problem—emissions of chlorofluorocarbons (CFCs)—to begin healing Earth's thinning ozone layer within a few years. Production of ozone-depleting CFCs was reduced by 87 percent between 1987 and 1997; after a lag of a few years, the lower emissions levels should allow stratospheric levels of ozone to accumulate, making the "ozone hole" progressively smaller, starting in just a few years. This experience is both encouraging and cautionary. It shows that concerted international cooperation—in this case, the quick crafting and signing of the Montreal Protocol in the 1980s—is possible, and to great effect. It is also sobering: most global environmental challenges are much more complex than ozone depletion, which is caused by a limited set of substances, most of which have ready, economical substitutes. Tackling climate change, deforestation, and water scarcity will require far more ingenuity and diplomatic skill.[31]

Caring for People

"Human beings are at the center of concerns for sustainable development," asserts the 1992 Rio Declaration—an indication of the importance of social issues for development. Here the world has seen some progress: important advances were made in health, education, and other social arenas in the 1990s. Yet the gains were spotty, and some, especially the decline in some infectious diseases, are fragile and could easily be reversed. Moreover, backsliding on important health issues was found in prosperous countries. This mediocre record is an indictment of national priorities that too often are not directed at the most important needs of the human family.[32]

In many developing nations, infectious diseases continue to pose major public health problems. It was not supposed to be this way: health officials in the 1970s expected infectious disease to be a minor problem by century's end, even in the poorest countries. Their attention, they thought, would turn to treatment of "diseases of affluence," such as heart disease and cancer. Instead, 20 familiar infectious diseases—including tuberculosis (TB), malaria, and cholera—re-emerged or spread in the last quarter of the twentieth century. And at least 30 previously unknown deadly diseases—from HIV to hepatitis C and Ebola—surfaced in the same period.[33]

Yet elimination of infectious disease is possible. Pneumonia, TB, diarrhea, malaria, measles, and HIV/AIDS account for 90 percent of infectious disease deaths, and all are preventable. Deaths for three of these six were reduced over the decade (see Table 1–1), and two factors seemed to be especially important in each case: affordable treatments, and the political will to put them to use.[34]

Child deaths due to diarrhea, for example, were reduced by half between 1990 and 2000, meeting the goal set at the World Summit for Children in 1990. Improvements in nutritional status and in access to safe water, along with greater practice of breastfeeding, played roles in this success. Many researchers give the greatest credit to oral rehydration therapy, the practice of administering an inexpensive solution of water, salt, and carbohydrates to children

Table 1–1. Progress and Problems in the Fight Against Leading Infectious Diseases

Disease	Deaths Worldwide 1990	Deaths Worldwide 2000	Spread of Resistance to Drugs
	(million)		
Lower respiratory infection	4.29	3.87	Data from lab samples indicate that 70 percent of chest infections are resistant to at least one of the first-line antimicrobials.
HIV/AIDS	0.31	2.94	Resistance to AZT and to protease inhibitors beginning to appear. Resistance to one protease inhibitor may quickly lead to resistance to the entire family of drugs, which were developed at great cost over many years.
Diarrheal diseases	2.95	2.12	Multidrug resistance is a growing problem. Ten years ago, for example, an epidemic of shigella (a form of dysentery) was easily controlled with co-trimoxazole. Today the drug is largely ineffective against shigella; only one viable medicine remains, and it also faces growing resistance.
Tuberculosis	2.04	1.66	1–2 percent of TB cases worldwide are now resistant to all anti-TB drugs. In Israel, Italy, and Mexico, the figure is 6 percent.
Malaria	0.86	1.08	Resistance to chloroquine, the first-line treatment, is widespread in 80 percent of countries where malaria is a major killer. Second- and third-line treatments also show increasing resistance.
Measles	1.06	0.78	Measles is effectively treated with a vaccine, but secondary problems associated with measles, such as pneumonia, are often resistant to antibiotics.

SOURCE: See endnote 34.

with diarrhea in order to replace vital supplies of water and nutrients. The practice was widely promoted by health agencies in the 1980s and 1990s, led by the World Health Organization (WHO), and adoption rates increased dramatically over those 20 years.[35]

Similarly, the decline in deaths from TB is in part the result of an inexpensive treatment program known as DOTS (for Directly Observed Treatment, Short-course). Aggressive promotion of the treatment by WHO doubled the share of the world's people with access to the program between 1995 and 1998. Today 22 countries, accounting for 80 percent of the disease's incidence, have adopted the program, and it is working. In India, which accounts for nearly a third of the world's TB cases, death rates among patients treated in a DOTS program were only 4 percent—about one seventh the rate in areas not covered by DOTS.[36]

In contrast, deaths from HIV/AIDS jumped more than sixfold worldwide in the past decade, from just over a half-million in 1990 to more than 3 million in 2000. (See Figure 1–2.) Nearly all of these deaths occurred in the developing world, and nearly four out of five of them were in sub-Saharan Africa, where drug treatments are largely unaffordable. About 1 percent of the world's adults are now infected with HIV/AIDS, but the rate is eight times higher in sub-Saharan Africa. There, the United Nations estimates that in seven countries, adult infection rates are 20 percent or higher.[37]

The toll of HIV/AIDS is enormous. UNDP asserted in 2000 that some 20 countries have experienced "reversals of human development since 1990 as a result of HIV/AIDS." Because the disease claims people in the prime of life, it has a devastating social and economic impact. AIDS reduces the ratio of healthy workers to dependents (children and the elderly), increasing the care-taking burden of the survivors. The number of orphans and children who have lost their mother, for example, is projected to double, to more than 26 million, by 2010. At the same time, the disease is a drag on national economies: the United Nations estimates that annual economic growth per person is falling by 0.5–1.2 percent each year in half the countries of sub-Saharan Africa because of AIDS. Add to these burdens the loss of civil servants (especially teachers), the diversion of scarce health care resources to treatment of the disease, and the dashed dreams of children who must leave school to help at home—and the devastating impact of the disease becomes even clearer.[38]

The harmful environmental effects of today's economies are partly to blame for the persistence and spread of infectious diseases. Pollution and degradation are directly responsible for about a quarter of all preventable ill health in the world today.

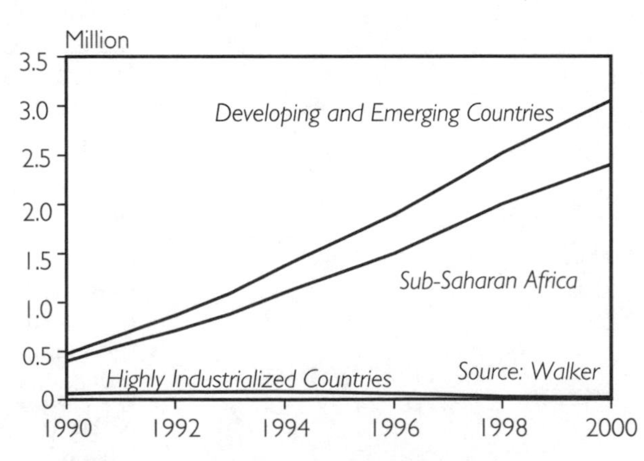

Figure 1–2. Deaths from AIDS in Selected Regions, 1990–2000

Climate change extends the range of mosquitoes and allows them to move to higher elevations, spreading malaria. Warmer temperatures also increase the incidence of algal blooms, which expands the habitat available to the microbes that cause cholera. And scarce or polluted water supplies and lack of sanitation are responsible for more than 10 million deaths each year.[39]

Indeed, economic activity and its environmental side effects may be causing infectious disease to bite once again in wealthy nations. After 60 years of near-continuous decline in deaths from infectious disease in the United States, the trend turned upward again in 1980, and deaths have nearly doubled since then. This has captured the attention of U.S. intelligence authorities: a 2000 report for the Central Intelligence Agency blamed increased trade and travel, new patterns of land use, microbial resistance to drugs, and climate change for the increase in U.S. infections, and described infectious disease as a new security threat for the country, since infections of Americans increasingly originate outside U.S. borders.[40]

Health care systems driven by profit-making are bound to overlook the health needs of those with little market muscle. A 1999 pharmacological study reported that only 13 out of 1,223 medicines commercialized by multinational drug companies between 1975 and 1997 were designed to treat tropical diseases. The great need for new medications to fight infectious disease is largely unaddressed because drug companies see few paying customers. Meanwhile, the market for cures for toenail fungus, obesity, baldness, face wrinkles, and impotence runs into the billions of dollars.[41]

Even the modest gains against infectious disease are now threatened by the growing ineffectiveness of anti-microbial medicines. People in search of pain-free lives pressure doctors to overprescribe drugs—by an estimated 50 percent in the United States and Canada. In developing countries, the reverse is often the problem. Patients underuse medicines when they cannot afford a proper dosage or when they fail to adhere to a full course of treatment. A 1997 study of patients in Viet Nam showed that more than 70 percent were given too few antibiotics. Either way, the number of microbes resistant to the drug multiplies. The result is a more depleted yet more expensive arsenal of antibiotics. The emergence of multi-drug-resistant TB, for example, has meant that a $20 medication must now be replaced with drugs a hundred times more expensive.[42]

Meanwhile, "diseases of affluence," such as cancer, diabetes, and heart disease, rose globally in the 1990s, many registering increases even in the developing world. (See Table 1–2.) Some of the increase is, paradoxically, a sign of successful development: as life expectancies increase, diseases associated with old age become more common. But the surge is also related to lifestyle characteristics increasingly common as the world industrializes, including poor eating, lack of exercise, and smoking. In Europe

Table 1–2. Deaths Worldwide from Leading Chronic Diseases, 1990 and 2000

Chronic Disease	1990	2000
	(million)	
Ischaemic heart disease	6.3	6.9
Major cancers	5.0	6.1
Cerebrovascular disease (stroke)	4.3	5.1
Diabetes mellitus	0.6	0.8

SOURCE: See endnote 43.

and North America, for example, more than 30 percent of all cancers are associated with dietary habits. And cancers caused by smoking are expected to increase in developing countries in coming decades, as more people there take up the habit.[43]

One health trend that indicates industrial countries are worse off than in the recent past—even if they are wealthier today—is the rapid rise in adult-onset diabetes. This disease is strongly associated with being overweight, a condition that is especially serious in industrial nations. In Europe, about a third of all adults are overweight; in the United States, the figure is 61 percent. And in both areas, obesity (the extreme condition of overweight) rose dramatically in the 1990s—by 10–40 percent in most European countries and by 50 percent in the United States. The problem, in turn, is directly related to policies that make fatty and sugary foods cheap and plentiful (see Chapter 3), and lifestyles that require less and less physical exertion.[44]

Global drug sales offer further evidence of a trend toward poor development. Nearly 90 percent of drugs sold globally in 2000 were sold in industrial nations, in part because prosperous people can afford drugs, but also because modern industrial lifestyles create expensive, unhealthy conditions. Five of the top 10 classes of drugs sold worldwide, accounting for some 18 percent of global sales, were for drugs that address heartburn, obesity, heart disease, stroke, and other conditions relating in part to tasty but poor-quality foods. In 2000, the world market for these drugs was more than $56 billion. (See Table 1–3.)[45]

Wealthy and poor nations alike understand the drag on national development created by poor health. Draft findings of WHO's Commission on Macroeconomics and Health, for example, show that Africa's gross domestic product (GDP) would be up to $100 billion greater today if malaria had been eliminated years ago. HIV infection of more than 20 percent of the adult population of a country translates into an annual decline in GDP of 1 percent. And the Red Cross is placing greater emphasis on fighting disease, since disasters are "built

Table 1–3. Global Sales of Selected Pharmaceuticals, by Category, 2000

Global Rank	Class of Pharmaceutical	Used to Treat	Global Sales, 2000	Share of Global Sales
			(billion dollars)	(percent)
1	Heartburn medicines	Indigestion, gastroesophageal reflux disease	17.4	5.5
2	Blood lipid (fat) reducers	Cardiovascular disease	15.9	5.0
4	Calcium antagonists, plain	High blood pressure and angina; treatment of stroke and coronary heart disease	9.8	3.1
6	ACE inhibitors, plain	High blood pressure, hypertension	7.3	2.3
10	Oral antidiabetics	Diabetes	5.9	1.9
	Total		56.3	17.8

SOURCE: IMS Health, at <www.imshealth.com/public/structure/navcontent>, viewed 30 October 2001.

on the shaky ground of poor public health," in the words of one official. Wealthy countries are not exempt from this tendency, either. Studies show that overweight accounts for 2–8 percent of health care costs in several industrial countries. In the United States, obesity was estimated in the late 1990s to account for some 12 percent of health expenditures. Resources devoted to these preventable diseases are resources that cannot be spent on other pressing issues of national development.[46]

As with health, developing countries made modest progress in expanding education for people in the 1990s. The number of children not enrolled in school dropped from 127 million in 1990 to 113 million in 1998. National governments increased the share of their budgets devoted to primary education in every region except Central Asia and Central and Western Africa. The number of students per teacher declined slightly in most regions between 1990 and 1996. And adult illiteracy rates fell, even in regions of greatest concern: India, for example, brought its rate down by 10 percentage points between 1991 and 1997.[47]

Still, nearly one in six adults today cannot read or write, a problem with strong implications for a nation's development. Education raises productivity, innovation, and output—important ingredients for economic prosperity—and tends to reduce economic inequality. Education is important for population stabilization as well, since educated women tend to marry later and bear fewer children. (See Chapter 6.) And each additional year spent by mothers in primary school has been shown to lower the risk of premature child death by some 8 percent. Not surprisingly, some 99 percent of illiterate people are found in developing countries; in the least developed ones, nearly half of adults cannot read or write. The rate of illiteracy of women, whose social advancement is key to development, is almost twice as high as that of men in developing countries.[48]

Much work remains to be done in providing education for all. In Latin America, for example, a quarter of children entering primary school do not continue past the fifth grade. And in nearly half of Latin American countries, at least 10 percent of children in primary school are repeating grades. These high rates of dropout and repetition suggest an increased need to focus on educational quality as well as access. Overlooking quality issues can be expensive: in the 1980s, children in Latin America required 1.7 years, on average, to be promoted to the next grade, a delay that cost primary and secondary schools $5.2 billion.[49]

Despite the challenges, the formula for educational success is increasingly understood. In a study of several nations and of the Indian state of Kerala, UNICEF found that countries with strong educational systems typically achieved universal primary enrollment early in their development process, gave emphasis to primary education without tuition or fees, and improved educational quality while minimizing costs per student, dropout rates, and repetition of grades. The study also highlighted the benefits for girls' enrollment of having female teachers, and the advantages of instruction in a child's mother language.[50]

Pioneering a New Economic Model

Most economies in the 1990s continued to use materials and fuel intensively and to depend exclusively on gross national product (GNP) to measure national well-being. It might have been otherwise: the end of the cold war created rare historical space to remake the world's political and economic landscape, to invest, for example, in new development initiatives, from poverty alleviation to mass transit. Instead, western nations seized the moment to further globalize the existing economic model, sometimes at the expense of local economies and cultures. And in the process of embracing free markets, many centrally planned economies weakened their commitment to health care and education—two key components of the UNDP definition of development.

Microfinance institutions make small but critical interventions in the lives of the poor to expand their options for a better life.

Despite the bias toward business as usual, however, signs of an emerging shift toward sustainable economics were evident as the decade unfolded. (See Box 1–1.) Imaginative thinkers in government, business, and academia found creative ways to redirect financial tools, engage the economic power of the poor, and rethink production and consumption. These initiatives were tiny in the context of the global economy, but are featured here because their vitality—and in many cases, their rapid growth—commend them as practical tools for sustainability.[51]

Governments and private individuals alike began to flex their financial muscle in favor of sustainability in the 1990s. Several European nations, for example, began to shift taxes from income to environmental "bads" such as pollution and fossil fuel use in search of a double dividend: degradation would be reduced as polluting became more expensive, and employment would rise as social security and other levies paid by employers were cut, lessening the cost of hiring new workers. Sweden led the way in 1991, followed by two multicountry waves of reforms in the mid- and late 1990s involving nine countries in total. The amount of taxes shifted was small—environmental taxes still account for only 3 percent of all taxes worldwide—but initial results are encouraging. Sweden estimates that a third of its 40-percent decline in sulfur emissions between 1989 and 1995 resulted from tax shifting, for example. The effect on employment has not been documented, but computer simulations consistently suggest that it is positive.[52]

Subsidies were also harnessed in the cause of sustainability, with impressive results. Organic agricultural area, for example, grew some forty-two-fold in Europe between 1985 and 2000 and now accounts for some 3 percent of agricultural area in the European Union, in part because of subsidies to farmers as they move from traditional to organic farming. Likewise, subsidies helped to boost global electricity generation from wind turbines tenfold between 1990 and 2000; wind now supplies 1 percent of the world's electricity. Conversely, Belgium, France, Japan, Spain, and the United Kingdom all slashed or eliminated subsidies on coal production, and collectively halved their use of this carbon-intensive energy source. By expanding access to healthy food and clean air, these policies increase the likelihood that people can live longer and healthier lives, a key developmental goal.[53]

Meanwhile, some private investors began to leverage their wealth for sustainable development through participation in socially responsible investment (SRI) portfolios. The number of these programs in the United States tripled between 1995 and 1999 and were valued at $2.16 trillion as the decade closed—accounting for 12 percent of all professionally managed funds.

BOX 1–1. DEVELOPMENT VERSUS GROWTH

For rich and poor alike, advances were made over the decade in distinguishing development from a simple growth in income. Institutions such as the World Bank acknowledged that poverty is not simply a lack of income but a lack of access to food, clean water, education, and other services that have a marked impact on opportunities for the poor. At the same time, UNDP devised indicators such as the Human Development Index, which combines life expectancy, access to education, and living standards to produce a yardstick of national well-being.

In prosperous societies, income indicators are insufficient for a different reason. Studies have shown that happiness does not necessarily track with GNP—in the United States, for example, the share of people describing themselves as "very happy" declined from 35 percent in 1957 to 30 percent in the mid-1990s, despite a doubling of income per person. And the growing toll of wasted resources in prosperous nations—material that is discarded, time spent in traffic gridlock, and health damaged by overeating, among many others—are not included or are counted as benefits under GNP accounting rules.

SOURCE: See endnote 51.

SRI programs allow investors to avoid supporting firms with, for example, poor environmental or social records. The operations of many, however, are still fairly basic. Environmental screens used by some investment firms might avoid companies that deal with nuclear power, yet invest in a host of other companies that pollute heavily. Nevertheless, the movement is an encouraging grassroots expression of a desire for a greener and more just world. As investors become more savvy about SRI options, they might choose more rigorously screened funds that could eventually "green" capital markets.[54]

One of the most promising economic advances of the 1990s came from the developing world. Microfinance, the extension of small-scale credit and other financial services to the poor, came into its own in the past decade, 20 years after its birth in rural Bangladesh. The world's oldest and largest microfinance organization, the Grameen Bank, doubled the number of Bangladeshi villages it serves in the 1990s, to 40,000, and nearly tripled its clientele, to more than 2.3 million borrowers. Its success has also been exported: similar programs were established in 58 countries in the 1990s. Researchers are just starting to collect global data on microfinance, but these too show robust growth. The NGO Microcredit Summit reported a 48-percent increase in microfinance clientele just between 1998 and 2000, to 31 million participants (Grameen included). Nearly two thirds were classified as the "poorest of the poor"—the bottom half of those living below their nation's poverty line—and a large share were women. In the Grameen Bank, 94 percent of borrowers are women.[55]

Microfinance institutions (MFIs) make small but critical interventions in the lives of the poor to expand their options for a better life. Managing loans of as little as $50 and savings deposits as small as $5, MFIs help entrepreneurs, often home-based, to generate greater income, perhaps facilitating bulk purchase of supplies to lower a basketweaver's costs or allowing a farmer to store a harvest until market conditions fetch a better price. MFIs are not a panacea to end poverty—they are not helpful to the homeless, destitute, or others whose lives are highly unstable, and they do not end the need for social safety nets. But by targeting those on the economic margin—especially women, who account for 70 percent of the world's poor and who tend to use a higher share of earnings for family needs than men do—MFIs could become an important grassroots weapon in the fight against poverty. Indeed, if the Microcredit Summit Campaign succeeds in its efforts to reach 100 million microfinance client families by 2005—a figure that represents probably 40 percent of the world's 1.2 billion people living in absolute poverty—its impact on poverty could be substantial.[56]

Finally, governments, industry, and nonprofits spearheaded several ingenious changes in the way goods are made and used, with an eye to creating more sustainable economies. On the production side, "industrial ecology" embraces a range of practices to reduce dramatically the appetite of modern economies for energy and materials while preserving a high quality of life. Such ambitious reductions—90 percent is a goal often proposed for industrial countries—require more than increased factory efficiency or a redoubled effort by families to recycle. Instead, it requires a rethinking of industrial systems—another way to rethink the way we do development.[57]

Many of the imaginative initiatives of industrial ecology have been tried only in pilot projects. Yet some encouraging successes are worthy of note. "Zero-waste" factories, for example, radically reduce waste either by making production more efficient or by selling byproducts to others who can use them productively. In 1996, Canberra in Australia became the first city to mandate a goal of "zero waste" by 2010. Toronto has followed suit, as have about 45 percent of local governments in New Zealand. In addition, at least 29 countries—20 in Europe and 8 in Asia—have enacted packaging "take-back" laws that require companies to recycle or reuse packaging discarded by consumers. Another 9 countries require manufacturers to take back electronic equipment, and the European Union (EU) now requires take-back of automobiles. (See Table 1–4.) These initiatives are an important step toward comprehensive, economy-wide recycling, a key component of a sustainable world.[58]

Eco-industrial parks build on the zero-waste concept by bringing together factories that can use each other's wastes. The oldest and most famous example is the complex of industries in Kalundborg, Denmark, which includes a cement factory, a fish farm, a power plant, an oil refinery, a manufacturer of gypsum wallboard, a producer of insulin, and local farmers. Each produces a byproduct—once considered waste—that is an input to the production of another. Although Kalundborg started more than a quarter-century ago, the idea gained broad attention only in the 1990s. Today, according to the National Center for Eco-Industrial Development at Cornell University, more than 25 eco-industrial parks have been started around the world. This represents an infinitesimal share of the world's industrial capacity, of course, but indicates that the concept is alive and workable.[59]

Companies also increasingly design products for recycling or remanufacturing, which saves materials and energy. Appliance and automobile companies in Europe, for example, are designing products for easy disassembly and labeling components to indicate their chemical or metallic makeup. And Xerox now designs most of its copiers to be remanufactured, rather than discarded, at the end of their useful lives. Xerox reported in 2001 that 95 percent of the equipment returned to it in 2000 was

Table 1–4. Key Legislative Responses in the 1990s in Favor of Reuse and Recycling of Materials

Initiative	Description
German Ordinance on Packaging Waste, 1993	Requires manufacturers and distributors to collect product packaging and arrange for its reuse or recycling, or to join DSD, an organization that runs a package waste collection system in parallel with municipal waste collection. Consumers can also leave secondary packaging behind in retail stores.
European Directive on Packaging and Packaging Waste, 1994	Requires EU member states to recover 50–65 percent of all packaging waste, 25–45 percent of which must be recycled.
Japanese Packaging Recycling Law, 1997	Requires businesses to take back glass, plastic, paper, steel and aluminum cans, bottles, boxes, and other packaging. Material that is not readily recyclable must be collected, sorted, transported, and recycled at the manufacturer's expense.
European Landfill Directive, 1999	Biodegradable municipal waste flows to landfills must be reduced to 75 percent of 1995 levels by 2006, and to 35 percent by 2016. Prohibited wastes include liquid, explosive, corrosive, rustable and highly flammable waste, infectious hospital and clinical wastes, and whole tires.
European End of Life Vehicles Directive, 2000	By 2006, car manufacturers must recover and reuse 85 percent of the weight of "end-of-life" vehicles, and by 2015, 95 percent. Costs are to be borne largely by the manufacturer. In addition, the directive restricts the use of lead, mercury, cadmium, and hexavalent chromium.
Japanese Appliance Law, 2001	End-of-life televisions, refrigerators, washing machines, and air conditioners must be returned to retailers or local collection authorities, at the consumer's expense. At least 55 percent, by weight, of air conditioners and televisions and at least 50 percent of refrigerators and washing machines must be recycled.
European Directive on Waste from Electronic and Electrical Equipment, in draft	Recovery and recycling rates for computers, tools, toys, medical equipment, and other electronic and electrical equipment would be set at 85 percent recovery and 70 percent recycling under pending legislation. A companion Directive would prohibit the use of several heavy metals in these products.

SOURCE: See endnote 58.

reused or recycled. Such "design for disassembly" initiatives portend a major expansion of the reuse and recycle mindset that is key to a sustainable economy.[60]

Consumers also participate in the industrial ecology revolution: companies and nonprofits have worked to nudge consumer choice in a more sustainable direction in the 1990s by selling services, rather than goods, to meet people's needs. Recognizing that a service is often less energy- and materials-intensive than producing goods for each consumer—and that services may better provide what consumers are really looking for—these innovators began to reshape the idea of consumer choice. Xerox, for instance, began to sell copying services rather than copy machines in the

1990s—leasing machines to customers and also maintaining them. Clients' copy needs continued to be met as always, but with reduced waste flows and materials use, since Xerox now had a strong incentive to remanufacture machines at the end of the lease rather than throw them away.[61]

One of the great successes in applying this concept in the 1990s is found in "car sharing," a kind of neighborhood-based car subscription service. Subscribers pay a flat fee to join a car-sharing organization and are billed monthly for the time a car is used and the distance it is driven. They have access to the fleet of cars in the city's car-sharing network, one of which is typically stationed in their neighborhood. Most car sharers use public transportation, cycling, or walking as their principal mode of transport; they turn to cars only when they need to haul purchases, do a string of errands, or get to a place that is poorly served by their normal transportation mode.[62]

By confining cars to their best use—as a flexible option when alternatives are inadequate—car sharing helps build healthier cities. Studies in Europe have demonstrated that car owners who become car sharers cut their energy use for transportation by about half, and that each shared car eliminates four private cars from congested roads. And because cars are materials-intensive yet spend upwards of 90 percent of their lives sitting idle, shifting trips from private cars to other modes of transportation will create a far more materials-efficient transportation system.[63]

Car sharing has taken off as a viable transportation alternative in Europe, and fledgling initiatives are taking hold in more than a dozen cities in North America, including Boston, Portland, San Francisco, Seattle, Toronto, and Vancouver. The total number of subscribers is still small—an estimated 140,000 worldwide in 2000—but growth has been very rapid. No less an automobile aficionado than Bill Ford, chairman and now chief executive officer of the Ford Motor Company, understands the idea of selling transportation services rather than cars. "The day will come when the notion of car ownership becomes antiquated. If you live in a city, you don't need to own a car," he told a British newspaper in November 2000.[64]

Looking Ahead

Over the past decade, people and organizations in the nooks and crannies of the world's economies have begun to embrace the natural environment, to address the urgent needs of the poor, and to restructure production and consumption. Successes are small, to be sure. But just as surely they can be rapidly expanded, with enough focus and will. As the world witnessed in September 2001, the U.S. government scrapped old priorities overnight and vigorously pursued others it found more urgent. Within two days of the attacks on New York and Washington, Congress approved $40 billion to combat terrorism. Additional government assistance for airline relief and for an economic stimulus package brought attack-related spending to well over $100 billion—none of which was in the budget before September 11. With similar focus and will, the global community can ensure that a third U.N. conference on environment and development in 2012, if held, would find a sustainable world well under construction.[65]

Finding focus involves, above all, developing a clear set of achievable objectives. The Millennium Declaration issued by the United Nations in 2000 is a good place to start; it lists a series of laudable goals for human development to be achieved by

2015. Adding to these a set of environmental targets produces an ambitious—but still focused and achievable—work agenda that would greatly advance the cause of sustainable development. (See Table 1–5.) "Tend to people, mend their world" might summarize a workable strategy for sustainability over the next decade.

These complex social and environmental objectives can be pursued in a dizzying variety of ways, and it would be easy to lose the forest for the trees as various actors pursue them. But a few principles could help maximize the mileage from the global community's efforts at sustainability, and could keep those efforts on track.

A first principle is to encourage the involvement and ability of women in building a sustainable world. Investments targeted at women have multiple payoffs, which increases their likelihood of success. We have already seen that a peso or a rupee in the hand of a poor woman is more likely to be used for family needs, especially for nutrition and health, than it is in the hand of a man, making income-generating opportunities for woman especially valuable. Moreover, investments in the health of women are important for the healthy development of children, since women are a child's first source of nutrition. And ensuring educational opportunities for women is

Table 1–5. Goals for Sustainable Development by 2015

Source	Goals
Environmental Goals	Meet, and then extend, the Kyoto Protocol goals for reducing emissions of greenhouse gases. End progressive shrinking of global area of natural forests. Develop and meet national air quality standards based on WHO guidelines. Halve the rate of soil erosion. End overpumping of aquifers.
Millennium Declaration	Halve the share of the world's people living in extreme poverty, suffering from hunger, and lacking access to clean drinking water. Reduce maternal mortality by three quarters. Reduce mortality rates for children under 5 by two thirds. Achieve of universal completion of primary school and gender equality in access to education. Halt, then reverse, the spread of HIV/AIDS, malaria, and other major diseases.
Economic Goals	Establish and implement systems of national accounts that internalize environmental costs. Eliminate subsidies that encourage the extraction and use of virgin materials and fossil fuels. Encourage fourfold to tenfold reductions in materials use in industrial countries. Encourage an ethic of sufficiency in consumption.

SOURCE: Millennium Declaration from U.N. Development Programme, *Human Development Report 2001* (New York: Oxford University Press, 2001), pp. 21, 24.

a key to population stabilization, which in most low-income countries would facilitate development. (See Chapter 6.)

At the same time, women in prosperous countries are particularly valuable allies of environmentalism. Polls consistently show that women embrace the values shift toward sustainability more quickly than men do. Thus strategies to pursue sustainability in industrial nations are more likely to be successful if they appeal to women. Whether in wealthy or poor nations, a woman-centered strategy for sustainability harnesses the energy of one of the movement's most powerful resources and allies.[66]

Guiding consumer choices toward green products and socially responsible investing can steer huge chunks of a national economy in a sustainable direction.

A second helpful principle is to focus on whole systems. Major advances in sustainability over the next decade will not be attained through incremental efficiency gains. Indeed, engineers and activists who achieved numerous reductions in energy and materials use in the 1990s were successful because they defined production and consumption issues more broadly than their predecessors had. Sometimes stepping back and looking at the system-wide picture is not only efficient, but a money-saver as well. When New York City began to investigate options for protecting its high-quality drinking water, for instance, it found that construction of water treatment facilities would cost at least $4 billion, with another $200–300 million annually in operating costs. But investing in conservation of the upper watershed that supplies the city with drinking water would achieve the same aim for about $1.4 billion. So that is what the city did, acquiring land near waterways that supply the city, and taking other measures to protect the water supply. Only because the city viewed preserving drinking water not simply as a narrow technological challenge but a broader environmental one was it possible to consider this natural option.[67]

Linkages among problems give clues to how broadly we might draw the boundaries of a system. As countries tackle infectious disease, for example, many find that an important immediate cause is unclean water. Water may be unclean, in turn, because sewage flows openly in the streets, allowing pathogens to enter the water supply. This dire situation exists because governments lack the resources or the will to provide sewer lines and treatment plants for the entire population. If this causal chain is conceived as a single system, we might fight infectious disease not just with medicine, but also with inexpensive, sanitary composting toilets that require no water at all. Draw the system boundaries a bit further, and it becomes clear that the composting toilets are potentially a source of fertilizer for nearby farmers, reducing their need for chemical fertilizer, which in turn reduces water pollution from fertilizer runoff. None of these solutions come readily to mind if the problem is simply defined as infectious disease. The old adage "If you're stumped by a problem, make it bigger" is a neat pitch for systems thinking.

Out of a focus on systems comes a third principle: harness powerful tools. Systems change is best done with system-sized equipment. Steering taxing and government spending in service of sustainability, as a dozen European nations have begun to do, could change incentives economy-wide and prompt rapid change, if adopted broadly enough. Redirecting government and corporate procurement policies toward recycled goods could be just the boost needed to expand the market for recycled materials and make recycling economically viable. And guiding consumer choices toward green products and socially responsible investing can steer huge chunks of a national economy in a sustainable direction. Governments and NGOs will need to enlist big levers like these over the next decade if major progress on sustainability is to be made.

Persuading governments, corporations, the electorate, or centers of power to weigh in on behalf of sustainability requires an understanding of what makes people change their attitudes and behavior. If it is true that people in economically secure countries can afford to be more receptive to values of sustainability than people on the economic margins, strategies of persuasion will look very different in rich and poor countries. If a society-wide change in values requires generational change—not just greater persuasion of the current generation of leaders—understanding where contemporary society is in that transition will be important. If it is true that people have a particular interest in health issues, and that they respond readily to people-centered stories, communications strategies should be sensitive to this. Only by using intelligent change strategies will politicians, citizens, and business people develop the political will for larger-scale change.[68]

In the decade since the historic Rio conference, the challenge of putting the world's economies on a sustainable track has advanced only slightly—but importantly. Trends are still headed largely in the wrong direction, but a shift in global consciousness is clearly discernible. Efforts this decade to expand and build on that consciousness can increase the momentum in favor of sustainability. But those efforts must be made; at this stage, progress in this struggle is not inevitable. The good news, though, is that as sustainability builds steam, succeeding efforts become easier. With proper focus and sufficient will, debate at an Earth Summit in 2032 might center not on how to achieve sustainable development, but instead on what development means in a world of plenty for all.

Chapter 2

Moving the Climate Change Agenda Forward

Seth Dunn and Christopher Flavin

As the World Summit on Sustainable Development nears, international climate negotiations may seem to echo those of 10 years ago. Just as in the run-up to the 1992 Earth Summit in Rio de Janeiro, a U.S. Bush administration refuses to embrace mandatory commitments to counter climate change, while European negotiators push for binding national targets for reducing greenhouse gas emissions. Developing-nation delegates criticize their wealthy neighbors for a lack of leadership, and demand financial and technical aid for their own efforts to address the problem. Environmental groups warn that failure to reach agreement could cause irreversible damage to the global environment, while industry trade associations counter that a binding treaty would constrain the world economy.

But behind the curtain of this now-familiar drama, important developments have reshaped the debate over climate change during the past decade. In Rio, considerable scientific uncertainty existed about whether human activity had begun to alter Earth's climate, and whether the projected impacts would actually occur. The perceived costs of reducing greenhouse gas emissions were almost uniformly high. The potential of cleaner, more efficient technologies to move the world toward a greenhouse-benign energy system was just beginning to be recognized. Many businesses were opposed to any international agreement, with some using tobacco-industry tactics of questioning the underlying science. As a result, the United States was able to wield its political clout and water down the U.N. Framework Convention on Climate Change (UN FCCC) that was agreed to in Rio.

Ten years later, there is now broad scientific consensus that human-induced climate change is under way and accelerating, with a number of projected impacts of warming already occurring. The debate over the economics of climate change is maturing, with greater recognition that innovative policies can substantially bring down the cost of lowering emissions. Wind and solar power, fuel cells, and other "alternative" energy technologies have entered the marketplace and begun to create multibillion-dollar industries. A growing number of corporations are moving beyond denial to acceptance and action on climate change, some seeking competitive advantage by anticipating rather than responding to future policy changes. And after years of struggle, the international community is showing signs that it may yet have the political will to bring into force the contentious 1997 Kyoto Protocol—with or without the United States.

Indeed, the political landscape of climate change has been altered in subtle but significant ways since 1992. This is in part because many industrial countries have quietly begun to experiment with policies to reduce emissions, while the limited experience of several developing countries suggests that economic development can be decoupled from emissions growth without harming the economy. More dramatically, the Bush administration's abrupt announcement in March 2001 that it would not sign the Kyoto Protocol has had the unintended consequence of galvanizing international determination to reach a global agreement.[1]

Historians writing about the rescue of the Kyoto Protocol may come to view the Bush administration's rejection of the pact as a turning point, recharging negotiations that had been bogged down for over three years. The unilateral U.S. move backfired not only with Europe but also with Japan, Canada, Australia, and other nations that had previously been closely aligned with the American negotiating position. In Bonn, Germany, in July 2001—to the surprise of numerous observers and participants—representatives from 178 nations finalized many of the protocol's key rules while U.S. negotiators stood by and watched.[2]

It is unclear how the tragic terrorist attacks of September 11, 2001, will affect the future pace of climate negotiations. The Bush administration's unilateral approach to foreign policy during its first eight months in office has been modified by a broad multilateral effort to cope with the problem of international terrorism. But it remains to be seen whether the U.S. government will encounter the diplomatic need to demonstrate greater multilateralism on other emerging global threats—such as climate change.

The international community may have the political will to bring into force the Kyoto Protocol—with or without the United States.

Even if the United States does not return to the climate negotiations table in the near future, domestic pressure to eventually do so is likely to rise as other countries move forward to finalize the pact. Meanwhile, the September 11th attacks have renewed debate over energy security and the world economy's disproportionate dependence on Middle Eastern oil. It cannot be lost on climate negotiators that the same oil that has increased vulnerability to terrorism and international conflict has also made the world more vulnerable to climate change. Both reinforce the case for accelerating the transition to a more efficient energy system that is based on carbon-free, indigenous resources.

This chapter assesses how the climate change issue has developed in the decade since Rio. It outlines how the science, technology, economics, policy, business, and politics of the issue have evolved in often gradual but sometimes sudden ways. And it argues that these developments, taken together, make the climate challenge in 2002 fundamentally different from that in

1992. Recognizing this reality is essential for pushing the climate agenda ahead at the Johannesburg Summit—and for achieving far greater progress during the climate convention's second decade.

Science Evolving

Since its creation in 1988 by the World Meteorological Organization and the U.N. Environment Programme (UNEP), the Intergovernmental Panel on Climate Change (IPCC) has established itself as the most authoritative source of information on this subject. Drawing on a network of hundreds of experts around the world, the panel engages in a meticulous process of collecting, synthesizing, and peer-reviewing an enormous body of literature spanning dozens of fields that relate to climatic change. In each of three IPCC assessment reports (released in 1990, 1995, and 2001), the mandate has been to assess available scientific information on climate change, its potential impacts, and possible response strategies. The first two assessments provided the basis for negotiating the 1992 Rio treaty and the 1997 Kyoto Protocol. (See Box 2–1 for a description of the treaties.) The findings of the third assessment report have set the stage for the current round of climate negotiations.[3]

One clear finding is that carbon dioxide (CO_2), which is released into the atmosphere from the burning of fossil fuels, is the single most important greenhouse gas in contributing to the "anthropogenic forcing of climate change," or the warming of Earth's surface. The share of CO_2 in warming is expected to rise from slightly more than half today to around three quarters by 2100. Other important greenhouse gases, emitted mainly from agricultural and industrial practices, include methane, nitrous oxide, sulfur hexafluoride, hydrofluorocarbons, and perfluorocarbons.[4]

An important question during the 1990s was whether the warming that has already occurred—an increase of 0.3–0.6 degrees Celsius in average global surface temperature since the late nineteenth century—could be attributed to human activities. At the time of the first IPCC report, scientists could not determine whether human-induced climate change was under way or whether the warming was due to natural variability, such as sunspots and volcanic eruptions. Over the next few years, however, they made considerable progress in distinguishing between natural and human influences. By accounting for the release of sulfate aerosols, which have a cooling effect, they found a better match between simulations of climate change and actual changes. This led the IPCC to assert in its second report that "the observed warming trend is unlikely to be entirely natural in origin" and that "the balance of evidence suggests a discernible human influence on global climate."[5]

In the five years since the release of the second assessment, new studies of past and current climates and better analysis and comparison of data sets have further improved our understanding of climate change. The third IPCC assessment report notes that "an increasing body of observations gives a collective picture of a warming world and other changes in the climate system," including widespread decreases in snow cover and ice extent and a rise in sea level of 0.1–0.2 meters during the twentieth century. The panel concluded that the 1990s were likely the warmest decade—and 1998 the warmest year—since instrumental recordtaking began in the 1860s. (See Figure 2–1.) Based on measurements in the northern hemisphere, the average global surface temperature rose more during the twentieth century than during any other century in the last 1,000 years.[6]

Unprecedented increases in global temperatures have occurred in tandem with record levels of greenhouse gas concentrations and emissions. Since 1750, atmospheric CO_2 concentrations have increased by 31 percent, with more than half this increase occurring in the last 50 years. Current concentrations are the highest in the

BOX 2–1. RIO TO JOHANNESBURG: 10 YEARS OF CLIMATE CHANGE NEGOTIATIONS

The U.N. Framework Convention on Climate Change, which was signed at the 1992 Earth Summit and entered into force in March 1994, established the objective of stabilizing atmospheric concentrations of greenhouse gases at levels that will avoid "dangerous anthropogenic interference with global climate" and allow economic development to proceed. The treaty recognizes several basic principles:

- that scientific uncertainty must not be used to avoid precautionary action;
- that nations have "common but differentiated responsibilities"; and
- that industrial nations, with the greatest historical contribution to climate change, must take the lead in addressing the problem.

The agreement commits all signatory nations to addressing climate change, adapting to its effects, and reporting on the actions they are taking to implement the convention. It also requires industrial countries and economies in transition to formulate and submit regular reports on their climate policies and their greenhouse gas inventories. It commits these nations to aim for a voluntary goal of returning emissions to 1990 levels by the year 2000 and to provide technical and financial assistance to other nations. Today 181 nations and the European Union (EU) are party to the UN FCCC.

In 1995, signatories to the UN FCCC concluded that its commitments were inadequate, and launched talks on a legally binding protocol to the convention. These negotiations culminated in the 1997 Kyoto Protocol, which collectively committed industrial and former Eastern bloc nations—termed Annex B nations—to reducing their greenhouse gas emissions by 5.2 percent below 1990 levels during 2008–12. The agreement includes several measures designed to lessen the difficulty of meeting the target, such as "flexibility mechanisms" that allow the trading of emissions permits, the use of forests and other carbon "sinks," and the earning of credits through a Clean Development Mechanism or joint implementation projects (carbon-saving initiatives that take place in developing or Annex B nations, respectively). The protocol also commits developing countries to further their existing commitments to monitor and address their emissions.

In 1998, governments agreed to a plan of action and timeline for finalizing the rules on the protocol's implementation. At negotiations in The Hague, Netherlands, in late 2000, disagreement between the United States and the EU over several key provisions led to a breakdown in the talks. Following a U.S. withdrawal from the negotiating process in March 2001, 178 nations reached agreement in July in Bonn, Germany, on several key elements of the protocol's rules. Many details of the Bonn agreement concerned compromises on emissions trading, sinks, and compliance that allow additional flexibility in meeting the Kyoto targets. Governments also established a special fund to help developing nations adapt to the impacts of climate change. Outstanding issues were deferred to negotiations in Marrakesh, Morocco, from October 29 to November 9, 2001.

SOURCE: See endnote 3.

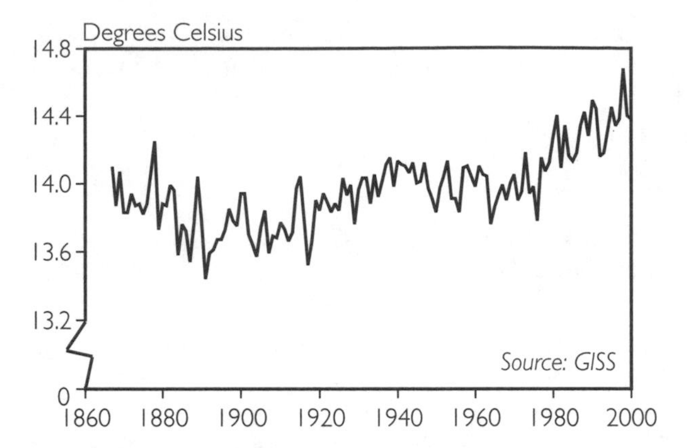

Figure 2–1. Global Average Temperature at Earth's Surface, 1867–2000

last 420,000 years, and probably in the last 20 million years. CO_2 levels are increasing at an unparalleled rate. About three quarters of the human-caused carbon emissions of the past 20 years are due to fossil fuel burning, with the remainder coming from deforestation and other forms of land use change. (See Figure 2–2.) Based on this evidence, the IPCC concluded that while natural factors have made small contributions to the warming of the past century, "there is new and stronger evidence that most of the warming observed over the last 50 years is attributable to human activities."[7]

According to the third IPCC report, emissions of carbon from fossil fuel burning are expected to be the dominant influence on future CO_2 levels, which are projected to range from 540 to 970 parts per million volume (ppmv) by 2100. Global average temperature is due to increase by 1.4–5.8 degrees Celsius between 1990 and 2100. This rate of warming is much larger than that experienced in the last century, and is likely to be without precedent in the last 10,000 years. Average sea level is projected to rise by 9–88 centimeters. Also projected are a continued decrease in snow cover and sea ice, and a more widespread retreat of glaciers and ice caps. Even after greenhouse gas concentrations are stabilized, climate change will persist for many centuries, with surface temperature and sea level continuing to rise in response to past emissions.[8]

Scientists are more confident in assessing the observed trends in weather extremes. The IPCC found it "likely" or "very likely" that the latter half of the twentieth century saw higher minimum and maximum temperatures and a higher heat index over most land areas, as well as more intense precipitation events over many mid- to high-latitude land areas in the northern hemisphere. All these changes, moreover, are "very likely" to continue during this century.[9]

A great deal has also been learned over the last decade regarding the risk of damage from projected climate change. There is evidence that regional climate changes have already affected a wide range of physical and biological systems. These changes include glacier shrinkage, permafrost thawing, later freezing and earlier buildup of ice on rivers and lakes, lengthening of mid- to high-latitude growing seasons, shifts of plant and animal ranges, declines of plant and animal populations, and earlier flowering of trees, emergence of insects, and egg-laying by birds.[10]

Scientists have uncovered new knowledge about the vulnerability of various systems. Several natural systems are recognized as especially at risk of irreversible damage, including glaciers, coral reefs and atolls, mangroves, boreal and tropical forests, polar and alpine ecosystems, prairie wetlands, and remnant native grasslands. Climate change will increase existing risks of extinction of the more vulnerable species and the loss of biodiversity, with the extent of the damage increasing with the rate and magnitude of change.[11]

More research is being conducted into the sensitivity of human systems, mainly water resources, agriculture, forestry, coastal zones and marine systems, human settlements, energy, industry, insurance and other financial services, and human health. Projected adverse impacts include:

- a reduction in potential crop yields in most tropical and subtropical regions for most temperature increases;
- decreased water availability for populations in many water-scarce regions, notably in the subtropics;
- an increase in the number of people exposed to vector-borne and water-borne diseases (such as malaria and cholera) and an increase in heat stress mortality; and
- a widespread increase in the risk of flooding for tens of millions of people, due to both increased heavy precipitation events and sea level rise.

Projected changes in climate extremes—droughts, floods, heat waves, avalanches, and windstorms—could have major consequences, as the frequency and severity of these events are expected to increase.[12]

The potential for large-scale, irreversible impacts has received more study, as they pose risks that have not yet been reliably quantified. Examples include a significant slowing of the ocean circulation system that conveys warm water to the North Atlantic, major reductions in the Greenland and West Antarctic Ice Sheets, accelerated warming due to carbon releases from terrestrial ecosystems, and the release of carbon from permafrost regions and methane from hydrates in coastal sediments. If these changes occur, their impact will be widespread and sustained. Slowing of the oceanic circulation would reduce warming over parts of Europe. Loss of either the West Antarctic or the Greenland Ice Sheet could raise global sea level up to 3 meters over the next 1,000 years, which would submerge many islands and inundate extensive coastal areas. Added carbon and methane releases would further amplify warming.[13]

Adaptation to climate change has also garnered growing attention. But this costs money, and the most vulnerable countries have the fewest resources and the least ability to adapt. The IPCC concluded rather forcefully that "the effects of climate change are expected to be greatest in developing countries in terms of loss of life and relative effects on investment and the economy." Regional assessments reveal major vulnerabilities around the globe—a National Research Council study points to serious adverse impacts in the United States—but those who will be hit hardest have contributed least to the problem.[14]

How much climate change occurs will

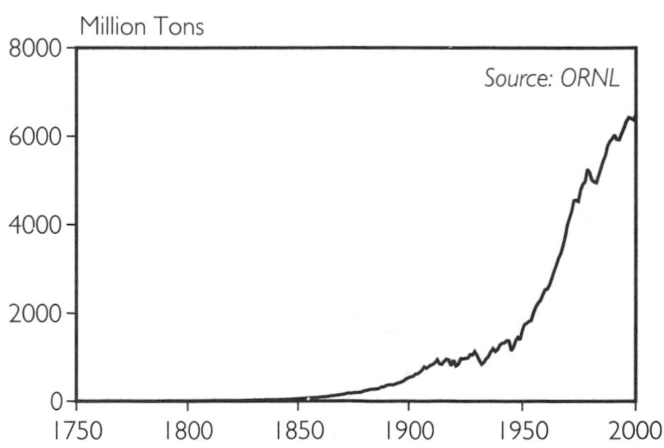

Figure 2–2. Global Carbon Emissions from Fossil Fuel Combustion, 1751–2000

depend on how high CO_2 concentrations rise, which in turn will be determined by trends in carbon emissions from fossil fuel burning. Stabilizing greenhouse gas concentrations at 450 ppmv, for example, requires that annual carbon emissions drop well below current levels within the next several decades, then to around 2 billion tons by 2100, and ultimately to less than 1 billion tons. This entails a cut of roughly 70–80 percent in global carbon emissions—much larger than the Kyoto cuts under negotiation.[15]

New Views on Technology and Economics

Lowering global carbon emissions will require major changes in existing patterns of energy resource development. Fortunately, the potential of new technologies and policies to slow climate change has grown dramatically since Rio. Since its 1995 assessment, the IPCC reports, "significant progress relevant to greenhouse gas emissions reduction has been made and has been faster than anticipated." Advances are taking place in a wide range of technologies that are in varying stages of development. These include the market introduction of wind turbines, the elimination of industrial byproduct gases, the emergence of highly efficient hybrid-electric cars, and the advance of fuel cell technology.[16]

What is the potential for reducing emissions in the relatively near future? Summarizing hundreds of studies, the IPCC concludes that global emissions could be reduced well below 2000 levels between 2010 and 2020. Specifically, the panel estimates that emissions could be reduced by 1.9–2.6 billion tons of carbon equivalent by 2010, and then by 3.6–5.5 billion tons by 2020. (At the moment, emissions are projected to reach 11.5–14 billion tons by 2010 and 12–16 billion tons by 2020.) The panel also found that half of these reductions could be achieved by 2020 in a cost-effective fashion.[17]

These low-cost opportunities lie primarily in the hundreds of technologies and practices that promote efficient energy use in buildings, transportation, and manufacturing. In addition, natural gas is expected to play an important role in reducing emissions in tandem with power plant efficiency improvements and greater use of cogeneration (the combined use of heat and power). Important contributions can also be made by low-carbon energy systems, such as biomass from forestry and agricultural byproducts, landfill methane, wind and solar power, hydropower, and other renewable sources of energy. Agriculture and industry can reduce other greenhouse gases: Methane and nitrous oxide emissions can be cut from livestock fermentation, rice paddies, nitrogen fertilizer use, and animal wastes, while process changes and the use of alternative compounds can minimize the emissions of fluorinated gases.[18]

Using these available or near-ready technologies, most models suggest that atmospheric CO_2 levels could be stabilized at 450–550 ppmv, if not lower, over the next 100 years. Bringing this about, however, would require major socioeconomic and institutional changes. These reductions imply an accelerated decoupling of economic development and carbon emissions, as measured in the carbon intensity of the global economy. (See Figure 2–3.) It also suggests that the supply and conversion of energy can no longer be dominated by low-priced fossil fuels.[19]

What are the costs and benefits of cutting emissions? Analyses vary widely, given different methodologies and underlying

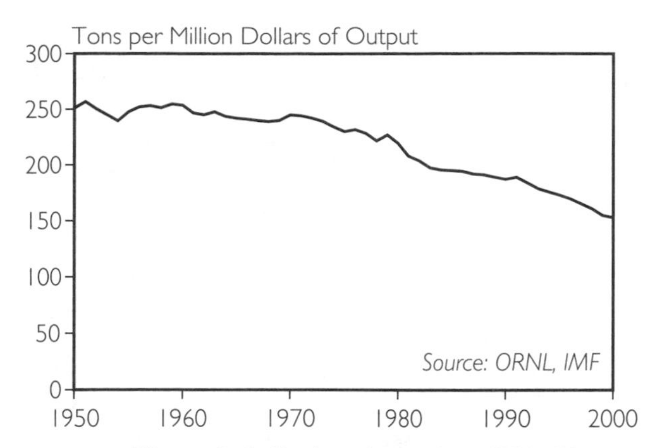

Figure 2–3. Carbon Intensity of World Economy, 1950–2000

assumptions. Estimates depend, for example, on whether the revenue of carbon taxes is recycled back into the economy through reductions in other taxes; whether the benefits of avoided climate change—including side benefits such as energy savings, reduced local and regional air pollution, energy security, and employment—are factored in; and whether the external costs of climate change are incorporated into market prices. Other assumptions shaping models of the economics of climate change include demographic, economic, and technological trends; the level and timing of the agreed-to target; and the degree of reliance on various implementation measures, such as emissions trading.[20]

There is a consensus among experts that some greenhouse gas emissions can be limited at no cost—or even a net benefit—to society through "no regrets" policies that address imperfections in the market. A lack of information, for instance, can prevent consumers and businesses from adopting efficient technologies that lower overall energy costs. If carbon taxes or auctioned emissions permits are used to finance reduced wage and labor taxes, the benefits become larger. In many cases, the side or ancillary benefits of reducing carbon emissions—lower air pollution, new jobs, reduced oil imports—balance out the costs of the policies themselves. For example, reducing carbon emissions can also lower emissions of particulates, ozone, and nitrogen and sulfur oxides—which can have significant human health benefits.[21]

Recent government studies support the notion that there is significant potential for low-cost or no-cost emissions cuts. A U.S. Department of Energy study estimates that the nation could meet the majority of its Kyoto target at no net cost, primarily by removing market barriers to the adoption of existing energy-efficient and renewable energy technologies. These policies would also reduce air pollution, petroleum dependence, and inefficiencies in energy use, leading to economic benefits that are comparable to overall costs. Similarly, a report from the Climate Change Programme of the European Commission indicates that the European Union can achieve its Kyoto target through cost-effective measures that amount to no more than $18 per ton of carbon dioxide, accounting for about 0.6 percent of the region's gross domestic product (GDP). These measures, primarily involving enhanced energy efficiency, have the potential to achieve more than double the emissions cut required of the EU under the protocol.[22]

How much would it cost industrial and former Eastern bloc nations (Annex B countries) to implement the Kyoto Protocol? That depends on how much trading is involved and what domestic measures are taken. Without emissions trading between these countries, most global studies show reductions in projected GDP of about 0.2–2 percent in 2010 for different regions. With full emissions trading, however, the reductions would be just 0.1–1.1 percent of projected GDP—amounts that would likely be lost in the "noise" of natural variations of the economy. The models also do not factor in the use of carbon sinks or non-CO_2 greenhouse gases to meet targets, the Clean Development Mechanism, side benefits, or revenue recycling.[23]

However the costs and benefits add up, they will be spread unevenly among different sectors of the economy.

Economies in transition, which are included in the Annex B grouping, represent a special case. For most of these, the effects range from an increase of several percent of GDP to negligible—reflecting enormous opportunities for improving energy efficiency. If energy efficiency is indeed improved drastically, emissions in 2010 could be well below the amounts assigned to them under the Kyoto treaty. In such instances, models show an increase in GDP, due to revenues these countries obtain from selling their trading surpluses.[24]

What would it cost to reduce emissions more aggressively? Conventional economic models typically suggest that costs will rise as the level at which greenhouse gas concentrations are stabilized drops (from 750 to 550 ppmv, or from 550 to 450 ppmv). But these models ignore the potential of ambitious targets to bring about deep technological change by spurring industry to make large rather than incremental innovations. "Induced technological change" is an emerging field of research in climate change economics, but most models do not account for it. Those that do suggest that certain policy regimes could lead to stabilization of CO_2 concentrations and GDP growth.[25]

Efforts to improve climate-related modeling have resulted in the "integrated assessment" model, which attempts to synthesize climate science, policy, and economic research—and is becoming increasingly influential in policy circles. These models are useful for assessing policies, coordinating issues, and comparing climate and non-climate policies. But a recent study from the Pew Center on Global Climate Change observes that most integrated assessment models are based on economic theories with simplifications that do not always apply to climate policy. In particular, they make unrealistic assumptions about how market forces drive technological innovation, the behavior of firms, intergenerational equity, and climate "surprises." Such assumptions tend to drive up the estimated cost of dealing with climate change.[26]

However the costs and benefits add up, they will be spread unevenly among different sectors of the economy. Generally speaking, it is easier to identify the sectors that are likely to face economic costs than it is to pinpoint those that may benefit. In addition, the costs are more immediate, more concentrated, and more certain—even if the benefits prove to be greater. Coal, possibly oil and natural gas, and certain energy-intensive sectors—such as steel production—are most likely to suffer an economic disadvantage. Others, including the renewable energy industry, are expected to benefit over the long term from price changes and the availability of financial and other resources that might otherwise have been committed to carbon-intensive energy sectors.[27]

Appropriate measures can help cushion some of the costs to various sectors. The removal of fossil fuel subsidies could increase total societal benefits by improving economic efficiency, while trading can cut the net economic cost of meeting the targets. Some policies, such as exempting carbon-intensive industries from these taxes, will redistribute the costs but also increase the total expense to society. And the revenues from a carbon tax can be used to compensate low-income groups who would otherwise suffer.[28]

Other countries will be affected by the actions taken by those facing initial emissions constraints. For oil-exporting developing countries, estimated impacts are as high as a 25-percent reduction of projected oil revenues by 2010. But these studies do not consider policies other than trading—which could lower the impact on oil exporters—and thus tend to overstate both the costs to these countries and the overall costs. Such nations can further reduce the impact by removing subsidies for fossil fuels, restructuring energy taxes according to carbon content, increasing natural gas use, and diversifying their economies.[29]

Other developing countries face both costs and benefits. They may suffer the effects of reduced demand for exports and the higher price of imports. At the same time, however, they may benefit from the transfer of environmentally sound technologies and know-how. No country is likely to experience the same net effect, and it is hard to identify winners and losers. As for "carbon leakage"—the possibility that carbon-intensive industries will simply relocate to developing countries in response to changing prices—the estimates range from a 5- to a 20-percent increase in non-Annex B emissions. But these models do not account for the transfer of environmentally sound technologies and skills, which could lower and in the longer term more than offset the environmental or economic costs of any leakage.[30]

Climate Policy: Theory and Practice

In order to tap various opportunities for reducing greenhouse gas emissions, governments will need to overcome the many technical, economic, political, social, behavioral, and institutional barriers to change. The options vary widely by region and sector, as well as over time, with poor people facing particularly limited options for adopting technologies or changing behavior. In industrial countries, the major barriers relate primarily to social and behavioral resistance; in economies in transition, they center on subsidized energy prices; in developing countries, they hinge largely on greater access to information and advanced technologies, financial resources, and training. But every country can find opportunities to surmount some combination of these barriers.[31]

Evidence to date suggests that national responses to climate change can be more effective if they are deployed as a portfolio of policy instruments that either limit or reduce greenhouse gas emissions. These might include:

- carbon/energy taxes,
- tradeable permits,
- removal of subsidies to carbon energy sources,
- provision of subsidies and tax incentives for carbon-free sources,
- refund systems,
- technology or performance standards,
- energy mix requirements,
- product bans,
- voluntary agreements, and
- investment in research and development.

Although there is no one policy of choice, market-based instruments show signs of being cost-effective. Energy efficiency standards have been widely used and could be effective in a number of countries. Voluntary agreements with industry have become more frequently relied upon, in some instances as a precursor to more stringent measures. Other measures include influencing consumer and producer behavior through information campaigns, environmental labeling, green marketing, and incentives. Government and private R&D are essential for advances in technologies that will lower costs further.[32]

Another lesson from the early history of climate policy is that it can be more effective when integrated with the "non-climate objectives" of national and sectoral policies and translated into broader strategies for long-term technological and social change aimed at sustainable development. Just as climate policies achieve side benefits, non-climate policies can yield climate benefits. For example, emissions could be reduced significantly through socioeconomic policies such as energy infrastructure development, pricing, and tax policies. Transferring climate-friendly technologies to small- and medium-sized enterprises is another case in point. Accounting for the side benefits of these policies can also lower the political and institutional barriers to actions pertaining to climate.[33]

Coordinating actions is another way to reduce costs and avoid conflicts with international trade. Taxes, standards, and subsidy removal can all be coordinated or harmonized, though steps to do so have thus far been limited. As for the timing of policies, the IPCC has reaffirmed the finding of its 1995 report: earlier action to mitigate climate change provides greater flexibility in moving toward the stabilization of atmospheric greenhouse gas levels. Economic models completed since the second assessment suggest that a gradual transition toward a less carbon-emitting energy system would minimize the premature retirement of power plants, factories, and other forms of capital stock. It would provide time for technology development and avoid an untimely lock-in to early versions of low-emission technologies that are developing rapidly. And greater near-term action would decrease the environmental and human risks associated with rapid climate changes, allow for a later tightening of targets, and address concerns about the effectiveness and equity of the climate regime.[34]

Despite the strengthening case for climate policy, the record of the past decade has been mixed. Global carbon emissions from fossil fuel combustion rose by 9.1 percent between 1990 and 2000. Cumulative global carbon emissions between 1990 and 2000, slightly over 68 billion tons, reflects a 15-percent increase over the 59 billion tons emitted worldwide between 1980 and 1990.[35]

As for the Kyoto Protocol's commitment of Annex B countries to reduce greenhouse gas emissions by 5.2 percent between 1990 and 2008–12, this group of nations reduced carbon emissions just 1.7 percent between 1990 and 2000. (See Table 2–1.) In other countries, meanwhile, carbon emissions rose by 28.7 percent. Annex B countries still account for the majority—58 percent—of global carbon emissions.[36]

The United States remained the largest national source of carbon emissions, as its share of the global total grew from 22 percent in 1990 to 24 percent in 2000. The Kyoto Protocol commits the nation to a 7-percent cut in greenhouse gas emissions between 1990 and 2008–12. But between 1990 and 2000, the United States increased carbon output by 18.1 percent, or 235 mil-

Table 2–1. Kyoto Emissions Targets, First Commitment Period (2008–12)

Country/ Region	Target 1990–2008/12[1]	Actual Emissions 1990–2000[2]
		(percent)
United States	– 7	+18.1
European Union	– 8	– 1.4
Japan	– 6	+10.7
Canada	– 6	+12.8
Australia	+ 8	+28.8
Russia	0	–30.7
All Annex B countries	– 5.2	– 1.7

[1]Basket of six greenhouse gases. [2]Carbon only.
SOURCE: See endnote 36.

lion tons. (See Figure 2–4.) The difference between U.S. emissions in 2000 and in 1990 is roughly equivalent to the combined annual carbon emissions of Brazil, Indonesia, and South Africa. U.S. per capita emissions, about 5 tons, are the highest in the world.[37]

U.S. carbon emissions stand at more than double those of the second leading emitter, China, whose carbon output increased by 7.7 percent between 1990 and 2000. This figure for China includes a sharp 19.8-percent decline since 1996, due to improved efficiency and a 30-percent reduction in coal use. (Some scientists have questioned the coal estimate, which may be revised downward). Chinese per capita carbon emissions, about 0.68 tons, are one seventh those of the United States and well under the global average.[38]

Russia, which agreed in Kyoto to maintain its greenhouse gas emissions at 1990 levels in 2008–12, experienced a 30.7-percent drop in carbon output between 1990 and 2000, mainly due to sharp declines in natural gas and coal consumption associated with economic slowdown and the closure of inefficient industries. Japan committed to a 6-percent emissions reduction, but actually expanded carbon output between 1990 and 2000 by 10.7 percent, with consumption of coal rising by 22.9 percent. India registered a 67-percent increase between 1990 and 2000, primarily due to a 54-percent rise in coal use. India's per capita emissions, though, are at 0.3 tons—well below the global average of 1.1 and the lowest of the major emitters.[39]

Through 2000, EU carbon emissions had dropped by 1.4 percent, although all member states maintained per capita carbon emissions above the global average. In Germany, carbon emissions fell by 19 percent, owing to a dramatic 36.2-percent decline in coal use associated with factory closings in the former East Germany and a gradual removal of coal production supports. The United Kingdom, which removed coal subsidies more sharply, experienced a 5-percent decline in emissions in line with a 41.9-percent fall in coal consumption.[40]

While only a few leading emitters—the United Kingdom, Germany, Russia—are on course to meet their Kyoto goals, national

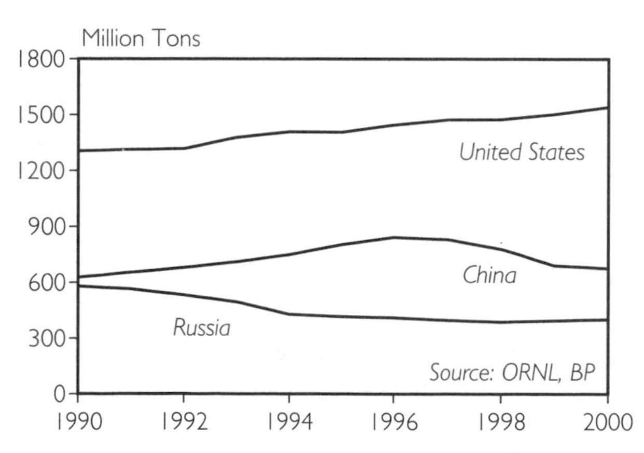

Figure 2–4. Carbon Emissions in the United States, China, and Russia, 1990–2000

governments of industrial countries are increasing their activity in the area of climate policy. The International Energy Agency (IEA) has identified more than 300 separate measures that its members undertook during 1999 to address climate change. The agency put these actions in five general categories: fiscal policy, market policy, regulatory policy, R&D policy, and policy processes. The IEA study noted that "good practice" climate policies would:

- maximize both economic efficiency and environmental protection;
- be politically feasible;
- minimize red tape and overhead; and
- have positive effects on other areas, such as competition, trade, and social welfare.

Based on these principles, it is possible to identify several good practices to date. (See Table 2–2.)[41]

While there is no "silver bullet" climate policy that can be applied across all countries, getting the prices right through subsidy reform and tax policy is crucial. According to the Organisation for Economic Co-operation and Development (OECD), a combination of fossil fuel subsidy removal and carbon taxes would cut carbon emissions in OECD member countries by 15 percent between 1995 and 2020. Market approaches and a mix of policies—voluntary agreements, standards, incentives, R&D—are needed, as are monitoring and assessment, good institutions, and international cooperation. Even with these criteria, however, climate policies face the barriers of perceived high cost and limited political will to act.[42]

Climate-related fiscal policies are increasingly popular, with nearly all industrial countries adopting such measures in 1999—most of them modest in size. These

Table 2–2. Climate Change Policies and Good Practices

Categories	Policies	"Good Practices" to Date
Fiscal	Ecotaxes	Denmark, Norway levies
	Tax credits, exemptions	U.S. wind/biomass tax credit
	Subsidy reform	U.K. coal subsidy removal
Market	Emissions trading	Netherlands, U.K. programs
	Green certificates	Denmark renewable certification program
Regulatory	Mandates/standards	Germany electricity feed law
		U.S. appliance efficiency standards
	Voluntary agreements	Netherlands, Germany covenants
		EU agreement with car manufacturers
	Labeling	U.S. Energy Star program
R&D	Funding and incentives	Japan renewable energy funding
	Technology development	Japan efficiency, renewables programs
Policy Processes	Advice/aid in implementation	Switzerland car-sharing program
	Outreach	France energy audits
		Canada multistakeholder consultations
	Strategic planning	EU studies of community-wide strategy

SOURCE: See endnote 41.

measures are appealing because they tend to reduce greenhouse gas emissions while stimulating national economies. Belgium, Portugal, and the United Kingdom have eliminated their coal subsidies since 1992. Subsidies are also being added to promote more-efficient vehicles and renewable energy in power generation, with the most successful example to date being the German electricity feed law—which has spurred the wind power business and been replicated in several other European nations. Nineteen industrial nations are planning more than 60 tax policy changes that will affect emissions, although only 11 of these are defined as carbon or emissions taxes. The most effective carbon taxes are in Scandinavia: Norway's levy, adopted in 1991, has lowered carbon emissions from power plants by 21 percent. One reason such taxes have been adopted slowly or contain several exemptions is that their impact on fairness and competitiveness is often overstated by industry.[43]

Interest in market-based mechanisms has also risen due to their expected cost-effectiveness and the success of the U.S. sulfur emissions trading program. Four countries have adopted greenhouse gas emissions trading proposals, with another nine, along with the European Union and the World Bank, considering their adoption or promotion. But only a few countries, such as Denmark, have begun to tackle the challenge of allocating the emissions within sectors. The Danish plan limits carbon emissions from electricity generation, with power companies given a quota, fined for exceeding it, and permitted to sell or bank unused quotas. Denmark has also pioneered a "green certificate" system that obliges electric utilities to supply customers with a percentage or quota of renewable electricity, allowing the companies to trade quotas among themselves.[44]

The most significant activity in this category has been the U.K. government's announcement in August 2001 of the world's first economy-wide emissions trading scheme, which will provide up to $312 million between 2003 and 2008 to encourage British firms to sign up to emissions reduction targets. The government estimates the program, scheduled to begin in April 2002, could be cutting 2 million tons of carbon per year by 2010, while giving industry a global competitive advantage by generating new job and investment opportunities.[45]

A third area of growing activity is voluntary agreements, which arise from negotiations between government and business or industry associations. These are attractive because they face less political resistance from industry, require little overhead, and can be complemented by fiscal and regulatory measures. Some 21 voluntary agreements were initiated in 1999, including 4 for power generation, 2 for transport, and 11 for industry and manufacturing. With respect to stringency, they are characterized as strong (in the Netherlands), containing legally binding objectives and the threat of regulation for noncompliance; weak (in Canada), lacking penalties for noncompliance but having incentives for achieving the targets; or "co-operative" (in U.S. manufacturing), with incentives for developing and implementing new technology.[46]

While voluntary agreements are relatively new, some interesting results have emerged. In Germany, where the business community has committed to reducing greenhouse gas emissions 20 percent between 1990 and 2005, the manufacturing and electric power sectors had achieved reductions of 27 and 17 percent, respectively, by 2000. UNEP and the World Energy Council (WEC) have identified more than 600 voluntary projects to reduce greenhouse gas emissions that are just completed, under way, or planned by industry. They estimate that by 2005 these could achieve annual reductions of up to 2 billion tons of carbon dioxide—roughly 6 percent of global greenhouse gas emissions. But UNEP and WEC believe that even as industry activity grows, governments remain too reactive.[47]

While these studies suggest growing engagement by industrial-nation governments in dealing with climate change, the IEA concludes that "there remains considerable scope for further improvements." Enacted and proposed policies might not be sufficient for countries to meet their Kyoto targets, and further action may be necessary. Meanwhile, a number of municipalities are not waiting for leadership from their capitals. (See Box 2–2.) Several American states, in addition, have initiated voluntary programs for tracking and reducing emissions. And in August 2001, the governors of six New England states and premiers of five East Canadian provinces adopted a resolution to bring their region's greenhouse gas emissions to 1990 levels by 2010 and to 10 percent below those levels by 2020.[48]

Meanwhile, climate policy continues to take shape in the developing world, mainly for economic reasons. One of the U.S. government arguments against ratifying the Kyoto Protocol is that it "exempts" 80 percent of the world's population. While this objection ignores the fact that this 80 percent accounts for only 37 percent of the past century's carbon emissions, it is true that developing nations are not subject to the protocol's first round of binding commitments. Nevertheless, research from the World Resources Institute suggests that developing nations are already taking substantial steps to reduce emissions growth. China's remarkable carbon cuts are related to fossil fuel subsidy reforms and energy efficiency programs—policies that are being adopted elsewhere. Mexico, India, and the Philippines have set national goals to increase renewable energy and improve energy efficiency. Thailand and Brazil have successful demand-side energy management programs. And natural gas vehicles are being introduced in India and Argentina, where 10 percent of the automobile fleet runs on compressed natural gas.[49]

The Business of Climate Change

The corporate response to climate change has undergone a major shift over the past decade. In the early 1990s, when the scientific basis for climate change was less compelling, many corporations were skeptical about whether the threat existed at all or whether its impacts would be significant. By the run-up to the 1997 Kyoto conference, however, some companies had softened their stance on the science, while those still opposed to action had begun focusing on the potential economic impacts.[50]

Since Kyoto, the business landscape on climate change has diversified further. The high profile of that conference convinced executives to study the problem more closely. Many concluded that the issue was not going away, and that by integrating it into their corporate strategies they could minimize the costs and risks, while at the same time identifying market opportunities that the effort to slow climate change would inevitably open up. In 1999, attendees at the World Economic Summit in Davos voted climate change the most serious global problem facing companies in coming decades—and an issue on which the business community should play a leadership role. As Kimberly O'Neill Packard and Forest Reinhardt argue in the *Harvard Business Review*, "business leaders need to inform themselves about climate change and think systematically about its effects on their companies' strategies, asset values, and investments."[51]

Unlike much of the public, who still view climate change and climate policy as abstract and long-term, many industry executives see immediate economic stakes in the outcome of climate deliberations. Coal companies and other energy-intensive industries, facing short-term costs, have an obvious interest in slowing efforts to reduce the use of carbon-based fuels. Renewable energy companies, on the other hand, see

BOX 2–2. ARE CITIES MOVING FASTER THAN NATIONS ON CLIMATE?

By taking voluntary steps to reduce greenhouse gas emissions, local governments are helping strengthen the international effort to stabilize global climate. In the last decade, a pledge made by the city of Toronto to reduce its carbon emissions by 20 percent of the 1988 level by 2005 has been copied and modified by city governments worldwide.

In the early 1990s, 13 cities in Canada, the United States, Europe, and Turkey joined Toronto in drawing up plans to slash carbon emissions. City-to-city networking led by the Toronto-based International Council on Local Environmental Initiatives (ICLEI) helped to multiply this effort. In 1993, ICLEI launched a campaign to help more local governments devise their own plans to reduce emissions. As of October 2001, some 500 cities, responsible for an estimated 8 percent of global carbon emissions, had signed up.

Although the precise goals vary from place to place, some cities are aiming higher than their national governments are. In the industrial world, many local governments have committed to reducing emissions by 20 percent relative to a baseline year somewhere between 1990 and 1995; the target date for the reductions to be complete ranges from 2005 to 2010. The first cities to join the campaign have already measured progress. Some 110 cities and counties in the United States had eliminated 2.5 million tons of carbon by June 2001. By 1995, Toronto had cut its total carbon emissions to 7 percent below its 1990 level, and by 1996, Copenhagen reduced emissions by 22 percent from 1990.

More recently, the campaign has begun to help cities in rapidly industrializing economies fix inefficient buildings, transportation, and energy systems that not only release carbon dioxide but also waste money and create air pollution. For example, Cebu City, in the Philippines, is calibrating the engines of all its city-owned vehicles. Local officials expect improved engine efficiency to cut municipal fuel costs by 12 percent, roughly $60,000, annually, and to improve air quality. Building on this, Cebu City aims to cut carbon emissions to 15 percent below the 1994 level by 2010.

Although local governments are not party to the climate treaty, ICLEI sends city officials to key meetings. By endorsing strong targets and quantifying success stories in emissions reduction, these local authorities have received a fair amount of press coverage—raising public awareness that aggressive targets and timetables for reducing carbon emissions are achievable and beneficial.

—*Molly O'Meara Sheehan*

SOURCE: See endnote 48.

enormous profit potential. Automobile and energy companies derive considerable revenues from the status quo but see long-term market opportunities in greenhouse-benign technologies and fuels. The climate issue is also forcing many industries to seek competitive advantage through energy- and cost-saving opportunities within their own walls. Whether they exploit emerging energy markets, take part in emissions trading, manage the risks of future regulations, gain a technological edge over rivals, or enhance credibility and policy influence by demonstrating environmental leadership, companies are beginning to recognize their strategic interests in engaging more proactively on the climate issue.[52]

The variety of industry positions was on display during the July 2001 negotiations in Bonn. Some U.S. business groups praised the Bush administration for its rejection of what they view as an economically risky and unnecessary agreement. Other companies, especially European ones, criticized the U.S. government and urged others to maintain the Kyoto process. Etienne Davignon, Vice Chairman of Société Générale de Belgique and Co-chairman of the EU-Japan Business Dialogue Round Table, said, "We need a protocol; it's indispensable."[53]

Such views contrast with the vocal anti-Kyoto business lobby in the United States, which has been reinforced by the rhetoric of conservative think tanks such as the Cato Institute. Some groups have also been criticized for funding scientists whose work had led them to a skepticism about the existence or seriousness of climate change. The most open opponent of Kyoto, the Global Climate Coalition (GCC), included in its heyday some of the world's most powerful corporations and trade associations involved with fossil fuels. But the group's extreme behavior—which included attacks on climate scientists—backfired: BP, DuPont, Royal Dutch/Shell, Ford Motor Company, DaimlerChrysler, Texaco, and General Motors all withdrew from the GCC between 1997 and 2000. The group was subsequently limited to industry associations, allowing individual companies to conceal their support. The GCC has also altered its message to support voluntary, technology-based efforts as the centerpiece of any effort to address climate change.[54]

ExxonMobil, which continues to publicly and aggressively oppose the Kyoto Protocol, has become the focus of a campaign led by Greenpeace and other organizations to boycott its Esso gasoline stations in Europe. Some businesses that have left the GCC—DaimlerChrysler, Texaco, General Motors—support action on climate change in general but oppose the Kyoto Protocol in particular. Auto and energy multinationals have also supported energy policies that run counter to climate change objectives, most visibly in the United States, through the Bush administration's proposed energy strategy and its emphasis on fossil fuel extraction and combustion.[55]

Even as they support policies that would prolong a carbon-intensive energy path, energy companies are hedging their bets by diversifying their portfolios to include renewable energy and hydrogen—in recognition that today's multibillion-dollar markets in these fuels could become hundred-billion-dollar markets in coming decades. Solar, wind, and other forms of renewable energy represent the largest growth areas (in terms of percentage) of the energy industry over the past decade. Phillip Watts, Chairman of the Royal Dutch/Shell Group, argues that—based on his company's long-term scenarios to

2050—future energy needs could be met in radically different ways, including a revolutionary shift to a hydrogen economy with natural gas used as a bridging fuel. The group has created two new businesses—Shell Renewables, a core business, and Shell Hydrogen—to explore these opportunities. In the automobile industry, DaimlerChrysler and other major manufacturers are racing to introduce the first commercial fuel cell vehicles between 2003 and 2005, with mass production expected to begin toward the end of the decade.[56]

BP offers an intriguing case study of an energy company's response to the climate challenge. Speaking at Stanford University in 1997, CEO John Browne announced that "the time to consider the policy dimensions of climate change is not when the link between greenhouse gases and climate change is conclusively proven but when the possibility cannot be discounted and is taken seriously by the society of which we are part. We in BP have reached that point." At the Yale School of Management in 1998, Browne committed his company to reduce greenhouse gas emissions from its operations by 10 percent from 1990 levels by the year 2010. As of 2001, the company was already halfway toward this goal, having traded the equivalent of 4 million tons of CO_2 through an internal trading program. BP has also made its solar business into one of the world's largest, established a hydrogen division, and launched advertising campaigns that use "Beyond Petroleum" as the theme and solicit public opinions on energy issues.[57]

Yet BP's $100-million annual investment in clean energy equals only about 1 percent of the company's overall expenditures of $12.5 billion. While this positions the company to gain market share in a growing industry, it does little to reduce vulnerability to policies that reduce demand for carbon-intensive products. Such vulnerability is faced by the entire fossil fuel business, which is now making capital investments that may be rendered unsound by future climate policies. For BP and others, plans to expand oil and gas exploration and production activities aggressively will increase their "carbon risk exposure"—which could cause them to lose a significant percentage of their market capitalization. The investment strategy firm Innovest believes these risks "strike to the very heart of the company's strategic direction and could, on a more practical basis, influence future earnings and shareholder value." In the most extreme scenario, BP would see its earnings erode by as much as 5 percent over the next 20 years.[58]

BP's $100-million annual investment in clean energy equals only about 1 percent of its overall expenditures of $12.5 billion.

While the financial world has been slow to take such factors into account, a number of leading asset management and insurance firms—Swiss Re, Munich Re, Deutsche Bank, Gerling, Nikko—are calling for greater integration of climate change into future investment and underwriting activities. A UNEP initiative with financial leaders estimates that climate change impacts could cost around $300 billion annually by 2050. Prior to the Bonn meeting, the initiative's bankers and insurers—predicting a new investment dynamic as capital shifts "from carbon fuels toward renewable energy, efficiency programs, and advanced public transit systems"—called for national and international market mechanisms to address climate change. But other industries that stand to be negatively affected—from tourism to forest products to agriculture—

have been less aware or supportive of the policy process.[59]

A growing element of corporate engagement on climate change is participation in the design of a potential multibillion-dollar emissions trading market that may develop under the provisions of the Kyoto Protocol. According to the global energy brokerage firm NatSource, 55 million tons of greenhouse gases have already been traded on a pilot basis since 1996. The current size of the market is estimated by the World Bank at about $100 million, but it has been projected to reach $250–500 billion by the end of this decade. The brokerage firm Cantor Fitzgerald and consulting giant PricewaterhouseCoopers have teamed up to create co2e.com, an online hub to help companies manage the transition to carbon commerce. (Carlton Bartels, CEO of co2e.com, and two colleagues died during the World Trade Center disaster, but the firm has resumed operations.)[60]

A leader in the charge toward carbon trading is Richard Sandor, who helped pioneer the sulfur emissions market that has reduced sulfur emissions by 29 percent since 1990. Sandor, who believes that greenhouse gas allowances will become "the biggest commodities market in the world," is working with 33 organizations—including BP, DuPont, and Ford—in the U.S. Midwest to design a Chicago Climate Exchange that will test out the trading of carbon at the regional level, much like corn is traded on the Chicago Board of Trade. The voluntary market will begin trading credits on a pilot basis in 2002, with a near-term goal of reducing the emissions of participants—which account for a fifth of the region's emissions—by 5 percent below 1999 levels over five years.[61]

The prospect of trading has led a growing number of companies to begin monitoring and verifying emissions, and to announce goals for reducing them. (See Table 2–3.) Shell has met its initial target of reducing emissions by 10 percent below 1990 levels by 2002, having achieved an 11-percent cut by 2000. The firm has accomplished this partly through improved energy efficiency, reduced gas flaring, and an internal permit system that has traded more than a million tons of carbon dioxide equivalent per year. DuPont has a goal of cutting greenhouse gas emissions 65 percent from 1990 levels by 2010—and has already managed a 50-percent cut, primarily through improved nylon manufacturing methods.[62]

Nongovernmental groups are partnering with corporations to help them address greenhouse gases. The U.S. nonprofit Environment Defense has joined with nine multinational corporations, including Canadian aluminum company Alcan and the Mexican oil company Pemex, to set targets for reducing their emissions by over 80 million tons of carbon dioxide equivalent by 2010. The World Wildlife Fund and the Center for Energy and Climate Solutions have partnered with several multinationals, including Nike and Johnson & Johnson, to reduce emissions through efficiency and fuel switching. The World Resources Institute and World Business Council for Sustainable Development have developed a common international standard for the corporate accounting and reporting of greenhouse gas emissions.[63]

Business opportunities from trading and new technologies are also prompting the formation of corporate coalitions aimed at promoting cost-effective climate policies rather than blocking national and international action. Thirty-seven companies, including Boeing, Enron, Hewlett-Packard, IBM, Intel, United Technologies, and

Table 2–3. Greenhouse Gas (GHG) Emissions Targets, Selected Companies

Company	Target(s)
ABB	Reduce GHG emissions by 1 percent each year through 2005
Alcan	Reduce GHG emissions by 500,000 tons from 2001 to 2005
Alcoa	Reduce GHG emissions by 25 percent from 1990 to 2010
Baxter International	Reduce energy use and associated GHG emissions by 30 percent per unit of product value from 1996 to 2005
BP	Reduce GHG emissions by 10 percent from 1990 to 2010
Dow Chemical	Reduce energy use per pound of production by 20 percent between 2000 and 2005
DuPont	Reduce GHG emissions by 65 percent from 1990 to 2010 Hold total energy use flat at 1990 levels Derive 10 percent of global energy use from renewable resources by 2010
Entergy	Stabilize CO_2 emissions from U.S. generating facilities at 2000 levels through 2005
Federation of Electric Power Companies of Japan	Reduce CO_2 emissions from electricity generation by 20 percent from 1990 to 2010
IBM	Reduce CO_2 emissions from fuel and electricity use by an average annual 4 percent of baseline from 1998 to 2004
Intel	Reduce PFC emissions by 10 percent from 1990 to 2010
Interface	Reduce nonrenewable energy use per unit of production by 15 percent from 1996 to 2005
Johnson & Johnson	Reduce GHG emissions by 7 percent from 1990 to 2010
Nike	Reduce CO_2 emissions by 13 percent from 1998 to 2005
Ontario Power	Stabilize CO_2 emissions at 1990 levels in 2000 and beyond
Pechiney	Reduce GHG emissions by 15 percent from 1998 to 2008–12
Shell International	Reduce GHG emissions by 10 percent from 1990 to 2002
STMicroelectronics	Achieve zero CO_2 emissions by 2010
Suncor	Reduce GHG emissions by 6 percent from 1990 to 2010
Toyota	Reduce CO_2 emissions by 5 percent from 1990 to 2005, by 10 percent from 1990 to 2010
TransAlta	Achieve zero net GHG emissions from Canadian operations by 2024
United Technologies	Reduce energy consumption as a percentage of sales by 25 percent from 1997 to 2007

SOURCE: See endnote 62.

Whirlpool, have joined the Business Environmental Leadership Council. This group, an initiative of the Pew Center on Global Climate Change, is based on several principles, including that "The Kyoto agreement represents a first step in the international process to address climate change" and that "Businesses can and should take concrete steps now in the U.S. and abroad to assess opportunities for emission reductions, establish and meet emission reduction objectives, and invest in new, more efficient products, practices and technologies." Sixteen member companies have set emissions targets, and several more are in the process of creating them.[64]

Other business groups with proactive climate positions include U.S. and European Business Councils for Sustainable Energy and the Social Venture Network. The latter, representing over 100 small companies, proclaimed in a June 2001 ad in the *New York Times* and other major newspapers that "We must all act. We need U.S. leadership now." Most recently, a group of 150 companies, mostly European and Japanese, has organized under the name "e-mission 55"; counting Deutsche Telekom and leading insurer Gerling Group among its members, it has called on governments to bring the Kyoto Protocol into force by 2002.[65]

As company views evolve, they are revealing a trans-Atlantic divide. While the fallback position of most European companies is now to openly support the Kyoto Protocol, in the United States most companies support the principle of climate protection but remain silent on the protocol itself—perhaps for fear of alienating the administration. This has created a rift between Ford and its Volvo Car unit in Sweden: the former publicly opposes Kyoto because it has different standards for industrial and developing nations, while the latter supports the pact. Coca-Cola belongs to the U.S. Council for International Business, which opposes the protocol; but a representative of Coke's Spanish subsidiary says, "We are in line with the general idea of the Kyoto Protocol ...It's the price of entry [to an emissions trading system]."[66]

Increasingly, U.S. companies are concerned that the Bush administration's disengagement will insulate them from pressures to innovate and from the opportunity to trade emissions. The U.S. Council for International Business, which opposes binding restrictions, recognizes that there cannot be an entirely free-market approach and that government must establish rules and methods for trading. Some companies believe the U.S. stance will hurt the economy by giving competitors a headstart in developing new technologies. Thomas Jacob of DuPont fears that delaying climate-related decisions will cost industry more in the long run and "could threaten America's economic supremacy." Firms such as American Electric Power, Cinergy, Enron, and Entergy are pushing the government to embrace some form of limits on greenhouse gas emissions; others, such as Southern Company and Peabody Energy, continue to oppose them. Multinationals like BP and Shell, meanwhile, see little sense in having their overseas operations covered by an agreement while their U.S. plants are not. As one executive with a large international energy company told the *New York Times*, "What businesses want is policy certainty. Bush has injected only turbulence."[67]

Such turbulence may increase before it lessens, but the long-term direction of industry is becoming clearer. If the Kyoto Protocol enters into force, this will prompt the private sector to invest billions of dollars in lower-emissions technologies and practices. As firms are better able to recognize the opportunities as well as the risks of climate change, and to understand that further emissions cuts are inevitable, their resistance is likely to lessen, changing the political dynamic. Companies will also become more active as they understand that clear guidelines are needed for emissions trading to take off. On the whole, the business community may become less a foot-dragger and more an activist: working to shape the rules of the carbon-constrained marketplace now taking shape.[68]

The Political Weather Vane

It is unclear how the international politics of climate change will evolve in the months leading up to the World Summit on Sustainable Development. With terrorism and an economic downturn topping the global policy agenda, government leaders have been distracted from the climate change negotiations. In the wake of the compromises achieved in the Bonn agreement, however, blocking progress has become diplomatically more difficult—and the benefits of joining the regime are increasingly seen to outweigh the potential costs. As of late 2001, there was growing momentum to bring the Kyoto Protocol into force by the end of 2002, independent of U.S. actions. (See Box 2–3.) The question is whether the protocol's remaining details can be worked out in sufficient time to allow national governments to ratify the pact and make it a binding international agreement.[69]

The Kyoto Protocol's "rulebook" of procedures and institutions was to be finalized in Morocco in late October and early November 2001. Issues to be addressed at Marrakesh included the election of the Executive Board of the Clean Development Mechanism and the establishment of a system for coordinating the various funds for developing countries. The Bonn agreement also did not fully address the extent to which nuclear power could be used to achieve emission reduction targets, the specific rules governing emissions trading between countries, or the accounting procedures for measuring carbon sinks. Decisions on these issues were to be finalized in Marrakesh and formally adopted as a package, together with the decisions made in Bonn—clearing the way for ratification.[70]

One of the most politically sensitive issues surrounding ratification is the potential impact of the protocol on international competitiveness. As the Marrakesh talks opened, the Bush administration argued that the Kyoto accord would make U.S. industry less competitive by forcing companies to adopt costly technologies. Meanwhile, Japan's Ministry of Economy, Trade, and Industry and domestic industries are worried that Japanese firms will be placed at a disadvantage if they—but not U.S. firms—are obligated to reduce emissions. Several U.S., European, and Canadian firms have repeated this concern about being put at a competitive disadvantage by shouldering the financial burden of emissions constraints that their U.S. competitors do not face.[71]

But recent studies suggest that the international competitiveness of the EU and Japan would in fact not be substantially weakened by implementing the Kyoto Protocol without the United States. In fact, they are likely to benefit in the short term due to the absence of the U.S. companies from the trading market, which would lower the price of an emissions permit substantially. Japan's National Institute for Environmental Studies, which is affiliated with the Ministry of Environment, estimates that without U.S. buyers, emissions permit prices would drop from $69 to $23 per ton of carbon, allowing Japan to meet the Kyoto target with a minimal impact on global competitiveness: a reduction in GDP of a negligible 0.07 percent relative to projections.[72]

Even this projection may be pessimistic, for Japanese industry has a history of responding proactively to regulation through technological change. If the nation's industry responds passively to energy taxes, then Japanese GDP would in fact fall. But a more realistic scenario would be for industry, especially automakers, to respond with technological innovation that results in improved productivity, reduced energy consumption, and ultimately lower prices. This is precisely how Japanese carmakers, particularly Honda, reacted to the U.S. clean air legislation and oil shocks of the 1970s and 1980s—by reducing vehicle emissions and enhancing car quality. Today, Japanese carmakers have a 25-percent share of the U.S. car market. (Honda and Toyota have been the first to market hybrid-electric vehicles, which nearly double the average fuel economy of passenger cars, in Japan and the United States. Over 50,000 Toyota hybrids are on the road in Japan; Honda has sold more than 5,000 of its version in the United States.)[73]

According to a study by the Japanese consultancy Shonan Econometrics, proactive implementation of the protocol by business could translate into a 0.9 percent increase in Japan's GDP, or an increase of $47.3 billion. Other nations could also benefit from the spillover effects of Japan implementing Kyoto: Southeast Asia's GDP would increase by $11.5 billion, and that of Western Europe by $13.9 billion. The study concludes that "Japan could greatly benefit its own economy by taking the initiative and going ahead with ratification of the Kyoto Protocol. . . . To Japan, ratification could very well serve as an excellent springboard to break out of its economic slump."[74]

What about Europe? Studies of the costs and benefits of EU ratification of the Kyoto Protocol without the United States also suggest an overall gain. According to the

BOX 2–3. HOW CAN THE KYOTO PROTOCOL ENTER INTO FORCE?

The Kyoto Protocol does not become an instrument of international law until it is ratified by 55 countries, representing 55 percent of the emissions of industrial and former Eastern bloc nations in 1990. As of October 2001, 40 countries had ratified the agreement. The majority of these are developing nations, including Argentina, Mexico, Senegal, and many small island states such as Trinidad and Tobago. If the United States chooses not to ratify, entry into force may require ratification by the European Union, Russia, Japan, Canada, and Australia (see Table)—all of whose governments have stated their intention of ratifying the protocol by the 2002 Johannesburg Summit.

Country	Share of Annex I Country 1990 Carbon Emissions
	(percent)
United States	36.1
European Union	24.2
Russia	17.4
Japan	8.5
Poland	3.0
Other European nations	5.2
Canada	3.3
Australia	2.1
New Zealand	0.2
Total	100

SOURCE: See endnote 69.

Dutch consultancy ECOFYS, the EU could achieve 85–95 percent of its Kyoto target without harming the competitiveness of its economies, with smart policies being able to offset the remaining competitive impacts. The cost of meeting the Kyoto goal could be as small as 0.06 percent of the GDP in 2010. Since climate policies reduce other air pollutants, financial savings on end-of-pipe technologies that reduce acid rain would also be realized. The study concludes that unilateral implementation of the Kyoto Protocol by the EU could give European industry a head start in developing innovative technologies that reduce greenhouse gases. Not implementing the protocol, on the other hand, may lead to substantial increases in mitigation costs in the longer term. Professor Kornelis Blok, coauthor of the report, contends that "If the U.S. does not ratify Kyoto and the EU and Japan do, they will gain a competitive advantage."[75]

Back in the United States, a high-level task force has been slow to produce a proposed alternative to the protocol. The political expediency of terming the treaty unworkable has found intellectual support among economists who use conventional models that project high costs of compliance, and who call for a go-slow approach. The U.S. government has also selectively drawn on the arguments of commentators such as David Victor, a Senior Fellow with the U.S.-based Council on Foreign Relations and long-time observer of the international climate process. In *The Collapse of the Kyoto Protocol and the Struggle to Slow Global Warming*, Victor contends that the treaty is unlikely to enter into force, and that its failure will offer the opportunity to create a more realistic alternative. Arguing that a worldwide trading system is impractical, he calls instead for a focus on national policies such as emissions taxes—an approach that the U.S. government has repeatedly resisted.[76]

Such critiques of the Kyoto Protocol are useful in identifying the challenges that lie ahead, such as determining effective ways to monitor emissions and ensure compliance with the treaty. But these are manageable issues, not fatal flaws—and any alternatives that have thus far been offered tend to be far less politically feasible than the regime now being created. As Michael Grubb of Imperial College notes in a critique of Victor's book, "there are answers, and ironically, some may be easier to find while the present U.S. administration is withdrawn from the Kyoto negotiations." This may prove a prescient observation, should the international community succeed in bringing the treaty into force with the United States on the sidelines.[77]

Indeed, insider and expert opinion on Kyoto calls to mind Winston Churchill's description of democracy as the worst possible form of government—except for all the alternatives. Jan Pronk, the Dutch environment minister who led the Bonn talks to their successful conclusion, claims that "Kyoto is the only game in town." In a study for Climate Strategies, a climate expert network created by the Shell Foundation, Benito Muller and colleagues at the Oxford Institute for Energy Studies examine other alternatives being considered by the U.S. government, such as emissions intensity targets and price caps on emissions trading. They conclude that these are not likely to be acceptable to the international community and that, given its reasons for disengagement, the United States is unlikely to come up with a credible, viable alternative to Kyoto.[78]

If Kyoto is rescued, can it eventually bring the world's leading emitter back in? Another Climate Strategies study, led by

Michael Grubb, argues that in the absence of credible alternatives, the Kyoto Protocol remains the best way to achieve global action on climate change. The report contends that the EU should lead an international effort, joining with Japan and Russia, to bring the protocol into force, and that the United States will then rejoin. This, the authors argue, will:

- provide a long-term structure for controlling emissions and strengthen the international framework for continuing action;
- demonstrate industrial-country leadership, making it easier to bring other nations on board at a later date; and
- bring certainty to the private sector, fostering the technological development and spread of energy-efficient and low-carbon technologies.[79]

In the absence of credible alternatives, the Kyoto Protocol remains the best way to achieve global action on climate change.

Even with the United States initially on the sidelines, the study adds, commercial pressures will push U.S. private-sector investment toward lower-carbon technologies. And by demonstrating the economic benefits of various climate policies, EU nations can influence the implementation of similar domestic policies in the United States, whether or not the government has signed on. In other words, "keeping Kyoto" could be the best way to prompt action both within and without the United States.[80]

This debate should not distract decision-makers from the greater long-term challenge of making major emissions reductions, and engaging the United States and the developing world in this effort, over coming decades. As Edward Parson of Harvard University's Kennedy School of Government points out, lessons from the effort to slow the loss of the ozone layer can help—yet are being overlooked. The 1987 Montreal Protocol on Substances That Deplete the Ozone Layer, signed in the United States by the Reagan administration and now endorsed by 177 countries, was successful not because of its initial targets but because it was flexible enough to allow industry and government to accept firm goals in the first place, and allowed the world to move forward in a common direction.[81]

Over the course of 15 years, the Montreal Protocol's goals have been revised five times, in accord with expert advice and advances in science and technology. This adaptive approach prompted an intensive effort by private industry to reduce the use of ozone-depleting chemicals and develop substitutes and to identify commercial opportunities for phasing out these chemicals. By including financial incentives and technological support through a Multilateral Fund, the protocol gradually and equitably phased in binding commitments for China, India, and other developing countries. Since 1987, use of ozone-depleting chemicals has declined 90 percent globally at a modest cost. Applying this experience to the climate change process, Parson recommends a focus on providing incentives for industry to innovate, and believes that domestic politics may force the United States to become more engaged on the issue.[82]

A more enlightened U.S. approach appears to be emerging not from the White House but from the Senate, where bipartisan support exists for domestic action and international engagement on the issue. In August 2001, the Senate Foreign Relations Committee unanimously called on the

United States to continue participating in international negotiations in a manner that is consistent with the economic interests of the United States and that includes developing nations—either through the Kyoto agreement or through an alternative binding pact. Congress has also considered several climate technology initiatives, the addition of carbon dioxide to existing clean air legislation, "cap-and-trade" programs for controlling domestic carbon emissions, and "early action" programs to ensure that companies that lower emissions ahead of regulations receive credit for doing so. These initiatives could keep the United States in parallel with Kyoto, and serve as a useful reminder that the climate battle will ultimately be won or lost at home, through implementation of domestic policies. (See also Chapter 8.)[83]

But the international framework remains essential, and the United States appears to have lost its prior influence over nations like Australia, Canada, Japan, and New Zealand in the climate negotiations. Indeed, the U.S. administration's sudden withdrawal created a strong backlash with the foreign ministries of these nations, with the inadvertent effect of pushing them toward agreement with the European Union—leaving the United States isolated. This has been especially true for Japan, for whom salvaging the Kyoto Protocol is an important matter of saving face, and who may finally be emerging as a major player in climate diplomacy. In October 2001, the Japanese government announced that its Diet would seek ratification of the protocol in early 2002; Prime Minister Koizumi has directed his Cabinet to work out the details, while several agencies have been instructed to prepare new domestic policies to implement the pact. Japanese officials plan to continue to work to persuade the United States to return to the agreement.[84]

Climate policy does not operate in a vacuum, and the international and domestic processes are likely to be affected in some manner by the terrorist attacks and signs of economic recession that emerged in late 2001. A shift in focus toward countering terrorism may lessen scrutiny of the global effort to address climate change. Concerns about a recession or energy security, meanwhile, may inhibit attempts to agree on, much less implement, a treaty with uncertain economic implications.

At the same time, there is growing awareness, even in traditional foreign policy circles, that climate change shares characteristics with terrorism: it is a new and looming threat to global security and human well-being of which experts have warned for more than a decade, it requires a response with short-term costs that are worth bearing, and it cuts across borders and thus merits greater international collaboration. Indeed, the breakthrough in Bonn was widely seen as a triumph of multilateralism over unilateralism—an approach since emphasized in the global response to terrorism. The imperative of cooperation has been stressed by Nobel laureate Joseph Stiglitz of Columbia University, who one day after receiving his award for economics in October 2001 called upon all governments to immediately adopt cost-effective climate policies and create an agenda for global collective action.[85]

The Johannesburg Summit provides an extraordinary opportunity to move the climate change agenda forward. Bringing the Kyoto Protocol into force before the Summit would be of critical symbolic importance, signaling to governments, businesses, and civil society that the international process to address one of our most pressing global environmental issues is gaining, not losing, momentum. Many observers doubt whether this will be accomplished. But many also doubted that negotiators would reach agreement in Kyoto, or that the Bonn meeting in 2001 would breathe new life into the process. As with the science, so with the politics of climate change: we can expect surprises in the future.

WORLD SUMMIT PRIORITIES ON CLIMATE CHANGE

- Bring the Kyoto Protocol into force before the World Summit.
- Account for climate change developments in reviewing *Agenda 21* implementation in the areas of atmosphere, energy, finance, industry, and technology.
- Reaffirm the importance of the IPCC Third Assessment Report as the authoritative starting point for policymakers seeking to implement the Kyoto Protocol.
- Set forth a blueprint for post-Johannesburg climate negotiations, emphasizing the need to re-engage the United States, consider a second period of emissions cuts, and expand the group of countries with emissions targets.
- Work to establish a voluntary Global Climate Compact, modeled after the Global Compact established in 2000 between the United Nations and the private sector, that challenges business leaders to commit to accelerated deployment of energy-efficient products, renewable energy, and hydrogen and fuel cell technologies.

Chapter 3

Farming in the Public Interest

Brian Halweil

The fragile hillsides and forest edges of Central America are home to some of the poorest and hungriest people in the western hemisphere. Hybrid corn seeds, synthetic fertilizers, and other technologies that helped raise food production elsewhere are not available or affordable to many people in these communities. Even where these technologies have been used, the results have not always been optimal—they have eliminated local crop varieties, polluted water supplies, depleted the soil, and left families in debt. In the 1970s, food production had stagnated throughout the region. With more and more villagers fleeing for nearby cities, residents of places like Guinope, Honduras, often referred to their home as a "dying town"—a common refrain throughout the region.[1]

By the 1980s and 1990s, however, towns throughout the region had in many ways been reborn. Take San Martin Jilotepeque, a town in the central highlands of Guatemala. In 1972, farmers there began to adopt a series of low-cost innovations to help improve the health of their land, including planting grass hedges to control erosion; rotating corn with beans, peas, and other legumes that add nitrogen to the soil; and covering the ground with vegetation year-round to reduce water and soil loss. Between 1972 and 1979, the amount of corn harvested from the average hectare jumped from 0.4 tons to 2.5 tons, without the use of any chemical fertilizer or pesticides.[2]

The initiative, led by the U.S.-based development organization World Neighbors, was designed not so much to introduce particular technologies as it was to boost the capacity of farmers to innovate, experiment, and "become the protagonists of their own development," according to Roland Bunch, coordinator of the effort. Even after World Neighbors left in 1979, the corn yield nearly doubled again in San Martin, to 4.5 tons by 1994—on a par with the average U.S. yield—because farmers continued to explore better ways to farm. Yields of beans, the other staple crop, grew ninefold—from 170 kilograms per hectare to 1,500—between 1972 and 1994. Villagers adapted many techniques for their own needs, and in many cases developed entirely new systems of production, from crop rotations to cheese-making to organic crop production.[3]

Emigration from San Martin to nearby cities dropped 90 percent as more productive farms paid higher wages and needed more workers. Nutrition, public health, and literacy all improved as additional food and income led to greater investment in health care and education. Soil quality improved, tree plantings increased, and water quality benefited from reduced agrochemical use. As the amount of organic matter in the soils improved, so did resistance to drought, making farmers less susceptible to climate variability—a growing problem, as deforestation in the region has made the rains more erratic. Farmers became more involved in local decisionmaking and civic responsibilities, and many of them got jobs as agricultural extension agents, spreading their knowledge to other communities throughout Central America.[4]

In fact, this experience of healthier soils, higher yields, higher wages, and improved prospects was carried to hundreds of towns throughout the region by thousands of farmers sharing knowledge with other farmers, a movement known as Campesino a Campesino. In Guinope, corn yields increased from 0.6 tons to 2.4 tons between 1981 to 1989, and reached nearly 3 tons by 1994. Elias Zelaya of nearby Pacayas, Honduras, says that "now, no one ever talks of leaving."[5]

Perhaps the most surprising benefit to the region came in 1998—two decades after the project ended—when Hurricane Mitch dumped 2 meters of rain on Central America, wiping out nearly all crops in some areas. Hillsides and the mostly poor people who inhabit them were hit hardest, but the thousands of farms throughout Guatemala, Honduras, and Nicaragua that were touched by Campesino a Campesino withstood the force of the storm much better than others, retaining more topsoil and more of their crop. Landslides were three times more severe—in terms of area affected—on conventional plots.[6]

This experience—in which farms anchor hillsides, store carbon, house biodiversity, provide stable rural incomes, and yield food without expensive or toxic inputs—stands in stark contrast to the type of farming that prevails in much of the world today. As currently practiced, agriculture delivers a great deal of food while it wears down ecosystems, while people go hungry, and while rural communities wither.

The growing costs of this "destructive" food system are pushing farmers, scientists, politicians, and consumers all over the world toward a very different model—what might be called a "regenerative" food system—that better serves the public interest. This new model is also sometimes called "multifunctional" or "agroecological." In this vision, farms function more like self-sufficient ecosystems and depend less on chemical inputs—a shift that will reconcile agriculture's tension with the environment while offering hope for poor farmers around the world. And rather than just driving tractors and spraying ammonia, farmers play an active role in agricultural research and decisionmaking, a change that will help revitalize rural communities. Finally, the food chain will also look different, with consumers buying more food directly from farmers and caring more about the source of their food.

These were the sorts of ambitious goals for agriculture that participants at the Earth Summit in Rio de Janeiro envisioned 10 years ago. Implementation of these goals in the last decade, however, has fallen far short. The upcoming World Summit in Johannesburg offers an important opportunity to regain the momentum. In a few cases, governments are supporting such a shift by closing the wide gap in women's access to agricultural resources, eliminating pesticide subsidies, reducing trade distortions, or using any number of other policies. The synergy between different farming practices or policies fortunately makes the task less daunting. Substitutes for agrochemical inputs, for instance, will not only be good for the ecological performance of agriculture, they will benefit poor farmers who cannot afford expensive inputs. In general, the same policies that will improve farmers' economic prospects, such as more secure land tenure and credit institutions, will also speed the spread of regenerative or ecological farming practices. The successes remain scattered, but they offer principles that can be applied widely to better meet the needs of the land and the people who depend on it.[7]

The Rise of Dysfunctional Farming

There is no doubt that modern farming has demonstrated great capacity to amass mountains of homogenous commodities—evidenced by large increases in production in many parts of the world and falling commodity prices over the past 50 years. (See Figure 3–1.) This is no small achievement. But the same farms have generally wrought a great deal of environmental and social dysfunction. Nations have used food output as the sole measuring stick of agricultural success for so long that it has become difficult to comprehend the price we pay for ignoring all other criteria.[8]

Recently a team of agricultural economists tried to put a price tag on the cost of modern farming in the United Kingdom, and they came up with a conservative figure of more than $2 billion each year. The estimate included the costs of removing pesticides and other agrochemicals from drinking water, the damage from soil erosion, and the medical costs of food poisoning and "mad cow" disease, but it did not include the more than $4 billion of government subsidies paid to farmers or the billions in health care costs due to poor food choices. Still, the figure equaled 90 percent of what British farmers earn each year. The study's lead author, Jules Pretty of the University of Essex, concluded that people in the United Kingdom pay three times for their food: once when they subsidize farmers, a second time when they pay to clean up the mess from polluting farm practices, and again when they buy food at the checkout counter. Costs like these—to ecosystems, rural communities, and society as a whole—are not confined to the United Kingdom.[9]

Many of these costs grew out of the "cheap food policies" of Europe and North America after World War II. (A more accurate name might be a "mass food policy," with an emphasis on quantity rather than quality that generates many hidden costs.) The policies included government support for domestic crop production in an effort not so much to prop up rural communities as to assure affordable food for working-class citizens. This encouraged overproduction by inspiring a single-minded focus on extracting as much crop as possible from a given plot of land, unintentionally marginalizing considerations of how food was produced and who benefited. Several decades later, the developing world embarked on a similar course that focused on production at any cost—the package of improved seeds,

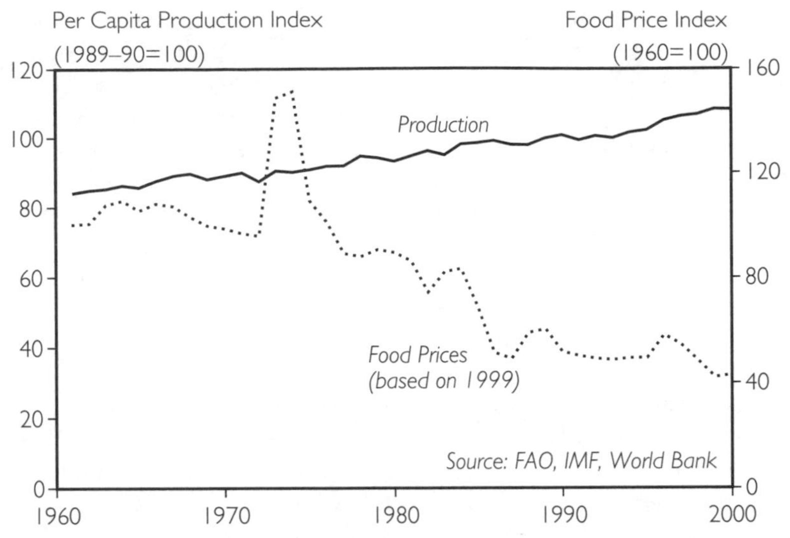

Figure 3–1. Per Capita Food Production and Agricultural Commodity Prices, 1961–2000

agrochemicals, and irrigation known as the Green Revolution. The intentions, as in the North, were good—more food to feed hungry mouths. But they neglected an important fact about hunger around the world: producing more food does not automatically eradicate hunger.[10]

At the same time, farmers gradually began contributing to some of today's most widespread ecological problems—including contamination of waterways, biodiversity decline, the spread of toxic chemicals, and climate change. Ecologist David Tilman at the University of Minnesota and colleagues recently suggested that in coming decades, industrial farming will rival climate change as a source of "massive, irreversible environmental impacts."[11]

That agriculture and the environment always seem to be at odds is more unfortunate than inevitable. From its roots more than 10,000 years ago, agriculture has always represented a transformation of nature. But in search of ever increasing levels of output, contemporary farmers have raised this transformation to a new plane. For example, whereas Chinese farmers were using some 10,000 varieties of wheat in 1949, that number had declined to 1,000 by the 1970s, and to about 300 varieties today. The 14 leading varieties occupy more than 40 percent of China's wheat fields. Of the 7,000 crop species that have been domesticated by humans, a mere 30 species provide an estimated 90 percent of global calorie intake—indeed, wheat, corn, and rice provide more than half—and occupy the vast majority of global crop area, a pattern that leaves farmers and the global food supply vulnerable to erratic weather or pest outbreaks.[12]

Wildlife populations generally decline when farmland replaces forests or other natural ecosystems, but the less diverse the farm, the smaller the homes and food sources for wildlife. A recent assessment noted a marked decline in the diversity of "landscape structures" in industrial nations over the last 50 years as farmers removed stonewalls, hedgerows, grass strips, ponds, windbreaks, and trees to accommodate the machinery used on larger and less diverse plots. The populations of nine species of farmland birds in the United Kingdom fell by more than half between 1970 and 1995.[13]

But monoculture does perhaps its greatest damage when it spills out of the field. Consider the U.S. Midwest, one of the most productive agricultural regions on the planet and a model for farm practices elsewhere. At any given point in time, over 80 percent of the cropland in states like Iowa, Illinois, and Indiana is planted in just two species: corn or soybeans. This necessitates heavy pesticide and fertilizer use, since monocultures invite pests and draw a lot of nutrients out of the soil.[14]

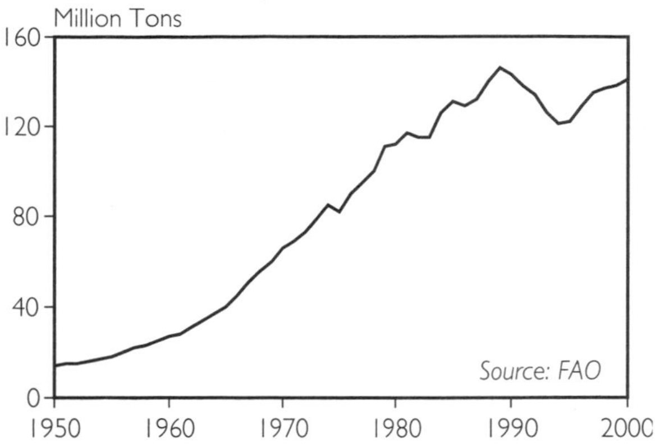

Figure 3–2. World Fertilizer Use, 1950–2000

Locally, chemical use raises levels of nitrates and pesticides (both hazardous to public health) in groundwater and reduces soil quality; several decades of heavy chemical fertilizer use has acidified many midwestern soils, a condition that leaches out key nutrients and compromises the long-term productivity of the region. Worldwide, farmers use 10 times more fertilizer today than in 1950, and spend roughly 17 times as much—adjusted for inflation—on pesticides. (See Figures 3–2 and 3–3.) Yet the effectiveness of these applications has plummeted—a tenfold increase in fertilizer use has coincided with just a threefold increase in food production, while the share of the harvest lost to pests remains largely the same as in 1950 despite the use of much greater quantities of pesticide.[15]

Since large monocultures do not make very efficient use of inputs, nutrient runoff from midwestern farms leaks into the Mississippi and ultimately concentrates in the Gulf of Mexico. Once there, the excess nutrients help to produce algae blooms that suffocate ocean life in a "dead zone" that at times covers over 18,000 square kilometers and has decimated local fisheries.[16]

The Midwest has a long history of this sort of food production, but the same pattern—and the associated problems—is emerging in most of the world's other major agricultural areas. Rivers, lakes, wetlands, and other bodies of water that drain farming regions have become repositories for excess agricultural nutrients, which alter the composition of species in water and on land, favoring some organisms and driving others into extinction. Deadly algae blooms and coral reef destruction related to farm pollution have become common in coastal areas on all continents, and dead zones have emerged in the Baltic and Black Seas that are even larger than the one in the Gulf of Mexico.[17]

Beyond outright pollution, industrial

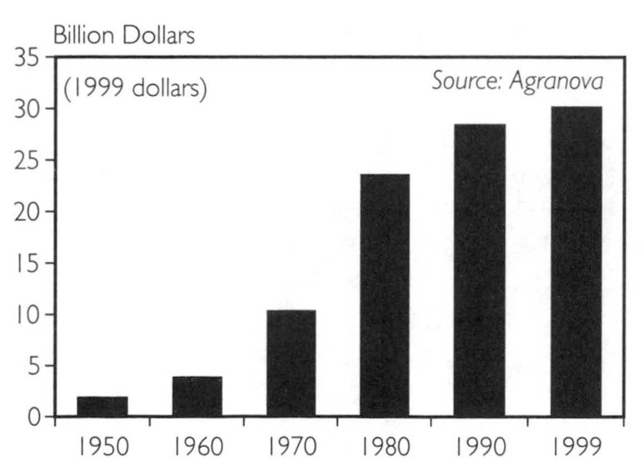

Figure 3–3. Global Pesticide Sales, 1950–99

farming has placed particularly heavy demands on water resources. Much of the growth in food production in the last half-century was built on the expansion of irrigated area, which grew from 100 million hectares in 1950 to 274 million in 1999. Today, the 17 percent of the world's cropland that is irrigated yields more than 40 percent of the world's food. Continued expansion of irrigated area with little regard for water conservation is no longer realistic in most of the world in the face of growing competition with nonfarm demands and mounting concern over the impact of large dams.[18]

Perhaps the strongest evidence that our food system is dysfunctional is the fact that farmers are the poorest people on the planet.

Northern China provides one dramatic case. In recent decades, continuous cropping of irrigated corn and wheat has spread in this breadbasket of the world's most populous nation, putting tremendous pressure on water resources. Today, water tables in northern China are falling 1–1.5 meters each year as farmers pump more water out than is replenished by rainfall. Water deficits are growing in all other irrigated regions as well, including the Indian subcontinent, the western United States, and across North Africa and the Middle East. Nearly 10 percent of the world's grain harvest is now produced by drawing down water supplies, like money from an overdrawn back account. Such overuse also means a drain on the water available to natural ecosystems: water use for irrigation threatens more than half of the nearly 1,000 major wetlands recognized as vitally important by the international community.[19]

Perhaps the strongest evidence of how dysfunctional our food system has become is the fact that farmers, as a group, are the poorest people on the planet.

Of the 1.2 billion people worldwide who earn a dollar a day or less, 75 percent work and live in rural areas. Across a wide range of nations in Africa, Latin America, and Asia, poverty is considerably more prevalent in rural areas than in cities. Even in the United States, the rural poverty rate is 23 percent higher, and in many areas farm families depend on food donations from social services agencies, church pantries, and soup kitchens. Rural indicators of income, health, education, and political participation continue to lag far behind urban indicators. Ironically, hunger is also concentrated in rural areas, worsened by poorer access to a safe water supply and sanitation. Although generally on the decline, hunger persists in much of the developing world for between 800 million and 1.1 billion people. And in sub-Saharan Africa, the share and the absolute number of hungry children have actually increased in the last two decades.[20]

From Brazil to Bangladesh, such dismal prospects have fueled a mass exodus from rural areas as the chance to make a living as a farmer disappears. This shift is not surprising, considering that most of the money in the food business now flows to cities and factories, not the farm. In 1950, for instance, American farmers captured over 50¢ of the average dollar that an American spent on food. By 1997 they were getting just 7¢. The vast majority of the money now goes to food processors, food marketers, and agricultural input suppliers—a pattern mirrored around the world. Part of this has to do with the fact that today people eat more packaged and prepared foods that bear little resemblance to the original crop harvested by the farmer. But it is also due to farming's growing reliance on expensive inputs and machinery, as well as the rise of agribusiness "cartels" that leave little room for farmers to make a profit. Global integration of food production has intensified these economic pressures that are fueling the exodus from the countryside.[21]

Hunger Amidst Plenty

In 1996, delegates at the World Food Summit in Rome committed to cutting world hunger in half by 2015—a significant retreat from a 1974 goal to eradicate hunger within a decade. In 2001, the U.N. Food and Agriculture Organization (FAO) declared that at the current pace, even the less ambitious goal would not be reached for more than 60 years, too late for many of the world's poor.[22]

How can a sizable chunk of the world remain hungry when food production soared in the last half-century, modestly outpacing population growth? The growth in global food production is in some ways a less important consideration than our capacity to boost yields where food is needed most, particularly where conventional farming techniques have failed.

Many of the world's hungry people have been unable to plug into the standard approach to raising food production—genetically uniform fields supported by chemical cocktails—because it was too expensive or inappropriate for local conditions, or because of a lack of land, market access, and other constraints. Since existing tools have not worked for these people, conventional agriculture has largely given up on them and assumed they will be better off moving to a city or doing something besides farming. It is estimated that almost 1.8 billion people in developing nations live in forests and woodlands, arid regions, steeply sloping hillsides, or other lands unsuitable for modern food production. These "marginal" or "less-favored" areas—the Sahel of Africa, the hills of the Andes, or the rainforests of Indonesia—now house the bulk of the rural poor and the world's hungry.[23]

Some agricultural scientists hope to find a solution to this problem at the genetic level—engineering crops to thrive in a wider range of environments—although to date the technology has proved largely irrelevant to the needs of the world's hungry. (See Box 3–1.) "Throughout the world, poor farmers are seldom limited by the genetic potential of the crop," notes Roland Bunch, who is now with COSECHA, an agricultural consulting group based in Honduras. A given corn variety might yield 5 tons per hectare under ideal conditions, but the same variety planted in depleted and drought-prone soils might yield less than 1 ton per hectare. According to Bunch, who has worked for decades with farmers in Africa, Latin America, and Asia, "ecological conditions, like soil fertility and water availability, are their major constraints"—constraints that cannot be readily overcome by genetic improvement, whether through biotechnology or more traditional means. To help the poor farmer, innovations must reduce these ecological constraints with low out-of-pocket costs, while building the resilience and stability of production and allowing sufficient flexibility so that they can be used in diverse ecological settings.[24]

The big gains for these farmers will come from taking advantage of "free" biological services, including nitrogen-fixing (or leguminous) plants such as beans or clover, the nutrient cycling abilities of soil microbes, and beneficial insects—an approach now widely referred to as agroecology. In many ways, this is the most sophisticated approach to farming because it depends on an intimate understanding of ecological interactions in the farm landscape. The best use of local resources and local knowledge substitutes for chemicals and techno-fixes. Instead of a package of inputs that is deployed in the same way everywhere, an agroecological approach depends on principles whose specific application varies by site.[25]

The importance of this approach for poor farmers has been confirmed by a recent sur-

BOX 3–1. A BIOTECH FIX FOR HUNGER?

Have you heard of Golden Rice? It's the yellow-tinted strain of a staple food that has been genetically engineered to contain beta carotene, and that could be a blessing for the hundreds of millions of people in the developing world who lack enough vitamin A to lead healthy lives. (Beta carotene is the precursor of vitamin A, an especially important nutrient for children.) Worldwide, 100–140 million children suffer some degree of vitamin A deficiency, a condition that can suppress the immune system, cause blindness, and, in extreme cases, even kill.

Unfortunately, the average person would have to consume an unreasonable amount of Golden Rice every day—some 9 kilograms of cooked rice, 12 times the normal intake—to get the necessary vitamins. Some nutritionists argue that it would make more sense to help poor people grow green vegetables, which produce more beta-carotene than Golden Rice does as well as various other nutrients completely lacking in rice, Golden or otherwise. Moreover, beta carotene can only be converted to vitamin A in the body of an already well nourished person. (Body fat and some other nutrients are necessary in the reaction from beta carotene to vitamin A.) Geneticist Richard Lewontin notes that "the developers of Golden Rice have not dealt with this problem in their publicity releases."

The much-touted promise and the sad reality of Golden Rice mirror the broader discussion of the role that genetic engineering may play in eradicating hunger. There is no doubt that biotechnology is an extremely powerful tool that holds some real potential for agriculture. But its current emphasis bears little relevance to the needs of poor farmers and the world's hungry.

The United States, Canada, and Argentina contain 98 percent of the global area of genetically engineered (or transgenic) crops. The biotechnology industry has funneled the vast majority of its investments into crops and traits designed for large-scale, mechanized farms of the First World—soybeans engineered to tolerate herbicide spraying or corn that churns out its own insecticides. This is not a big surprise, considering the technology is largely controlled by the private sector and defined by a landscape of patents and other proprietary obstacles. A report from the U.N. Development Programme recently acknowledged this commercial reality, but clung to the hope that biotech could play a large role if only given the chance.

If there is a role for biotechnology in improving the way we farm and in reducing hunger, it may be as an informational rather than an engineering tool. The ability to map and study the genetic code of agricultural plants—the field called genomics—can greatly enhance traditional breeding or improve our understanding of how plants respond to drought and disease. This role for biotechnology may ultimately prove less risky—and more palatable—than swapping genes between wholly unrelated species.

SOURCE: See endnote 24.

vey by University of Essex researchers Jules Pretty and Rachel Hine of over 200 agricultural projects in the developing world that depended on ecological approaches. They found that for all the projects—9 million farms, covering nearly 30 million hectares—yields increased an average of 93 percent, and substantially more in some cases. Most important, a majority of these projects succeeded in boosting production under adverse conditions, in marginal areas where everything else had failed.[26]

One particularly useful principle for raising production in these areas is the use of leguminous crops to boost soil fertility. Whereas First World farms face nutrient overload, nutrient shortages plague Third World farms. Annual rates of depletion for the principal plant nutrients (nitrogen, phosphorus, and potassium) range from 40–60 kilograms per hectare in Latin America to well above 60 kilograms per hectare in parts of Africa. In East Africa, an estimated 50,000 farmers who cannot afford chemical fertilizers are sowing several different leguminous tree crops (such as *sesbania* or *tephrosia* trees) during the fallow season as a way to boost the yield of the subsequent crop. Such "improved fallows" can often boost corn yields two- to fourfold in the following season, while also reducing pest pressures, yielding fuelwood and animal fodder, and improving soil health. The improved fallow system also lends itself to local adaptation—farmers can grow trees for differing lengths of time or can plant them with other crops—increasing the likelihood of success in a wide range of circumstances.[27]

Some might argue that doubling or tripling yields is less impressive when yields are starting from the low levels found in much of the developing world. But an ecological approach may in some cases hold significant untapped productivity that has been obscured by prevailing farm practices. Consider sustainable rice intensification (SRI), which confounds traditional rice growing in a number of ways. Rather than grow rice plants in clumps in flooded fields, SRI transplants seedlings at a much younger age, spaces individual plants widely, periodically waters the field, and aerates the soil throughout the season. These relatively simple changes mean that the plants develop much more extensive root systems, with additional strength to withstand drought and disease; flooding, it turns out, can suffocate and stunt the roots. Typical yields for SRI, which has been used by thousands of farmers in all major rice-growing regions, are 6–10 tons per hectare, several times the 2-ton average for rice grown in much of the world.[28]

Where the extra labor needed for certain agroecological techniques is not available, adoption of these techniques can be slowed. In certain parts of Asia, for instance, farmers have abandoned SRI techniques for this reason, despite the large potential increases in yields and profits. In such cases, farmers can emphasize innovations that require the least labor (adding legumes to the rotation, for instance) over practices that require more labor, such as regularly aerating the soil.[29]

Perhaps the most important testing ground for any attempt to eradicate hunger is so-called rain-fed areas—the agricultural equivalent of inner cities. These arid regions without irrigation are home to a disproportionate share of the world's hungry, and nearly half of the projects in the University of Essex survey took place in these areas. One project focused on boosting food security in the Sahelian countries of Africa—a region that includes Ethiopia, Mali, Niger, Senegal, and Somalia and that is characterized by erratic rainfall, low natural soil fertility, and high rates of desertification. The Sahel is also one of the most entrenched pockets of hunger on the planet: more than half the children in several nations are chronically malnourished. Nearly 16,000 farmers on 26,000 hectares used a combination of measures to check erosion and boost the fertility and water-holding capacity of the soil, which resulted in a sustained tripling of both millet and peanut yields compared with control farms. In drought years—when hunger tends to deepen—these practices also resulted in less severe and less frequent crop failures. In general, agroecological systems exhibit more stable levels of productivity over time than chemical-intensive systems—a sort of risk management that results from strengthening the ecological infrastructure of the farm.[30]

Since mainstream agricultural research has tended to neglect these arid areas—the conventional approach has been to focus on irrigated areas because they have generally offered more stable and higher crop production—many of the relevant strategies will have to grow out of local innovation. On at least 100,000 hectares in Niger and Burkina Faso, a farmer innovation called *tassas* (or *zaï* holes) has tripled yields on land that has generally been considered too infertile, dry, and cracked for most agricultural endeavors. *Tassas* are small pits dug in the soil, filled with manure, and then planted once they fill with rain. Households using this technique have shifted from not having enough food for half the year to producing a surplus of 153 kilograms annually. These small schemes cost less than large-scale irrigation projects, are easier to manage from the bottom up, build on traditional knowledge of climate and hydrology, and are often the only ways for the very poor to get irrigation.[31]

Clearly, for this sort of farming to thrive, farmers will need to control resource use and other decisions in a way that has not always been common. A lack of such involvement is one of the main reasons that many low-cost but high-yielding farming systems do not flourish in the first place, particularly since the success of any ecological farming technique depends on location-specific knowledge and adaptation.

For instance, improving the availability of water often depends on greater involvement of farmers groups and cooperation between farmers. In Sri Lanka, a pilot program began in 1981 to improve water management in the irrigation scheme of Gal Oya—the largest and most disorganized system in the nation—by giving local farmers organizations control over the timing and distribution of water releases. These managerial changes alone doubled water use efficiency, so that twice as much cropland could be served with available water. In combination with improvements in agricultural practices, the amount of rice produced per cubic meter of water released from the reservoir quadrupled. Based on these results, the government decided in 1988 to hand management of irrigation systems nationwide over to local water users groups.[32]

More recently, during a drought in 1997, when the government considered suspending rice production completely, Gal Oya farmers organizations were given the opportunity to proceed with the small amount of water available. They were able to cultivate the whole area, and harvested an average to better-than-average crop. This is the sort of increase in water productivity that Sandra Postel of the Global Water Policy Project says will be required to cope with limited world freshwater supplies in coming decades.[33]

Perhaps no other group commands as little control over agricultural decisions as women. In the developing world, women tend most of the fields, plant most of the seed, pull most of the weeds, haul most of the water for crops and family, harvest most of the food, and then cook most of it. Their role as nutritional gatekeeper has swelled as more men migrate to towns and cities—nearly 40 percent of households in rural India, for example, are headed by women. Yet rural development programs consistently overlook women, targeting extension, credit, and other services at men. Women own just 2 percent of land worldwide; even where they do, their ability to use it is hampered by limited access to agricultural infrastructure, credit, and extension. In Kenya, Malawi, Sierra Leone, Zambia, and Zimbabwe, for example, women do most of the farming, yet receive less than 10 percent of the credit awarded to smallholders and just 1 percent of the farm credit overall.[34]

Such discrimination means women are less able or willing to invest in the land. Liz Alden Wily, a political scientist in East Africa, argues that lack of secure landownership rights for women is the most significant obstacle to reducing hunger and poverty in sub-Saharan Africa. Yields of maize, beans, and cowpeas could be increased by over 20 percent in that region by giving women equal control over agricultural inputs and equal access to extension services. In Kenya, a new weeding technique raised crop yields by 56 percent on women's plots when the women controlled the output, but by only 15 percent on men's plots when women weeded but the proceeds went to men.[35]

Women are not the only marginalized group in this respect. In most nations, an elite minority owns most of the farmland, and largely determines how that land is used. Roughly 100 million farm families—about 500 million people—lack ownership or owner-like rights to the land they cultivate, including a near majority of agricultural populations in South and Southeast Asia, Central and South America, and Southern and Eastern Africa. (See Table 3–1.) They have little incentive to build up their soils, plant tree crops, or adopt many of the other agroecological techniques that require long-term investment, even when such practices may be in their best interest. In the case of improved fallows in East Africa, the poorest farmers have difficulty changing to the new system because they lack credit services and they need some cash flow until the fallow begins to pays off; obtaining credit often depends on owning some land for collateral.[36]

Rural development programs consistently overlook women, targeting extension, credit, and other services at men.

The participation of farmers in agricultural research can often make the difference between success and failure in reducing hunger. This requires building the capacity of local people to experiment, innovate, and better understand their ecological surroundings—a radical shift for many agricultural institutions, which still tend to regard farmers as a relatively marginal part of the agricultural R&D machine. The irony, according to agricultural sociologist Ann Waters-Bayer, who recently surveyed farmer innovations in sub-Saharan Africa, is that often the best place to look for solutions to the problems faced by farmers is in the fields of neighboring farmers who have been wrestling with the same problems for years.[37]

As noted earlier, in San Martin in Guatemala, crop yields continued to

Table 3–1. Land Distribution in Selected Countries and Worldwide

Country	Description
Zimbabwe	Some 70,000 whites (0.5 percent of the population) own 70 percent of the land; 4,000 whites own nearly one third of the farmland.
South Africa	Blacks, who account for 75 percent of the population, occupy 15 percent of the land.
Namibia	Some 4,000 whites (less than 1 percent of the population) own 44 percent of the territory.
Brazil	Just 3 percent of the population owns two thirds of the land.
India	Some 9 percent of the farm population owns 44 percent of the agricultural land.
United States	Only 16 percent of farmers control 56 percent of all the land.
Worldwide	In 28 of 44 nations surveyed by the International Labour Organisation, the top 10 percent of landowners controlled over 40 percent of the land.

SOURCE: See endnote 36.

increase dramatically long after World Neighbors staff left, and some 80–90 different agricultural innovations were documented in the villages studied, including two new nitrogen-fixing crops, two new species of grass used for erosion barriers, marigolds used for control of parasitic worms, and homemade sprinklers for irrigation. In Latin America, only 30 percent of households who have participated in farmer field schools still suffer food shortages, compared with 50–65 percent of their neighbors. Rural residents who are better prepared to cope with shifting conditions—whether climate change or global market change—are better able to feed themselves.[38]

The Nature of Farming

As agriculture has abandoned much of its original ecological complexity, it has become more of a drain on the global environment—worsening rather than lessening floods, emitting rather than storing carbon in soils, and destroying rather than hosting biodiversity. Since agriculture occupies such a large share of the world's land area—nearly 40 percent worldwide, and at least half in large nations like the United States and India—farming to build up rather than erode ecosystems offers widespread environmental benefits.[39]

Consider the relatively simple technique of planting two varieties of the same crop in one field. In China, when farmers have replaced the standard monoculture with two rice varieties, pest pressures have plummeted, allowing elimination of pesticide use. (A side benefit was a 20-percent boost in total yield, since more diversity allows greater use of field niches.) In settings like the U.S. Midwest, where single-species fields that stretch for kilometers are the norm, the addition of a winter rye crop to the normal corn and soy rotation cut the amount of nitrogen leaking from fields by at least half, with huge benefits to water quality. Farms that rely more on ecological processes within the field and less on chemical inputs will themselves begin to function more like the wetlands, forests, and grasslands they replaced—with the added benefit of producing food.[40]

Trees are one element that has generally not fit the stereotypical "modern" farm landscape. But by reintroducing trees and other perennials, farmers can reduce erosion, sequester carbon, retain water, and generally buffer agriculture against ecological extremes that accompany climate instability. Tree planting can also form part of the strategy to combat salinization, the dominant form of land degradation in irrigated regions. (Trees improve soil drainage and prevent water from pooling near the soil surface, where it evaporates and leaves salt residues.) In Algeria, the government has decided to convert a large chunk of its grainland—starting with 15 percent and eventually growing to 70 percent—to tree crops in an effort to stop the spread of the Sahara Desert and reduce the nation's considerable salinization problem.[41]

Reintegrating trees into the farm landscape is one part of a broader strategy to use agriculture to help conserve biodiversity. "Many people believe that biodiversity can be preserved simply by fencing it off," said World Conservation Union–IUCN scientist Jeffrey McNeely. "But agriculture and biodiversity are inextricably linked." Almost half of the areas currently protected for biodiversity are in regions where agriculture is a major land use. To avert widespread extinctions, a recent IUCN report recommends that farmers create spaces where wildlife can thrive in and around farms. Letting part of the farm go wild often has the added benefit of boosting production, as when grass hedges provide fodder for livestock or habitat for pollinators. In the Philippines and Indonesia, fishing communities have banned fishing in "no-take" reserves to provide breeding sanctuaries where fish populations can recover. A survey of the reserves found that in the first three years after they were established, fish number, size, and diversity in the surrounding areas all increased dramatically.[42]

For the farmer, improving environmental performance and raising the bottom line will often intersect. Consider the potential of "no-till" farming, which can simultaneously reduce costs, boost profits, and protect agriculture's most basic foundation—soil. The approach involves planting seeds in the stubble of the previous crop rather than plowing the soil each season, which can accelerate erosion. The growth of this technique in Latin America has been phenomenal: farmers are using no-till on 11 million hectares in Brazil, up from 1 million in 1991, and on 9.2 million hectares in Argentina, up from 100,000 hectares in 1990. In the Brazilian state of Paraná, where half of cultivated land uses no-till, costs for weeding, tillage, herbicides, and fertilizers have dropped dramatically, boosting profits by nearly $200 per hectare. And the technique has cut soil erosion by 90 percent, greatly reduced water pollution, and boosted soil organic matter—the form in which soils store carbon—pointing to a role for better soil management in efforts to mitigate climate change. (See Box 3–2.)[43]

Reducing agrochemical use and farm pollution will also be essential to biodiversity preservation on the farm. A recent report from the Soil Association in the United Kingdom tallied the findings of 23 comparative studies of the biodiversity benefits of organic and conventional farming. On the organic farms, it found:

- substantially greater levels of both abundance and diversity of species, including five times as many wild plants and several rare and declining species;
- 25 percent more birds at the field edge and 44 percent more in the field in the fall and winter;
- 1.6 times as many of the bugs that birds eat;
- three times as many non-pest butterflies;
- one to five times as many spiders; and

BOX 3–2. FARMERS BATTLE CLIMATE CHANGE

Last June it was not rainclouds that darkened the skies over Tintic Junction, Utah, but hordes of "Mormon crickets" eating their way across 600,000 hectares of farmland and more than $25 million of produce. That same month, provinces across China were hit by the worst locust plague in years, reaching a peak density of 3,000–10,000 insects per square meter in some areas. As in Utah, the affected areas in China had experienced warmer winter months than usual, followed by a prolonged dry spell, creating the perfect conditions for insects to breed and destroy cropland.

Farming is vulnerable to many of the spasms that are likely to accompany a changing climate. Yet according to FAO, local disasters such as hurricanes, flooding, or massive crop infestations pose less of a threat to food production than the steady changes in rainfall patterns and regional temperatures.

Fortunately, the same practices that help farms adapt to climate variations are the most potent weapons for mitigating the effects of climate change and reducing greenhouse gas emissions. For instance, building up a soil's stock of organic matter—the dark, spongy material that gives soils their rich smell—not only increases the amount of water the soil can hold (good for weathering droughts), it helps bind more nutrients (good for crop growth). Organic matter is also the form in which our soils store carbon dioxide—the principal greenhouse gas.

Whereas well-managed soils in temperate regions can accumulate 100 kilograms of carbon per hectare a year, and in the tropics 200–300 kilograms, farms planting green manures or using no-till methods can accumulate up to 1,000 kilograms of carbon in a year. Farms with trees planted strategically between crops will better withstand torrential downpours and parching droughts and will "lock up" even more carbon; the improved fallows being used in Africa typically store three times as much carbon as nearby croplands or grasslands do.

The systems that store more carbon are often considerably more profitable, and they might become even more so if farmers get paid to store carbon under the international climate treaty. In Chiapas, Mexico, farmers are already paid to shift from systems that involve regular forest clearing to agroforestry. The International Federation of Automobiles is funding the project as part of its commitment to reducing carbon emissions from sponsored sports car races.

SOURCE: See endnote 43.

- dramatic increases in life in the soil, including earthworms.[44]

The organic systems surveyed held the biodiversity advantage by including more diverse crop rotations, the year-round presence of ground cover, greater habitat (hedgerows, trees, wild vegetation) at field boundaries, no use of agrochemicals, and use of green manuring (leguminous crops that are worked into soil)—all practices that have been progressively abandoned as agriculture has industrialized over the last century. Moreover, the authors concluded that the spread of organic farming in the United Kingdom was an essential component of any attempt to reverse the well-documented decline in Britain's farmland wildlife, and could deliver much better results than other government wildlife conservation programs.[45]

Financial mechanisms can be among the most powerful drivers for changing the way we farm. Yet governments have rarely penalized farm pollution. Pesticides, chemical fertilizers, and animal feedlots are just a few things that might be taxed. Denmark, Norway, and Sweden already have substantial taxes on pesticides, with the goal of cutting usage by 25–50 percent in coming years, and the Netherlands taxes farms that generate excess manure. Dave Brubaker of the Center for a Livable Future at the Johns Hopkins University suggests a tax on industrial animal production, now the fastest-growing form of meat production worldwide and a major source of water pollution. Such a tax not only would make it more expensive to set up and run factory farms, and therefore make grazing and organic meat production more competitive, but it would also help slow the loss of smaller farms because the tax would be based on herd size.[46]

Combining financial incentives with education about reducing agrochemical use is also powerful. After attending farmer field schools on insect ecology and non-chemical pest control (and after higher taxes on certain pesticides), 2 million farmers in Viet Nam cut pesticide applications from 3.4 to just 1 per season. And following a two-year campaign to explain to rice farmers that spraying during the first 40 days after sowing is unnecessary because insect damage during this stage rarely reduces yields, nearly 80 percent of Mekong farmers had stopped early spraying and 30 percent now farm rice entirely without pesticides—with no drop in yields.[47]

Today, however, most agricultural policy acts as a powerful disincentive against shifting to cleaner methods of food production. A case in point is the more than $320 billion that governments of industrial nations spend each year to support agriculture. The lion's share of these subsidies are tied to the production of a handful of commodities—such as corn, soybeans, and beef. This arrangement helped create the less diverse system in the first place and inhibits the adoption of resource-conserving practices by making them less profitable. Farmers interested in diversifying out of the handful of crops that receive payments lose a significant source of income.[48]

But there is huge potential to use this money more creatively. Recent food safety crises in Europe and the United States, by exposing the public health fallout of current farm practices, have pushed the political momentum toward redirecting production payments to "green" or "stewardship" payments, which would support farmers who meet certain ecological goals. An important side effect of decoupling these payments from production of a specific commodity would be the boost given to rural communities, since the current structure funnels the vast majority of funds to the largest and most well off farms. But the powerful commodity lobbies—including trading and processing firms that reap benefits in the form of lower commodity prices—are not likely to take any loss of income lightly, and represent one of the strongest barriers to subsidy reform.[49]

Although most industrial nations more than doubled public expenditures on agricultural conservation programs between 1993 and 1998, these payments still represent just 2 percent of total agricultural budgets in these nations, a meager counterweight to the massive commodity payments that perpetuate dysfunctional farming. Moreover, prevailing conservation payments focus mainly on marginal lands or the edges of fields, not on land in production. For example, 85 percent of U.S. conservation payments, including the Conservation Reserve Program that pays farmers to protect erosion-prone lands, are for lands not in production.

Although beneficial, these programs do not affect farmers' practices on the majority of their land.[50]

In other words, "greening the edges" will not be sufficient to restore biodiversity in farm fields and to reduce most farm-related pollution, since "the middle" of the farm holds most of the potential for both negative and positive ecological impacts. Some agricultural economists have suggested that any payments to farmers should depend on a basic level of ecological compliance and that farmers who go beyond the minimum should receive more money. Nations can already do this without making major changes to existing policy, but generally they do not. For instance, European states can deny subsidies to farms that do not comply with environmental requirements, although only a few have done so. And the existing U.S. Farm Bill provides for several conservation programs that are perennially underfunded. Still, France is considering shifting 20 percent—up from just a few percent today—of all direct payments to farmers toward rural development and ecological farming programs in coming years. This would support France's Contrats Territoriales d'Exploitation (land management agreements), a new grassroots program that involves rural communities in deciding what changes to farming practices will most benefit local environmental needs but also farm profitability.[51]

Whatever the policy change, new incentives can elicit a dramatic response. Consider the effects of broad government support for organic farming in the European Union. Over 80 percent of the explosive growth in organic area there has occurred in the last six years, spurred by the 1993 Union-wide policy to support farmers in the first years of conversion from conventional to organic production. (See Figure 3–4.) Conversion has been highest in the nations with the highest transfer payments per hectare—Austria and Switzerland, where roughly 10 percent of the area is organic today.[52]

Without transition payments, many farmers may be unable to afford to shift to a different farm system: for many ecological farming practices, productivity and profitability are likely to drop for awhile as the ecological infrastructure of the farm (soil quality or insect predator populations, for instance) and the farmer's expertise improve. For example, where decades of pesticide use have wiped out a farm's insect populations, it may take several years to rebuild the diversity that helps control pests naturally once a farmer decides to move toward nonchemical pest control. Even though studies show that losses are often recouped by greater returns after the transition period, farmers generally see cost as a chief reason not to change. The transition is also complicated by the inertia of university professors and researchers, agricultural

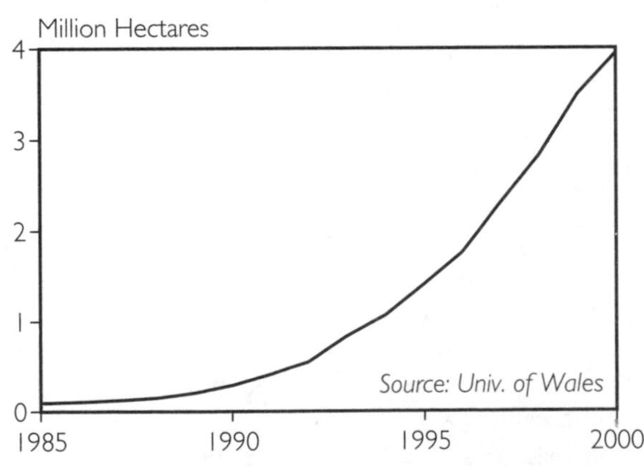

Figure 3–4. Certified Organic and In-conversion Land in European Union, 1985–2000

extension agents, and government agricultural officials, who are all often quite unfamiliar with new modes of farming.[53]

Why Care About Rural Areas?

Because the farm sector is such a small share of the economy in wealthy nations, it is easy to think that governments can ignore rural areas. Michael Lipton of the Poverty Research Unit at the University of Sussex describes a contradiction between the rhetoric of poverty reduction among international lenders, like the World Bank and aid agencies, and the large-scale neglect of rural areas—home to most of the world's poor. International aid to agriculture has declined two thirds in real terms since the 1980s; rural investments represented less than 10 percent of World Bank commitments in 2000.[54]

Such reductions send a worrisome signal to governments in Africa, Asia, and Latin America, who have already cut spending on education, credit, marketing assistance, and other essential support services in rural areas, partly as a result of austerity measures encouraged by international lenders like the International Monetary Fund and the World Bank. "As public commodity procurement boards—along with the provision of rural credit and extension and rural infrastructure maintenance—are privatized as part of structural adjustment policies," notes Rafael Mariano, chairman of a Filipino farmers union, "the new entities are under no obligation to service marginal rural areas and often result in even spottier coverage than the institutions they were intended to replace."[55]

Rampant consolidation at all layers of the food chain has further squeezed the role of farmers in the economy. (See Table 3–2.) In Canada, for example, just three companies control over 70 percent of fertilizer sales, five banks provide the vast majority of agricultural credit, two companies control over 70 percent of beef packing, four companies mill 80 percent of the wheat, and five companies dominate food retailing—a situation that means farmers pay more for inputs and get paid less for what they harvest. Although industry analysts often argue that such consolidation is necessary to deliver affordable food, it reduces choice and ultimately allows a few companies to control prices. To date, few nations have shown interest in enforcing existing antitrust laws in agriculture, with even less hope of such action at the global level.[56]

But the argument for investing in rural areas is quite strong, since rural people still constitute the majority of the population in developing nations. Moreover, rural investment generally spills over to the rest of the economy and to urban areas, becoming an indispensable engine of economic growth (not to mention the growing and more explicit role for farmers in taking care of the countryside). Rising farm production and farm incomes provide the base for growing urban industrial centers, which gradually draw people out of agriculture—the history of the industrial world over the last century or so. In West Africa, for example, each $1 of new farm income yields income increases in the local economy ranging from $1.96 in Niger to $2.88 in Burkina Faso. And the growing prosperity of millions of small farms in Japan, South Korea, and Taiwan following World War II is widely cited as the major stimulus to the dramatic economic boom those countries enjoyed. In contrast, many farmers today—far from choosing to move to the city—are driven off the land by desperate economic circumstances.[57]

Table 3–2. Concentration in Various Layers of Agribusiness

Business Sector	Description
Agrochemicals	Five companies control 65 percent of the global pesticide market.
Seeds	The top 10 seed firms control 30 percent of the global seed market; five companies control 75 percent of the global vegetable seed market.
Trade	The top five grain trading enterprises control more than 75 percent of the world market for cereals. A handful of transnational companies control about 90 percent of the global trade in coffee, cocoa, and pineapples; about 80 percent of the tea trade; 70 percent of the banana market; and more than 60 percent of the sugar trade.
Meat	One firm controls 60 percent of chicken purchases in Central America. In the United States, four companies control over 80 percent of beef packing, and five companies pack 75 percent of the pork.
Retail	Five retailers control 50 percent or more of all food purchases in France, Germany, and the United Kingdom; two firms control over 80 percent of Hong Kong's retail market; between 1994 and 1999, the share of the retail sector in Brazil controlled by the top 10 supermarkets grew from 23 percent to 44 percent.

SOURCE: See endnote 56.

Prosperous rural areas can thus take pressure off of urban infrastructure by reducing migration to cities. Research in Brazil has found that in just one month the total cost to the state of maintaining a given person in an urban shantytown, including water services, electricity, and infrastructure, can exceed the yearly cost of helping landless laborers get established on idle farmland. As a result, urban welfare groups there have joined with farmers, unions, and environmentalists to support the Landless Workers Movement, a grassroots coalition that pushes for land reform, as an alternative to the growth of city slums.[58]

Ecological farming systems might offer an even better alternative, since they generally require greater management and labor, usually a plus for rural communities. (No-till farming is a notable exception; it reduces labor needs, although the farmer will spend more time orchestrating the diverse rotation that helps keep weeds down once plowing stops.) In the Indian states of Maharashtra, Gujurat, and Tamil Nadu, since the introduction of water storage tanks, widespread tree planting, and other measures to boost water conservation, seasonal migration out of rural areas has declined sharply, as enough water is now available to farm in the dry season.[59]

Land reform has fallen off the development radar in recent years, even though lack of land rights is still one of the dominant constraints to improving rural lives. Analyses of poverty trends in India between 1958 and 1992 have shown that poverty fell the most in the states that implemented more land reform. China's move from collectivized management of farmland to relatively equitable household responsibility between 1977 and 1985 resulted in tremendous gains in food production, moving a huge chunk of the world's rural poor out of poverty. Especially where land is scarce or widespread redistribution is unlikely, even small amounts of land for a garden or home can mean better family nutrition, higher income and status, and access to credit.[60]

Collective action among farmers in the form of cooperatives, unions, or research collectives can be particularly important to bolstering power in the food chain. Economist Bina Agarwal notes that throughout South Asia, forming groups has been an essential strategy to boost the power of women in the struggle for land rights. Farmers cooperatives in Mozambique enjoy greater market access, improved transportation of produce, and better prices, including 22 percent more for maize and 93 percent more for groundnut than paid to individual farmers. In general, these groups help farmers take back some of the profit currently captured by the rest of agribusiness.[61]

In other cases, the central challenge for farmers will be finding new market opportunities, a pursuit that remains largely neglected. A recent survey in the developing world found that just 12–15 percent of agricultural projects included some focus on either marketing or increasing the value of the farm product through processing. (As with insecure land rights, farmers are much less likely to invest in their farms without access to markets.) The Association for Better Land Husbandry in Kenya has developed the Farmer's Own brand, which markets energy bars, cooking sauces, and other food items made from locally produced crops as a way to get farmers a higher price for their harvest. Among the most helpful initiatives would be basic credit and storage facilities that allow farmers to wait for the best time to sell their harvest and thus capture some of the windfall profits that usually go to local merchants, lenders, and brokers. Cash-strapped farmers often need to sell their produce at harvest time, when a market glut means that prices have bottomed out, and then buy the same crop back for their own consumption later in the season, when prices might be several times higher.[62]

The exact opportunities will vary with location, but the ability of a wide base of the rural population to make more money on or off the farm depends partly on closing the gap between rural and urban areas in schooling, literacy, health care, and other basic services. A recent analysis showed that of six types of public investment in rural areas in China—education, agricultural research, roads, telephones, electricity, and irrigation—education has the greatest impact on reducing poverty. Extra investment in rural areas generally enhances welfare more than it would in urban areas, since rural communities start from a much lower level of service; for example, an extra year of schooling for an urban child will likely mean costly college education, whereas for a rural child it means grammar school.[63]

Lack of land rights is still one of the dominant constraints to improving rural lives.

Many government officials and development economists view trade as an essential component of reducing rural poverty. Yet in most nations, market liberalization has tended to benefit larger farmers and agribusiness companies and to widen inequalities between these people and small, poor farmers. The United Nations surveyed 16 developing nations implementing the last phase of the General Agreement on Tariffs and Trade and concluded that "a common reported concern was with a general trend towards the concentration of farms," a process that tends to exacerbate rural poverty and unemployment. During the first seven years of the North American

Free Trade Agreement (NAFTA), all three participating nations saw commodity prices and farmer incomes plummet, as the companies that trade and process agricultural commodities reaped windfall profits. As farmers depend on markets that are farther and farther away, moving, storing, processing, and brokering of food begins to assume greater importance than production.[64]

Current international trade agreements actually restrict the ability of nations to protect and build domestic farm economies, forbidding domestic price support and tariffs on imported goods. (Politically powerful industrial nations have, nonetheless, boosted their own barriers to trade in recent years.) At the same time, these agreements leave considerable wiggle room on other forms of trade distortion, including the ability of wealthy nations to dump subsidized crops on the world market well below the cost of production—an economic weapon that can squash local food production by driving prices down and actually worsen poverty among those who depend on agriculture for their income. Michael Widfuhr of FIAN, an international hunger rights group, argues that trade agreements must create some space for nations to pursue domestic goals of eradicating hunger, maintaining a base of family farmers, or striving for some level of self-sufficiency—"a trading system where food sovereignty is the priority and fair trade prevails."[65]

Ethical Eating

"Eating is an agricultural act" is how farmer-poet Wendell Berry explains the fact that how we eat determines to a large extent how we farm. For the average eater, this implies a new identity—from a relatively apathetic purchaser to an active critic of the food system, ever curious about the origins and history of food. The depths of consumer preference have often been limited to subtleties of packaging, color, or flavor, but a new generation of eaters seems to hold much higher expectations for their food system. As one small example of emerging consumer power, the recent decision by Monsanto to permanently discontinue its genetically engineered Bt-potatoes was not prompted by a corporate change of heart, but rather by consumer and environmentalist pressure put on McDonald's and Frito-Lay, the major U.S. purchasers of potatoes.[66]

Interest in taking an active role in the food system will grow as consumers begin to understand their personal stakes in various types of farming. A series of well-publicized food safety crises—from the on-going mad cow crisis to the recent foot-and-mouth outbreak—has made this abundantly clear to Europeans in recent years. By undermining consumer confidence, these events opened the door to greater support for organic farming, regional food self-sufficiency, and pressure to shift the massive Common Agricultural Policy budget toward ecological goals. In Germany, the detection of the first mad cows in the nation's herd prompted the prime minister to replace the agriculture minister with an environmentalist, who quickly set a goal for increasing Germany's organic area from the current 2.6 percent to 20 percent by 2010 and declared "the end of intensive farming as we know it." Perhaps the most promising element of this U-turn on farm policy is the sense that politicians as well as consumers view the recent food scares not as isolated incidents but as symptoms of an agricultural system gone wrong.[67]

The self-interest component of food activism, however, runs much deeper than food safety. Greater freshness, nutritional value, and food quality are all potential payoffs. The argument for average citizens to take a more active role in their food system is also strengthened by the fact that in many industrial nations, farmers now get nearly half of their income from government payments. Consumers have a right to demand that farmers better serve the public interest.[68]

For instance, the public pays for the prevailing dysfunctional food system through increased medical costs associated with poor food choices. Don Wyse, an agronomist at the University of Minnesota, thinks that these health impacts offer an opportunity to enlist the urban majority as a political base for redirecting the food system. He notes that the prevailing corn-and-soybean system in the U.S. Midwest basically provides society with inexpensive meat and sugar—two products that contribute to the national obesity crisis. (Seventy percent of the corn and nearly all soybeans are used in industrial meat production, while high-fructose corn syrup has become the primary sweetener in the American diet.) The range of public health concerns associated with U.S. farm practices also includes the rapid emergence of antibiotic-resistant microbes due to the overuse of antibiotics in animal feed as well as other health risks associated with unhygienic animal farms.[69]

In some cases, farmers are already beginning to operate in more of a public service function—and the public is paying accordingly. For example, German water supply companies in Munich, Osnabrück, and Leipzig now pay neighboring farmers to go organic—a cheaper investment than removing farm chemicals from the water. In Washington State, a coalition of farmers, a consumer food cooperative, a local Indian tribe, and the Department of Fish and Wildlife are purchasing sections of farmland that border salmon breeding grounds and switching them to organic production in an effort to reduce water contamination in the spawning and nursery habitat. Australia's Landcare movement involves rural communities, both farmers and nonfarmers, in projects to reclaim soil, plant trees, clean rivers, and reconcile farming practices with local ecological health. The movement has grown from 200 community Landcare groups in 1990 to 4,250 today; one third of Australia's farmers now belong to a group.[70]

Such efforts are not restricted to wealthy nations. Coffee growers around San Salvador, El Salvador's capital, are being encouraged to reintroduce trees into their farm landscape in an effort to boost the city's water supply, which went into steep decline in recent years as farmers in the surrounding hillsides cut down trees. In one proposal, a share of residential waterbills in San Salvador will be earmarked for a farmer fund.[71]

Consumers can nurture a particular food system by seeking out foods produced with care to ecological and social consequences. Since people in the First World exercise power by virtue of their money, they can drive the market for organic produce or shade-grown coffee; for people in the Third World, the ethical choice might simply be how to get enough to eat. William Vorley of the International Institute for Environment and Development argues that the virtual monopoly in many national retail markets "makes retailers very sensitive to campaigns designed around ethics, safety or environment." He points to Christian Aid's Global Supermarket Campaign as a model of farmer and consumer groups joining forces to publicize corporate commitment to animal rights, family farms, or fair trade. Such "food activism" can often have a profound impact on the lives of farmers halfway around the world.[72]

Consider the growing fair trade movement, a partnership between First World consumers and Third World food producers that seeks to improve the often unfavorable conditions of trade. The typical fair trade arrangement guarantees that farmers receive a fair share of the retail profit (often several times more than they would receive from mainstream distributors) and that agricultural workers receive fair wages and enjoy labor rights. The product label also generally carries more information about the people and process involved in production than is typical, reinforcing consumers' interest in having an impact with their purchase. Worldwide, an estimated $400 million worth of fair-traded products are bought each year.[73]

In a global food market, one of the most significant selections a consumer can make is to buy locally grown food.

Perhaps more significant for the well-being of food producers in developing nations, fair trade standards often overlap with organic farming standards to demand that farmers use no pesticides. Although most of the world's pesticide use today occurs in the North, a lack of safety equipment or proper instructions means that most pesticide poisonings occur in the developing world. The World Health Organization estimates that every year 3 million people suffer from severe pesticide poisoning, matched by a greater number of unreported, mild cases that result in acute conditions such as skin irritation, nausea, diarrhea, and breathing problems. These poisonings result in as many as 20,000 unintentional deaths, in addition to an estimated 200,000 "pesticide suicides." (Suicides are more visible and therefore reported more frequently than unintended poisonings.) Many of the most popular export crops, from cut flowers to miniature vegetables, are also the most pesticide-intensive, reinforcing the potential benefits of fair trade.[74]

Globalization in some ways threatens to obscure this story behind our food, because of the inevitable breakdown in the crop's identity as it is processed and moved great distances, but also because prevailing trade agreements tend to emphasize the product rather than the process used to make it. Such agreements even threaten local sovereignty over public health or ecological standards, ceding many food quality decisions to international bodies that are non-democratic and dominated by industry representatives. Or consider the World Trade Organization's ruling that Europe must import hormone-treated beef from the United States and Canada or face retaliatory sanctions, even though European countries ban such practices on their own farms. Consumers have good reason to be skeptical of claims that they are among the primary beneficiaries of the trade agreements—seven years into NAFTA, for instance, inflation-adjusted prices for foods at the checkout counter in the United States, Canada, and Mexico are considerably higher, even as commodity prices bottomed out.[75]

Food policy expert Tim Lang doubts, however, that citizens around the world will allow the global integration of the food system to threaten their ability to know about their food. "At its apparent moment of triumph, the globalisation of the food supply is engendering a worldwide political opposition," Lang notes, "characterized by a set of countertrends that celebrates the local over the global, fresh over processed foods, and diversity over homogeneity." A Slow Food movement was founded in 1989 to celebrate the wisdom and pleasures of local cuisines—and, as its name implies, to retain alternatives to the proliferation of fast food—and now includes 65,000 members in 45 countries.[76]

In an increasingly global food market, one of the most significant selections a consumer can make is to buy locally grown food. In much of the world, farmers no longer sell food to their neighbors. Instead they sell it into a long and complex food chain of which they are a tiny part—and are paid accordingly. Apples in Boise's supermarkets are from China, even though there are apple farmers in Iowa; potatoes in Lima's supermarkets are from the United States, even though Peru boasts more varieties of potato than any other country. Buying food produced locally will help take some of the profits of food traders, brokers, shippers, and processors and put them back in the pocket of the farmer and the rural community.

An additional benefit of reconnecting farmers with consumers will be to take some of the distance out of the modern food chain, whose sprawl now means that transportation is one of the food system's biggest energy uses and sources of greenhouse gas emissions. Food eaten in the United Kingdom travels 50 percent more on average than two decades ago. The average distance traveled by food to reach one Chicago, Illinois, wholesale market has increased by 22 percent in the last two decades. While this might mean greater variety for the global eater, it also requires large amounts of energy, generates excess packaging and pollution, and can reduce food quality. In the United States, refrigerating, transporting, and storing food uses eight times as much energy as is provided by the food itself.[77]

"Eating local" can go a long way toward reducing this toll. In the Iowa Food System Project, a given basket of Iowa-grown foods traveled an average of 74 kilometers to reach its destination, compared with 2,577 kilometers if these foods had arrived from conventional national sources. The conventionally sourced meals also used 4–17 times more fuel than the local meals and released 5–17 times more carbon dioxide. Eating locally generally also means eating more fresh, whole foods, since many of the additives and processing that go into our food are the consequence of the time that commercial food spends in transit and storage. Shorter trips can have food safety advantages as well, since opportunities for contamination proliferate over long-distance hauls and long-term storage.[78]

Farmers markets, shopping in season, local-food labels, and other direct buying schemes are just some of the ways to support local food systems. Concerted efforts to get schools, hospitals, government agencies, and other institutions to set food procurement standards that favor local or regional farmers can also have powerful impacts. The benefits are often not just financial, but social and psychological, as the wider community begins to understand what it takes to produce the food it eats, and as relationships develop between food growers and food eaters.[79]

Finding as much of this common ground as possible will build the coalition for transforming our food system. Once stable farm communities are seen as beautiful landscapes that arrest the invasion of asphalt, then people who are sick of urban sprawl become an ally of farmers. When city folk realize that the cleanliness of their drinking water depends on the practices of the farmers in their watershed, then support for farmers no longer seems an unreasonable drain on public coffers. And when governments and aid agencies understand that alleviating poverty in the countryside assures more prosperity for the whole nation, then redistributing land or shoring up rural banks become urban priorities. The agricultural sector has operated alone in the political sphere for too long. Food is too essential to keep other parties away from the table.

WORLD SUMMIT PRIORITIES ON AGRICULTURE

- ➢ Shift agricultural subsidies to support for ecological farming practices.
- ➢ Tax pesticides, synthetic fertilizers, and factory farms.
- ➢ Redistribute land and guarantee secure ownership rights to both women and men.
- ➢ Eliminate export subsidies and food dumping.
- ➢ Assure women equal rights and support in agriculture.

Chapter 4

Reducing Our Toxic Burden

Anne Platt McGinn

In early December 2000, just three weeks after global talks on climate change reached a deadlock at the Hague, delegates negotiating a new global toxic chemicals treaty finalized a text that environmentalists and chemical industry representatives alike embraced. The treaty's primary goals are to ban 10 intentionally produced persistent organic pollutants (POPs) worldwide and to reduce emissions of two industrial byproducts, with the aim of eventually eliminating them. POPs are long-lived toxics that cause biological havoc as they bioaccumulate—collect and concentrate—in the food chain. The nine pesticides covered by the treaty had already been banned in at least 60 countries; one value of the treaty is that it sets up the process to expand that list.[1]

Signed in Stockholm in May 2001, the Convention on Persistent Organic Pollutants is one of the main environmental achievements in the decade since the 1992 Earth Summit in Rio. It outlines the key principles for a less toxic world, including the prevention of new toxic, persistent, bioaccumulative chemicals; the reduction of existing ones; substitution with less dangerous materials; and the great care needed with respect to all chemicals. Recent experiences in many industrial sectors and communities have shown that alternatives to toxics are available that not only protect human and environmental health but also improve the economic bottom line. They include unleaded gasoline, organic agriculture, bio-based industrial materials, and an overall reduction in consumption.[2]

Part of what is preventing these and other safer choices from becoming standard practice is the challenge of reframing how we think about toxic chemicals. In effect, we have based our collective well-being on a great deal of scientific ignorance and answers to the wrong questions. Instead of asking if a particular chemical is essential, we currently assume a certain amount of danger. The burden of proof for existing chemicals and many new ones now rests with public authorities and scientists who must prove something is harmful after it has been released and people can be exposed to it, rather than with chemical proponents who must prove a compound is safe over the long term. As structured, our current system puts the focus on which risks are acceptable rather than which are necessary and unavoidable. And what is considered acceptable changes over time, even within a few years, as scientific understanding evolves and society's values change.[3]

Officials at the Earth Summit were mindful of the need to protect people from accidental and routine exposure to thousands of hazardous chemicals. But the chemicals chapter of *Agenda 21*, the blueprint for change adopted at the conference, failed to address this adequately: it called on nations to promote chemical safety and information sharing, but offered little in the way of specific requirements to rid the planet of the most harmful compounds. The POPs treaty therefore represents an important milestone in international environmental law, not least because it applies to toxic chemicals management the "precautionary principle"—the rule that even in the face of scientific uncertainty, the prudent stance is to restrict or even prohibit an activity that may cause long-term or irreversible harm. (*Agenda 21* adopted a less controversial position: the chemicals chapter suggested that countries adopt a precautionary approach to risk reduction where deemed appropriate.)[4]

Since Rio, serious and previously unexpected human health effects have emerged concerning, for example, damage to the body's key communications systems: the nervous system that sends messages through electric pulses and the endocrine system that sends messages chemically, through hormones. Moreover, irreversible health problems have recently been shown to occur at exposure levels below what we normally think of as safe. This new and rapidly changing body of scientific evidence poses a serious challenge to our current way of dealing with toxic chemicals and supports widespread application of the precautionary principle.[5]

But before we can step off the toxics treadmill, we need to understand where these chemicals come from and what they are used for. The distinction between naturally occurring metals and humanmade persistent toxins is an important one. Metals such as lead and mercury are found in Earth's crust combined with other elements, typically sulfur. These toxic metals do not degrade, so if we continue to mine the ore and extract the metals or release them as byproducts, they come back to harm us. "Synthetic" toxins, on the other hand, are not found in nature and are not fundamental to life (although sometimes it may seem like they are because they are found in everything from plastic wrap to computer terminals). Synthetic toxins, such as all the intentionally produced POPs, were created either by trial and error, by deliberate intent, or, in some cases, by accident. By looking at what they are used for, we can begin to determine if they are absolutely necessary or not.[6]

Even when there is widespread agreement on which compounds need to go—toxic heavy metals and POPs, for example—people often find few viable and cost-effective alternatives. The issue is not simply one of banning "the bad guys." It involves developing and then adopting safer materials, processes, and products into our economy. While there is progress in this direction, the challenge remains enormous and the window of opportunity to change the way we use toxic chemicals and to prevent long-term environmental and health damage will not remain open for long.

The Chemical Economy

The chemical economy is one of the largest and most diverse industrial sectors in the world. Each year, tens of thousands of individual chemical compounds are produced and serve as the feedstock for countless industries, as the basic ingredients for virtually every consumer product manufactured today, and as the basis for such products as cleaning agents and pesticides. (See Table 4–1.) By 1998 (the most recent year with data), global sales of all chemicals totaled nearly $1.5 trillion, making the sector about twice as large as the global market for telecommunications equipment and services.[7]

Not surprisingly, the chemicals manufacturing sector has a major influence on the health of the global environment. In 1998, for example, the industry accounted for nearly 10 percent of world water use and 7 percent of world energy use. (Energy inputs, such as oil and natural gas, are used both as a source of fuel and as a feedstock material.) While this is considerably less than agriculture's thirst for water, the global chemicals manufacturing industry consumes 21 percent more water each year than all household water users.[8]

Quantifying the global toxic burden is difficult, given the incomplete picture of the life cycle of thousands of chemicals. Only a few countries measure toxic emissions, and these data are limited in scope. In 1999, for example, the U.S. chemicals manufacturing sector ranked third in terms of toxic emissions, behind metal mining and electric utilities, according to U.S. Toxics Release Inventory (TRI) data. Yet only large manufacturers are required to report, and the current list of 650 chemicals does not cover all toxic chemicals or sources, or emissions during use and disposal. According to the World Bank, the chemicals and plastics manufacturing sectors are among the most intensive in terms of toxic air pollutants. (See Figure 4–1.) (The global ship building and repair industry is the most intensive, emitting about five times more toxics to air than the chemical manufacturing sector.)[9]

Table 4–1. Global Chemical Output by Sector, Value, and Share of Total, 1996

Sector	Value (billion dollars)	Share of Total (percent)
Basic industrial chemicals	360	26
Pharmaceuticals	305	22
Plastics, resins, and synthetic resins	235	17
Soaps and toiletries	160	12
Other chemicals	131	10
Fertilizers and pesticides	90	7
Paints and varnishes	79	6
Total	1,360	100

SOURCE: Organisation for Economic Co-operation and Development, *OECD Environmental Outlook for the Chemicals Industry* (Paris: 2001), p. 112.

Moreover, the quantity of materials produced and used gives no indication of its potency. To bring in this year's agricultural harvest, for example, farmers worldwide will apply something on the order of 2.5 million tons of pesticides, the overwhelming majority of which are synthetic organic chemicals that are orders of magnitude more toxic than 50 years ago. Just as we have no concrete measures of our cumulative environmental burden of toxins, neither do we know the relative safety or danger of most chemicals in use. There are no basic health and environmental data for 71 percent of the

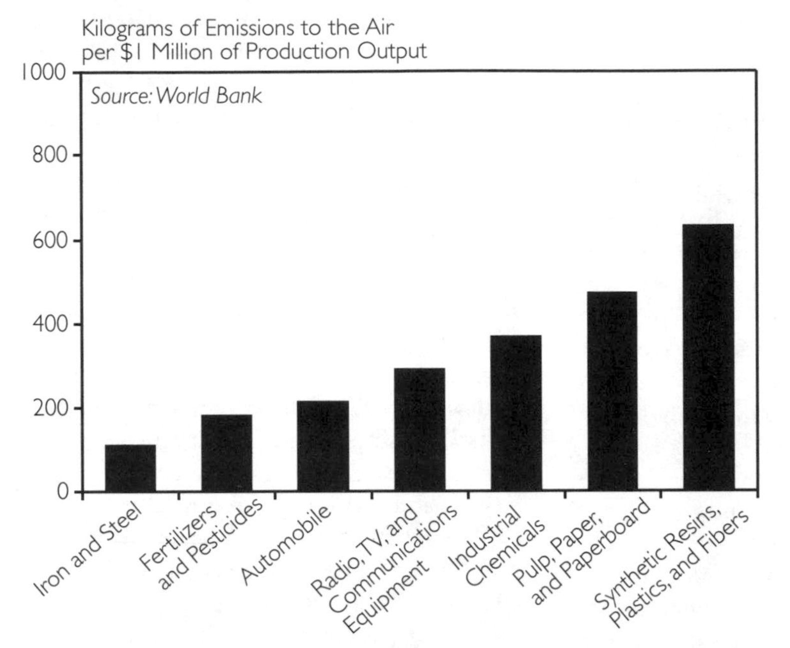

Figure 4–1. Toxic Intensities of Selected U.S. Manufacturing Sectors, Early 1990s

most widely used chemicals in the United States today, and less than 10 percent of new chemicals reviewed each year under premarket notifications having adequate test data on health effects. Meanwhile, chemical production keeps growing—it is expected to soon grow faster than the global economy. (See Figure 4–2.)[10]

Much of the expansion in chemicals production and use is now occurring in developing countries, in part because companies in traditional producing nations (primarily industrial countries) are shifting away from commodity chemicals, which are a mature market, toward speciality chemicals, which is a less cyclical business and has a higher profit margin. But several changes within developing regions are also contributing to the global realignment of the industry from North to South, including the growth in domestic demand, low labor costs, and expanding chemical-dependent sectors.[11]

Polyvinyl chloride (PVC) plastic provides a telling example. Every stage of its life cycle—from manufacture to disposal—creates dangerous chemicals, including some POPs, while toxic additives are used to stabilize the material and add flexibility. Nearly 25 million tons of PVC were produced in 1999. This material now has a constant presence in every channel of the global economy. Overall, production is accelerating, with much of the growth expected in Asia, where rapidly expanding cities are built with PVC building materials and filled with consumer goods made from PVC and other plastics.[12]

Similar trends are evident in the chemically intense pulp and paper sector. Some 40 percent of the world's pulp supply is bleached with chlorine compounds. A large share of these are based on elemental chlorine, a process that creates up to 35 tons of chlorinated byproducts a day per industrial-scale facility, as opposed to almost none for chlorine-free bleach methods based on hydrogen or oxygen. In 1998, the world volume of paper production was 294 million tons, more than a sixfold increase since 1950. It is expected to increase by another one third by 2010. Countries in Asia and Latin America are rapidly boosting their pulp production, eager to tap into lucrative

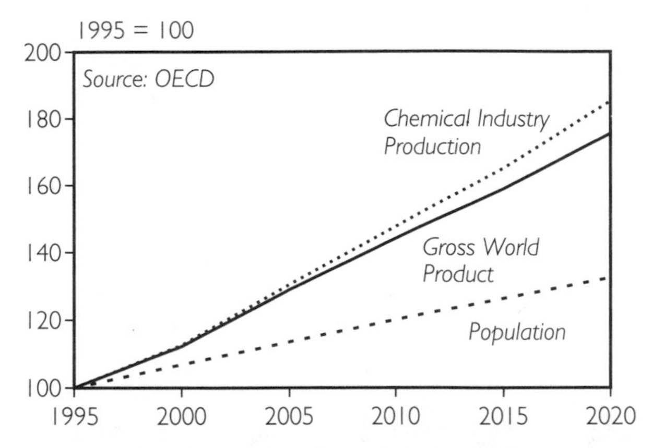

Figure 4–2. Projected Growth in World Economy, Population, and Chemical Production, 1995–2020

trade markets. In the next few years, Asia's paper and pulp output will likely surpass that of North America, making that region the world's top producer.[13]

Growth in these and other chemically intense industries promises to bring not only desperately needed jobs and export earnings, but also significant environmental liabilities. And as these activities expand in developing countries and economies in transition, which often have minimal capacity to monitor toxic contamination from persistent and mobile pollutants—let alone contain and reduce it effectively—global contamination could become much worse in the years ahead.[14]

In addition to releasing toxic compounds, industries producing PVC plastics and pulp and paper consume chemicals, and thus help propel the growing demand for existing and new chemicals. Part of the reason that these industries use so many chemicals is simply that all modern industrial production follows this pattern. But the demand of these industries for chemical inputs also results from deliberate—and successful—efforts by others to create markets for unwanted synthetic chemicals. Producers of materials such as petroleum have intentionally created markets for byproduct chemicals to reduce waste and make money. Each year, petroleum refineries create literally tons of highly toxic byproducts—including benzene, ethylene, and propylene. Over time, these were developed as chemical sources for secondary processing and manufacturing industries, most notably plastics manufacturing.[15]

Of course, recycling materials and closing the production loop are basic concepts in "industrial ecology," a new discipline that tries to model industrial processes on the efficiencies found in nature, in order to minimize waste and pollution. But in some cases, these principles have been applied to their extreme, essentially creating a justification for the continued production of toxic materials.[16]

Chlorine is the classic example of a chemical byproduct that was marketed as the basis for entirely new branches of industrial production. Because it is highly reactive, chlorine has a strong affinity for organic (carbon-based) compounds. (In nature, chlorine is almost never found alone in its elemental state—it normally binds with sodium or carbon.) Combined with an organic molecule, chlorine often imparts stability and persistence, making the resulting compound likely to bioaccumulate. Because of its versatility, chlorine is the basis for thousands of synthetic chemicals. About 60 percent of the final products in the chemical industry involve chlorinated chemicals at some stage of production. Initially generated as an unwanted byproduct of caustic soda (which is used in manufacturing pulp, paper, and soaps, among other things), chlorine has been hailed by W. Joseph Stearns of Dow Chemical as "the single most important ingredient in modern [industrial] chemistry."[17]

Many compounds—including the thousands that contain chlorine—are both innocuous and valuable for commerce and medicine. The challenge is to identify and regulate the most dangerous ones. At the moment, scientists do not even know how many dangerous ones exist. Estimates vary from dozens to hundreds. Despite the ubiquity of synthetic chemicals, many compounds have never been tested for basic health impacts, such as toxicity, let alone for bioaccumulative or persistent properties.[18]

There are, however, some clear choices for elimination among the thousands of chemicals on the market today. (See Figure 4–3.) Based on the degree of persistence and toxicity, high-priority chemicals include dioxins and furans (both POPs), chlorinated pesticides, and polychlorinated biphenyls (PCBs), along with mercury, lead, and a few other heavy metals. Other toxic compounds—including organic solvents and organophosphate pesticides—are not as harmful as POPs, but they are important from public health and ecological perspectives because of the harm they pose on their own or in reaction with other substances and because the lessons they offer for phasing out toxics.[19]

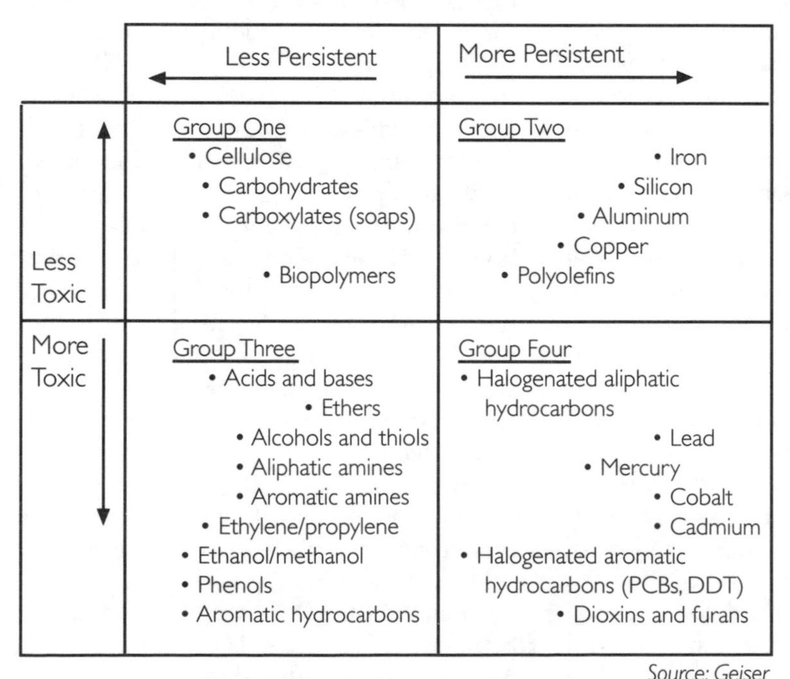

Less Persistent
More Persistent
Less Toxic
More Toxic
Group One
• Cellulose
• Carbohydrates
• Carboxylates (soaps)
• Biopolymers
Group Two
• Iron
• Silicon
• Aluminum
• Copper
• Polyolefins
Group Three
• Acids and bases
• Ethers
• Alcohols and thiols
• Aliphatic amines
• Aromatic amines
• Ethylene/propylene
• Ethanol/methanol
• Phenols
• Aromatic hydrocarbons
Group Four
• Halogenated aliphatic hydrocarbons
• Lead
• Mercury
• Cobalt
• Cadmium
• Halogenated aromatic hydrocarbons (PCBs, DDT)
• Dioxins and furans
Source: Geiser

Figure 4–3. Industrial Materials Groups

Old Metals, New Threats: Lead and Mercury

Metals are different from other toxic substances because they are naturally occurring, albeit trace elements in Earth's crust. They cannot be created or destroyed. Once emitted, they can reside in the environment for hundreds of years. Natural forces such as volcanoes, forest fires, and ocean tides cycle metals through the environment. But humans also play an important role and, in many cases, a larger role than nature. By influencing the rate of release and transport of metals through the environment and by altering their biochemical state, humanity has increased by several orders of magnitude the emissions of and its own exposure to toxic heavy metals. In particular, the stories of lead and mercury—two potent neurotoxins (compounds that harm the nervous system)—demonstrate the scale of contamination, the resulting human and environmental health problems, the difficulties of addressing such releases, and, especially in the case of lead, the enormous health and economic benefits of reducing usage.[20]

Emissions of lead date back at least 8,000 years, to the first lead-smelting furnace. During the nineteenth century, large-scale coal combustion released significant quantities of mercury (a common contaminant in coal) into the atmosphere, while the use of large quantities of mercury to amalgamate gold and silver dates back at least to the sixteenth century in Latin America. Despite our long history with these two elements, the twentieth century brought enormous change to the relationships. Metals consumption in the United States jumped sixteenfold between 1900 and 1998, compared with a tripling in the use of wood products. At their peak in the mid-1980s, global atmospheric releases from human activities exceeded natural sources by a factor of 28 to 1 for lead and 1.4 to 1 for mercury.[21]

The use of leaded gasoline throughout much of the last century boosted global lead levels to unprecedented heights. In 1924, three U.S. companies—General Motors Corporation, Du Pont Chemical, and Standard Oil—formed a separate company known as Ethyl Corporation solely for the purposes of producing and selling tetraethyl lead (TEL), a compound that reduced the audible "knocking" sound in cars during fuel combustion and was supposed to improve overall engine performance. Well before the additive was marketed, company and government officials knew of its dangers but assumed they could control its release in factories and protect workers. Moreover, because TEL dissipates easily, many assumed it would never cause any significant environmental or public health problems.[22]

Despite several initial setbacks, including a challenge by the U.S. Surgeon General in 1925, the Ethyl Corporation aggressively pushed TEL onto U.S. and eventually world gasoline markets. The company favored TEL because it could patent the compound—as opposed to ethanol, a more effective and less polluting compound, but one that anyone could make. Leaded gasoline went on to become the global standard for decades. Between 1926 and 1977, U.S production of TEL increased from 1,000 tons to 266,000 tons per year. With widespread use of leaded gas came a parallel rise in global contamination. In Japan, airborne lead emissions increased about a thousandfold from 1949 to 1970. Today, TEL is responsible for some 90 percent of airborne lead emissions in developing countries.[23]

Quite literally, the legacy of the Ethyl Corporation and other manufacturers that deal with lead is written in human blood: the average person today carries levels of lead that are 500–1,000 times higher than our preindustrial ancestors. Lead is now found in all living things and throughout the environment. (Unlike copper or iron, free lead was virtually nonexistent in the precivilization biosphere, which meant that humans and other species had no opportunity to evolve a natural defense to it.)[24]

But the story of TEL does not end at the tailpipe. In the process of solving a noise problem, burning TEL created a corrosive byproduct that ruins the engine. So in order to get the lead out of the engine and into the atmosphere as quickly as possible, scientists added another toxic compound, ethylene dibromide (EDB), to leaded gas. When EDB is burned it produces methyl bromide, a developmental toxin and potent ozone-depleting substance. Indeed, the World Meteorological Organization has identified automobile exhaust from leaded gasoline as one of the top three sources of methyl bromide.[25]

By the 1970s, countries as varied as Brazil, the Soviet Union, Thailand, and the United States began to phase out leaded gas, although often for reasons unrelated to the health effects of TEL and EDB. Brazil, for example, switched from gas to ethanol in an effort to reduce its dependence on foreign oil and save the national currency from collapse. The Soviet Union diverted high-octane, leaded fuel to the military during the cold war, leaving little choice for Russian consumers. And beginning in 1975, the United States required automobiles to have catalytic converters to reduce carbon monoxide and other hazardous air pollutants from vehicular emissions. As with older engines, leaded gas was incompatible with this new technology.[26]

The list of countries that have banned leaded gasoline continues to grow.

The list of countries that have banned leaded gasoline continues to grow. And although 100 or so countries still use leaded gas today, some have reduced the lead content and others have begun to introduce unleaded gasoline as an alternative. All told, some 80 percent of the gasoline sold today in the world is unleaded.[27]

As the markets for leaded gasoline declined, the Ethyl Corporation and other manufacturers faced significant profit losses. As early as the 1970s, the industry turned its attention to a manganese-based compound (MMT) that also had antiknock properties and enhanced gasoline octane. Although the U.S. Environmental Protection Agency (EPA) argued against its use until basic health tests were done, and although the American Automobile Association warned that its use would damage catalytic converters, in 1995 a U.S. federal court allowed Ethyl Corporation to introduce MMT, claiming it was not in EPA's jurisdiction to ban MMT on health grounds. (At high doses, manganese is extremely toxic and causes nervous disorders and symptoms of Parkinson's disease; at low, airborne doses, its effects are unknown.) Since 1977, MMT has been widely added to gas sold in Canada. Most U.S. companies now avoid it, however, because of public health concerns. As the story of tetraethyl lead in gasoline and the related bromide and manganese-based compounds illustrates, novel applications of chemicals can create new, unforeseeable problems, which then prompt chemical producers to offer "solutions" that in turn create their own problems.[28]

People have been exposed to and poisoned by lead in many other sources in addition to gasoline. Lead has been added to ceramic glazes, paints, electronics, batteries, and other products that emit it to varying degrees when they are burned or otherwise disposed of. Some applications are problematic during routine use: Lead in pipes leaches into water supplies, which happened as long ago as during Roman times, whereas lead-based paint can peel off walls, doors, and window frames and become a deadly meal of dust for curious children. Children are at special risk from mercury, lead, and other toxins because they "eat, drink and breathe three to four times as much per pound of body weight as adults do," according to Richard Jackson, Director of the U.S. Centers for Disease Control and Prevention's National Center for Environmental Health.[29]

These other uses are not insignificant. Worldwide, for example, tens of thousands of tons of lead (as well as other toxic metals) are added to PVC each year to stabilize it at high temperatures. In North America, lead is now only added to PVC wire and cables, but in Europe it is still used in rigid applications, such as pipes, where it can leach into water.[30]

While turning to unleaded gasoline, many countries have also improved waste incineration and wastewater treatment technologies and reduced the use of lead in paint, batteries, and other sources. Consequently, global lead emissions dropped two thirds from the mid-1980s to the mid-1990s. (See Table 4–2.) Although annual emissions have dropped, a huge reservoir of dispersed lead must still be dealt with. Global mercury emissions have followed a similar path in recent years, but the situation in developing countries is worsening.[31]

The primary human-based sources of mercury today are coal burning and solid waste disposal, both of which are increasing in many regions. (Another main source, the mercury cell method of industrial chlorine production, has been declining for many years.) Asia now accounts for about half of the world's annual mercury emissions from human activities, in large part because China and India burn about one third of the world's coal. Between 1990 and 1995, mercury emissions in Asia jumped 26 percent. Several hundred million Chinese regularly heat their homes and cook in unvented stoves, exposing family members to high doses of mercury as well as arsenic, fluorine, and other contaminants. Exposure to mercury and other toxics comes from polluted air and water, but in fact we absorb most persistent bioaccumulative toxics in our food. Mercury illustrates this point.[32]

In its inorganic state, mercury is a common but poorly absorbed compound. In its organic form, however, methyl mercury is both very toxic and easily absorbed by fish, birds, and humans. By unfortunate coincidence, bacteria commonly found in polluted waters readily convert inorganic mercury to its more dangerous organic state, bringing it directly into the aquatic food chain. What are often dismissed as inconsequential environmental discharges of inorganic mercury are easily transformed into methyl mercury and carried up the food chain, where the mercury is concentrated hundreds and thousands of times over. Some 2,200 tons of mercury are emitted

Table 4–2. Global Atmospheric Emissions of Lead and Mercury by Major Industrial Source, Mid-1990s, with Decline Since 1983

Source	Lead	Mercury
	(tons per year)	
Vehicular traffic	88,739	—
Stationary fossil fuel combustion	11,690	1,475
Nonferrous metal production	14,815	164
Iron and steel production	2,926	29
Cement production	268	133
Waste disposal	821	109
Other		325
Total emissions, mid-1990s	119,259	2,235
Change since 1983	– 64 percent	– 37 percent

SOURCE: Jozef M. Pacyna and Elisabeth G. Pacyna, "An Assessment of Global and Regional Emissions of Trace Metals to the Atmosphere from Anthropogenic Sources Worldwide," *Environmental Reviews* (in press).

from human activities each year, while as little as one seventieth of a teaspoon is enough to contaminate a 25-acre lake for a year.[33]

One indicator of the growing environmental burden of methyl mercury is the number of fish consumption advisories issued by governments. (An advisory is issued when officials find concentrations of a contaminant in local fish at a level that may pose a risk to the public or to groups at high risk, such as young children, the elderly, or the fetuses of pregnant women.) In the United States, the number of mercury advisories for noncommercial fish increased more than one and a half times between 1993 and 2000. Almost 80 percent of fish advisories issued by state officials now appear at least in part because of high levels of mercury. In February 2001, the U.S. Food and Drug Administration warned pregnant women not to eat any top marine predators, including swordfish and shark, because of mercury. Based on studies from the Faroe Islands and New Zealand, people who rely on fish for a large share of protein in their diets are especially at risk of mercury contamination. And in communities near gold mines, high mercury levels in the food chain have become a fact of life. (See Box 4–1.)[34]

People who rely on fish for a large share of protein in their diets are especially at risk of mercury contamination.

Mercury's impact on human health is well documented, unfortunately, because so many people have become ill after being exposed to it. As early as the eighteenth century, workers who used mercury to make felt hats from beaver pelts suffered from tremors, hallucinations, delirium, and other signs of mercury poisoning, which gained a reputation as "mad-hatter's disease." In the 1950s, large industrial discharges of mercury into Japan's Minimata Bay killed hundreds of people and left epidemiologists with a tragic record of the workings of this powerful neurotoxin. Children born after the initial incident suffered from cerebral palsy, mental retardation, and severe brain defects, and some adults became afflicted with a wide range of neurological disorders, including tremors, paralysis, blindness, and deafness. More recently, researchers have found that when low levels of methyl mercury strike at key points in fetal and childhood development—as opposed to repeated occupational exposure or large industrial releases—they can slow brain development significantly, prompting loss of cognitive skills and other effects.[35]

Demonstrating the links between the trends in the production of these toxins and the trends in human illness is difficult, but one thing is clear: efforts to reduce exposure to lead, a powerful neurotoxin, have paid off. Since 1976, blood lead levels of American adults have dropped, on average, more than 75 percent and those of children, more than 85 percent. This means that, on average, each American child born today has gained five IQ points over children born a generation ago, a gain that is quantified as being worth about $45,000 over the course of a lifetime (measured in terms of cognitive ability, memory, and educational achievement).[36]

But this good news is tempered by the reality that averages do not translate into equal gains for everyone. Research from places as varied as Mexico City, the Cape Province of South Africa, and Rhode Island shows that socioeconomic factors are important indicators of high blood lead levels, especially among children. Approximately one out of three inner-city

BOX 4–1. GOLD MINING'S TOXIC TRAIL

Since the early 1980s, when the price of gold reached its all-time high, hundreds of thousands of small-scale miners or *garimpeiros* have flocked to the rainforests of Brazil, Venezuela, Guyana, and neighboring countries in search of this precious metal. In the Amazon, as in southern Africa, the Philippines, and other gold mining regions, small-scale miners use the same age-old formula to extract gold from earth and rock. They pour mercury over crushed ore that they have dredged from riverbeds or mountainsides, believing the sediments may contain gold. They press out the excess mercury with their hands, and then burn the mixture in order to evaporate the rest of the heavy metal. The lucky few are left with a few grains of gold; almost all will have inhaled or absorbed some mercury in the process.

Not surprisingly, many miners and their families have extremely high levels of mercury in their bodies. Tests conducted on the Wayana Indians in French Guiana revealed that 57 percent of subjects had mercury concentrations two to three times higher than World Health Organization (WHO) standards. Studies from Venezuela and the Brazilian Amazon show similar results. Other residents of the region may be exposed to mercury by eating fish—an important part of the diet of most native peoples in the Amazon—containing mercury in its highly toxic form, methyl mercury.

It is believed that since the 1980s, Amazon *garimpeiros* have produced between 80 and 100 tons of gold annually. Mining this gold sends roughly 100 tons of mercury into the Amazon and another 100 tons into the atmosphere each year—accounting for about 8 percent of annual emissions of mercury from human activities. Metals mining is a leading polluter globally. In the United States, for example, it is responsible for nearly half of the toxins released by industry. In 1999, U.S. mines sent nearly 4 billion pounds of toxic pollutants such as mercury, lead, cadmium, and cyanide into the environment.

— *Payal Sampat*

SOURCE: See endnote 34.

African-American children today has elevated blood lead levels that are, on average, 80 percent higher than the U.S. figure for all children. (Lead poisoning persists in poor communities in part because the houses tend to be older and in disrepair, and frequently still have lead-based paint.)[37]

Other factors, such as proximity to highways and nutritional status, also contribute to the gross inequities in lead exposure and poisoning. Children living in rapidly expanding urban areas of China, for instance, have blood lead levels up to four times as high as the average level for American children in the 1970s, when it was at its peak. One in five children in Beijing carry more lead in their blood than is considered safe by the World Health Organization. In one district of the Chinese capital, 80 percent of children had readings above the unsafe level. Almost universally, lead exposure is worse in developing countries. People who live in Dhaka, Bangladesh, for example, breathe air that has the highest atmospheric lead levels in the world. And in Africa, much of the gasoline sold today contains among the highest levels of lead in the world.[38]

Although it has been 10 years since WHO described gasoline-based lead poisoning as "one of the world's worst environmental problems," this assessment remains true today. Given current rates of industrialization, the continuing use of leaded gasoline in some countries, rapid growth in vehicle production and road-building, and the persistence of lead in the environment, childhood lead poisoning and exposure among adults will continue to be an enormous global public health problem for many years to come. Almost universally, the urban poor will continue to bear the brunt of this health crisis.[39]

While most health professionals recognize the need for a global phaseout of leaded gas to improve public health, we have only begun to think in global terms with respect to mercury. The Governing Council of the U.N. Environment Programme (UNEP) recently called for an assessment of mercury to be completed by 2003. At the same time that we are gathering information, scientists are finding that the effects of mercury—like lead—will be with us for a long time. Gold mines operating in Nova Scotia from 1860 until 1945 produced some 3 million tons of tailings (mine waste), which include mercury as well as arsenic, cadmium, copper, iron, and thallium. Scientists recently tested lake sediments downstream of the mine and found that there is still "no evidence of [a] downturn" in contamination levels, despite the 50 years that have passed since the mines were closed.[40]

POPs and Precaution

Most chemicals are now tightly regulated under environmental laws, usually in terms of exposure limits for air, water, or soil. In contrast, regulatory approval to introduce chemicals is less stringent. "Like the science that informs it, the process of regulation has taken a reductionist approach; seeking chemical by chemical solutions; focusing on too few [biological] outcomes; neglecting additive, cumulative, and synergistic effects; and allowing balkanization of regulatory authority," according to Sheldon Krimsky, a professor of urban and environmental policy at Tufts University. It is no wonder that we are only beginning to discover how everyday chemicals, assumed to be relatively harmless—indeed, safe—are in fact jeopardizing our health and quite possibly that of generations to come. (See Table 4–3.)[41]

Consider PVC plastic: in addition to the problems associated with stabilizers such as lead, a majority of the additives that give this material its range of flexibilities belong to a group of compounds called phthalates. Because they are not chemically bonded to the resin (raw plastic), they can migrate to the surface and leak into the surrounding environment. Under particular conditions, some commonly used ones persist and bioaccumulate. In wildlife and laboratory animals, phthalates have been linked to a range of reproductive health problems, including reduced fertility rates, miscarriages, birth defects, abnormal sperm counts, and testicular damage, as well as liver and kidney cancer.[42]

Hospital patients receiving blood infusions have been shown to be at risk of exposure to a commonly used phthalate known as DEHP, which can leach directly out of intravenous tubes and into a patient's bloodstream. Adults who receive one or two transfusions are not believed to be in danger, but critically ill patients, such as premature babies, who require life-saving procedures are exposed to "very, very high doses," according to a researcher at Boston's Children's Hospital Medical Center. The U.S.-based National Toxicology Panel recently concluded, "there may be no margin of safety" with respect to DEHP.[43]

Recently, scientists at the U.S. Centers for Disease Control and Prevention detected

Table 4–3. Chemicals by Health Effects

Health Effects	Main Chemicals
Cancer	arsenic, benzene chromium, vinyl chloride *probable:* acrylonitrile, ethylene oxide, formaldehyde, nickel, perchloroethylene, PCBs, PAHs, metals, other endocrine disrupters
Cardiovascular diseases	arsenic, cadmium, cobalt, lead
Endocrine disruption	aldrin, aluminum, atrazine, cadmium, dichlorvos, dieldrin, dioxins, DDT, endosulfan, furans, lead, lindane, mercury, nonylphenols, phthalates (including DEHP), PCBs, styrene, tributyltin, vinyl acetate
Nervous system disorders/ cognitive impairment	aluminum, arsenic, benzene, ethylene oxide, lead, manganese, mercury, many organic solvents
Osteoporosis	aluminum, cadmium, lead, selenium
Reproductive effects (such as birth defects and miscarriages)	arsenic, benzene, benzidine, cadmium, chlorine, chloroform, chromium, DDT, ethylene oxide, formaldehyde, lead, mercury, nickel, perchloroethylene, PCBs, PAHs, phthalates, styrene, trichloroethylene, vinyl chloride

SOURCE: European Environment Agency, *Europe's Environment 1998* (Copenhagen: 1998), p. 122; Kenneth Geiser, *Materials Matter: Toward a Sustainable Materials Policy* (Cambridge, MA: The MIT Press, 2001), p. 130; Françoise Brucker-Davis, "Effects of Environmental Synthetic Chemicals on Thyroid Function," *Thyroid,* vol. 8, no. 9 (1998), pp. 829–31; "Agency Attacked Over Endocrine Disruptors Strategy," *ENDS Report,* March 2000, p. 39.

phthalate metabolites (breakdown products) in the urine of women of childbearing age. DBP, a phthalate used in perfumes, cosmetics, and other health care products marketed almost exclusively to women, was most commonly reported. Although this compound is not known to cause reproductive problems, others that are known offenders were also found in the general U.S. population, proving that exposure is far more common than previously suspected.[44]

The clearest and most undisputed body of evidence showing the ability of synthetic chemicals to disrupt the glands and hormones that make up the endocrine system comes from more than 100 species of mollusks (mussels, oysters, snails, and other shellfish), which have suffered worldwide population declines and, in some cases, complete disappearances because of the reproductive and hormone-disrupting effects of tributyltin (TBT). TBT, a form of organic tin, was first introduced in the mid-1960s as an additive in marine paint that was 10–100 times better than copper at fending off algae, barnacles, and other "fouling" organisms that cause structural damages to ships and slow them down in the water.[45]

Within a few years of the first use of these anti-fouling paints, shellfish in northern European waters began to develop an irreversible condition known as imposex, which leaves the species unable to breed normally. By 1981, scientists had established the link between reproductive toxicity and TBT paints, based on tests in and around marinas and harbors. Residues of TBT have been found in bottlenose dolphins and bluefin tuna, animals that are high on the aquatic food chain, showing that TBT is a bioaccumulative compound.[46]

Several countries have since banned TBT paints from vessels, particularly smaller, recreational boats that tend to spend more time in harbors and close to coastal areas. But this paint is still used on larger, ocean-going vessels. In October 2001, the International Maritime Organization adopted an international convention that will ban TBT and related compounds in marine paints.[47]

As this example suggests, endocrine disruption is potentially "a far more serious health problem than cancer," according to Dr. Terry Collins, a professor of chemistry and an expert in "green chemistry" (the scientific field that focuses on detoxification) at Carnegie Mellon University. There are at least four reasons for this. First, the animal or person often looks and appears healthy even while suffering the effects of reproductive, neurological, or immunological toxicity, so simple identification of the problem is difficult. Second, frequently there is a long time lag between exposure and effects, so it is difficult to predict—and prevent—such effects until it is often too late. Third, the effects of some chemicals, like TBT, cannot be predicted on the basis of the compound's chemical structure alone, making it difficult to screen chemicals and identify which ones may be endocrine disrupters. Fourth, many of our current regulatory limits are based on screening for cancer and other health effects from high doses. But because endocrine disruption can occur at low exposure levels, these compounds can slip below the regulatory radar screen and often are perfectly acceptable under our current regulatory definition of what is deemed safe for human health.[48]

Despite extensive counter-studies from industry-supported groups in the United States and Japan, a panel of scientific experts recently concluded that "estrogenic chemicals can cause biological effects at levels below those normally found safe," according to a report in *Science*. Lab tests even found damages to the reproductive organs and the neurological and immune systems that were absent at higher doses. Given mounting evidence of human reproductive and developmental problems—including declining sperm counts, rising rates of testicular cancer and other male reproductive disorders, increasing incidence of breast cancer, earlier ages of puberty among young girls—these findings regarding low doses in lab animals suggest that environmental factors, including exposure to endocrine-disrupting chemicals, may be to blame in causing such problems in people.[49]

As evidence of toxic and environmental damage mounts, the list of suspected POPs will grow and make the initial "dirty dozen"—10 pesticides plus dioxins and furans, the unintentional byproducts of combustion and other industrial and natural practices—look like easy targets. The challenge of pinpointing which compounds might be persistent organic pollutants and then proving they need to be banned is a task that quickly becomes complicated and costly. Adding to the challenge is the fact that long-term risks are not created solely by metals and POPs. Depending on the circumstances of their production and use, other chemicals may create long-term problems, even if they are not called POPs.

Chlorinated solvents, for example, are generally not persistent enough to qualify as POPs, yet many of them are quite toxic: they have been linked to miscarriages, infertility, kidney and liver cancer, and various immune system disorders. A recent study showed that women who regularly worked with organic solvents (such as factory workers, lab technicians, and graphic designers) had a thirteenfold higher chance of having a child with a major birth defect than did mothers in other occupations. Some chlorinated solvents are now effectively considered POPs by certain regional agreements, notably the 1992 OSPAR Convention for the Protection of the Marine Environment of the Northeast Atlantic. (While they may not be persistent, they may degrade into other toxic substances that are much more stable.)[50]

Another complication in identifying chemical culprits is that people are routinely exposed to mixtures of compounds that can react in unexpected ways. Researchers from the University of Wisconsin looked at the combined effects on mice of two pesticides and one fertilizer commonly used on U.S. farms—aldicarb, atrazine, and nitrate. Although one of these compounds alone did not trigger a significant change in the level of thyroid hormones, a similar concentration of a mixture of the three contaminants altered thyroid levels enough to trigger behavioral, endocrine, and immune changes.[51]

In formulating so-called safety thresholds, we invariably focus on—and get bogged down in a debate over—how much of a toxic material to use and release according to a highly politicized process of setting such limits. While the debates are usually based on the best available science, the science itself—because it is highly uncertain—becomes politicized and subject to delay as interested stakeholders question its methods, assumptions, and motives rather than weighing what is best for the economic bottom line of certain companies against what is needed to protect human and ecological health. Designing better regulations, while important, is an inadequate long-term response to persistent, bioaccumulative toxins. Because of the high stakes involved, these compounds require a new way of thinking about and producing materials, which is nothing short of a chemical revolution. Instead of asking ourselves how much harm we should allow, we should focus on preventing as much harm as possible.[52]

The Changing International Field

Prompted by rapidly emerging scientific evidence and heightened public awareness, the global community has moved far beyond the goals laid out in Rio for chemical safety. Indeed, we have begun to question—and, in some cases, reject—the long-held presumption of innocence for toxic chemicals and called for a higher standard of proof, a standard based on necessity and informed consent rather than convenience. With the Stockholm Convention on POPs now open for ratification and funding available on an interim basis, politicians, business leaders, health officials, environmentalists, and concerned citizens have an enormous opportunity to embrace the precautionary principle and rewrite the human relationship with toxic chemicals. While treaties alone will not get rid of toxic chemicals, they can help create a level playing field and spur the technical and financial transition that is needed to move the world away from these chemicals.[53]

The Stockholm Convention has many notable features, including provisions to "turn off the tap" on new and existing POPs; the option for countries to require—not simply promote—substitute materials, products, or processes; and a broad commitment to the precautionary principle. Parties to the treaty will examine any new pesticides and industrial chemicals "with the aim of preventing" additional persistent organic pollutants. Governments are also obligated to screen existing chemicals and reduce the use and release of those with the characteristics of a POP. Perhaps more profoundly, they must promote "the best available technology" and "best environmental practices" with respect to a number of major industrial sources, including oil refineries, paper and pulp mills, metal processing plants, and all types of waste incinerators. Although such technologies and practices have not yet been specified by the Conference of the Parties, these features will help change social behavior "down to the level of how municipalities deal with their trash," according to the treaty Chair, John Buccini.[54]

Richer countries have a special responsibility not to externalize their pollution costs via exports.

In an important compromise, the treaty allows countries to continue using DDT, one of the "dirty dozen" chemicals it addresses, in programs to control malaria-carrying mosquitoes or other disease vectors if a country files a request with the Secretariat, closely monitors such use, and reports regularly to a publicly available DDT registry. This is a notable improvement over the situation today, in which no one is responsible for tracking DDT. Twenty-six countries had requested such exemptions as of May 2001, but all parties to the treaty "must promote the research and development for alternatives to DDT," a significant obligation to ensure universal support for alternative methods of mosquito control. The Stockholm Convention also includes specific steps for implementing treaty requirements, including detailed mechanisms to ensure transparency and accountability as well as requirements for new and additional funding from industrial countries to help developing nations pay for required changes.[55]

Two other treaties—the 1998 Rotterdam Convention on the Prior Informed Consent Procedure (PIC) for Certain Hazardous Chemicals and Pesticides in International Trade and the 1989 Basel Convention on the Control of Transboundary Movements of Hazardous Wastes and their Disposal together with its 1995 amendment that bans the export of hazardous waste from rich to poorer countries—also have a big role to play in limiting the flow of toxic pesticides and wastes. In addition, they provide an opportunity for public access to information and greater transparency in the handling of hazardous materials, which too often occurs behind the scenes and is becoming a more pressing issue as disposal sites fill up and waste piles grow.[56]

On the surface, the PIC procedure pales in comparison to the far-reaching Stockholm Convention. Essentially, it is a reporting requirement that helps establish a global information exchange system on pesticides. It is intended to be an early warning system to prevent the proliferation of pesticides and encourage the adoption of alternatives. PIC was initiated on a voluntary basis at the global level in the 1989 revision of the International Code of Conduct on the Distribution and Use of Pesticides. At the 1992 Earth Summit, governments agreed that PIC should have the status of an international convention. And by 1998, prior informed consent had made the transition from voluntary tool to global legal instrument. Although it is not yet in force, most countries already abide by it.[57]

The PIC procedure requires exporting parties to share information globally on chemicals and pesticides each country has banned or restricted nationally. The Convention's Chemical Review Committee considers such products and decides whether to place them on a list that will be subject to the PIC procedure. Listed chemicals cannot be traded until recipient countries have been informed and have consented to the import. The sender is obligated to comply with that country's decision, and the decisions are made public so that other countries can track them and see how they were made. PIC gives potential destination countries the power to choose what they will or will not accept, along with a growing basis of information in order to make that decision.[58]

The 1995 amendment to the Basel Convention takes the PIC policy to another level. As with PIC, the amendment is not yet in force but countries have agreed voluntarily to abide by its prohibition on shipments of hazardous wastes from industrial to developing countries. A blanket ban such as this will not only make it easier to detect illegal shipments, it will, at least in theory, force industrial nations—typically the source of hazardous waste—to deal with treatment and disposal themselves rather than dumping their wastes on poorer countries. Worldwide, some 300–500 million tons of hazardous wastes are generated each year, according to UNEP estimates, with industrial countries accounting for 80–90 percent of the total. With the Basel Ban, the Basel Convention recognized that free trade in hazardous waste was not acceptable, and that richer countries have a special responsibility not to externalize their pollution costs via exports.[59]

Although the ban was passed by consensus and is supported almost universally in developing countries, a few industrial nations still oppose ratification. In August 2001, U.S. State Department officials argued that the Basel ban may prevent some legitimate recycling activities and could inhibit trade. (The United States signed the Basel Convention in 1989 but has not yet ratified it.)[60]

Like the Basel Convention itself, a central point of disagreement on the hazardous waste trade ban concerns the term "recyclable." Some argue that recycling wastes is preferable to using virgin materials, and may help encourage proper disposal, and therefore that developing countries should be allowed to accept hazardous wastes for recycling. Environmentalists argue that the recycling of hazardous waste via export is usually a polluting enterprise, as there are inevitably quantities of the material that remain as pollution and expose workers in the recipient country to health threats. Further, they argue that such export provides a major disincentive to preventing hazardous waste and avoiding the use of toxics in the first place. One of the fundamental goals of the Basel Convention is to minimize the generation of hazardous waste and therefore its trade. The Basel Ban is seen as a way of implementing the convention, starting first with the industrial countries that produce the most waste and have the most resources to reduce toxicity and quantities of waste dramatically.[61]

Behind the trade in hazardous wastes is a larger story involving the economics of unused materials and stockpiles. Like illegal drug trafficking, illegal movements of hazardous wastes are hard to detect, thought to be underreported, and difficult to control. Tracking hazardous wastes from "'cradle-to-grave' when the cradle is in one country and the grave in another is nearly impossible," according to a recent study on hazardous waste flows under the North American Free Trade Agreement.[62]

Noting these difficulties, global networks of activists, such as the Basel Action Network, have sprung up to work on these issues. In January 2001, for example, a 20-ton shipment of obsolete mercury left the now defunct HoltraChem facility in coastal Maine, bound for India. With an alert sent out from U.S. activists to colleagues in India, the union of port workers there successfully blocked the ship from unloading its cargo there. The ship was last seen in Port Said, Egypt, but activists are unsure where the mercury finally ended up. The remaining 110 tons of mercury from this facility are still sitting in Maine, awaiting their fate.[63]

While banning chemicals is increasingly an accepted tool for reducing toxic burden, dealing with toxic wastes in ways that do not exacerbate the problem is harder to do. Incineration and burning can create dioxins and furans and other harmful pollutants. Similarly, disposal of hazardous wastes on land and at sea has backfired, leaking toxic compounds into the environment, dispersing the problem to larger areas, and allowing toxics to interact in unpredictable ways to form new compounds. Recycling of hazardous wastes is also a serious problem. Recycling mercury, for example, reintroduces this toxic metal into products that almost always have safer substitutes.

The scale of the waste problem is enormous. Nearly every nation in Africa now shares the legacy of some 50 years of international development aid: more than 200,000 tons of abandoned pesticides, about one third of which are thought to be POPs. Such stockpiles are continually creating problems of their own—from water degradation to acute human exposures—through improper storage and misuse and subsequent exposure. The situation is equally grave in the former Soviet Union. The reality is that much of the world's unwanted pesticides are housed in places that are least able to deal with their disposal. Most of the 53 nations in Africa, for example, lack the institutional capacity to remedy the situation, much less the labs to do the testing and site analysis or the medical personnel to treat victims of exposure. Expensive high-tech waste disposal methods are not an option in countries that rely on waste imports for quick cash.[64]

While the waste problem is not new, it is becoming more pressing. The global toxic waste pile is growing rapidly: plastics waste, such as PVC from short-lived items, continues to pile up, and we are near the end of the useful life span of "long-lived" (20–30 years) PVC materials such as pipes, siding, and other construction materials. Electronic waste is also mounting due to rapid obsolescence of computers and other electronic equipment and the manufacturers' lack of attempts to reduce toxic inputs in their products. The present toxic waste challenge could take on the dimensions of a crisis during the next two decades as thousands of tons of PCBs and other POPs are phased out, as called for in the Stockholm Convention.[65]

Even though the yearly emissions of many toxic compounds are now declining and well below peak levels, what has accumulated over the last several decades in the environment is what ultimately matters in terms of public health. Persistent toxins in soil, water, and even bedrock can be reactivated by human or natural causes (as happened with arsenic poisoning from wells in Bangladesh), essentially keeping the threat alive. Further, many new compounds are invented and put on the market each year without proper testing as to their long-term impacts on the environment. Minimizing the generation of new toxic wastes and finding ways to detoxify or store current wastes are essential to protecting health.

In combination with the POPs treaty, PIC and the Basel Ban will help stimulate more responsible chemicals management and a better informed public. But having individual companies and countries report their activities to designated national authorities and banning particular activities still may not be enough to reduce the use and generation of toxics and to dispose of toxic wastes safely. What is needed is a market-driven impetus to refocus our efforts upstream toward prevention rather than the ultimately hopeless efforts at an end-of-pipe cure. With more accurate information about the chemicals available, nongovernmental organizations (NGOs) and the general public can help force this change through innovative market-based programs, community-based monitoring systems, and other tools.[66]

Environmental Democracy and Markets

In October 2001, the Aarhus Convention on Access to Information, Public Participation in Decision-making and Access to Justice came into effect, thanks to wide support from a number of economies in transition. (This regional agreement applies to 28 countries in Western and Eastern Europe but is open to other governments.) It encourages more citizen participation in environmental issues and greater public access to information previously limited to government authorities. U.N. Secretary-General Kofi Annan has called the Convention the "most ambitious venture in the area of 'environmental democracy'" since Rio.[67]

Establishing the public's legal right to know what they are being exposed to dates back at least to 1986, when following the 1984 Bhopal disaster the U.S. Congress passed the world's first community right-to-know law, over strong protests from industry officials. The Emergency Planning and Community Right-to-Know Act created a national database of toxic emissions and releases by manufacturing plants. Known as the Toxics Release Inventory, the data allow citizens, companies, and the media to publicize the worst polluters and to bring public attention to the issues of toxic waste management. This helped drive down releases of an original core group of 300 chemicals by 45 percent between 1988 and 1999. Despite some notable limitations, the TRI system is continually being improved. In April 2001, for instance, the U.S. EPA drastically lowered the reporting threshold for lead, from 25,000 pounds to 100 pounds. Accordingly, information on hundreds of thousands of pounds of lead emissions that were never previously reported will become public beginning in 2002.[68]

Such systems of tracking chemicals and emissions by industry are catching on elsewhere. Since Rio, eight industrial countries and two developing nations—Mexico and the Slovak Republic—have implemented systems like the U.S. right-to-know laws. Several others—including Argentina, the Czech Republic, Egypt, and five former Soviet bloc nations—are expected to adopt similar systems soon. Public right-to-know also extends to product labeling systems, which are now used in a variety of settings from PVC-free toys and mercury-free thermometers to organically grown cotton T-shirts and chlorine-free bleached paper. Simply by telling consumers what is in a product and how it was made, these systems give the public the power to refuse to buy particular toxics. In addition to monitoring emissions, registers and labeling systems will help develop national POPs inventories, as called for in the Stockholm Convention. And they help remove the wall of corporate secrecy, encourage greater public participation, and provide a check against government and corporate abuses.[69]

The vibrant and vocal NGO network that sprung up during the U.N. POPs treaty negotiations provides ample evidence that greater public access to information does set the stage for greater citizen involvement. The more than 250 NGOs represented in the International POPs Elimination Network outnumbered the number of countries participating in the U.N. treaty by almost two to one.[70]

Increased citizen awareness and participation, whether in international negotiations or our own backyards, often translate into growing political support for change. In Mozambique, for example, local activists and political leaders successfully blocked the construction of a Danish-funded incinerator. The country has since banned incineration as a method to get rid of stockpiled pesticides. For an alternative, the government can look to demonstration projects now under way in Slovenia and the Philippines to treat PCB wastes with non-burn technologies that do not emit additional toxic byproducts in the process.[71]

Chemical bans have also prompted proactive responses from the regulated industry. In late 2000, for example, the Swedish Parliament called for a national ban on all persistent and bioaccumulative chemicals by 2020. The law puts the onus on industry to prove that a chemical is safe (an important aspect of the precautionary principle) rather than on government to show it is dangerous. While this may seem to discourage innovation, it has in fact spurred new research as manufacturers whose livelihoods appear to depend on toxic substances like lead have moved in a new direction. Orrefors Kosta Bod, a world-famous Swedish crystal glass company that dates back several generations, is exploring the use of barium instead of lead to give its crystal a similar luster but a lighter feel and a much safer product. As a company spokesperson says, "We will have to educate our customers not to choose their glass by weight but only by its beauty."[72]

Similar sentiments concerning the importance of corporate education and public awareness-raising are heard elsewhere. In anticipation of a global ban on TBT (the antifouling marine paint), for example, the World Wide Fund for Nature is now working with a number of shipping and paint companies to organize a buyers' group for TBT-free paint. Several companies have already agreed to use safer paints by the end of 2002. Likewise, many toy manufacturers have pledged to phase out phthalate-softeners from toys and other items that children use in response to a ban in the European Union, growing public concern in the United States and elsewhere, and the fear of losing business worldwide.[73]

Taxes and fiscal policies can further support the progress made in parliaments and boardrooms. Since 1970, for example, the Netherlands has had great success in toxics reduction by charging households and companies for discharges of heavy metals. Originally intended to raise revenues, levies based on the quantities of toxics released—combined with a permitting system—proved to be effective incentives for companies to treat their own discharges or switch to cleaner processes. (See Figure 4–4.)[74]

Similar efforts have been undertaken with pesticides and gasoline. Sweden, for instance, has a pesticide tax that adds a 7.5-percent surcharge for every kilogram of active ingredient purchased. This was one of a set of government initiatives that helped Swedish farmers cut their pesticide use by 65 percent from 1986 to 1993. Many countries have reduced their consumption of leaded gas by taxing it at a higher rate than unleaded gas. Malaysia, for example, made unleaded gas 2.7 percent cheaper than leaded, which increased the share of unleaded to 60 percent of the total. Unleaded fuel was first available in 1991 in Singapore; by 1997, it accounted for 75 percent of the gas used there, thanks to differential gas taxes. Twenty industrial countries introduced differential taxes at the same time they implemented other policies, such as stricter emissions controls, thereby accelerating the shift from leaded to unleaded gas.[75]

Combining the influence of financial markets with the power of the news media has helped reduce pollution in a number of communities around the world. It is an especially powerful incentive in countries where monitoring is lax and enforcement is weak, so that polluters typically have little incentive to change their ways. In an experiment in Indonesia, for instance, government officials publicly graded factories using a color-coded system: black for those that made no attempt to manage wastes, red for significant violators, blue for those that met national standards, and green for those that went beyond what was required.

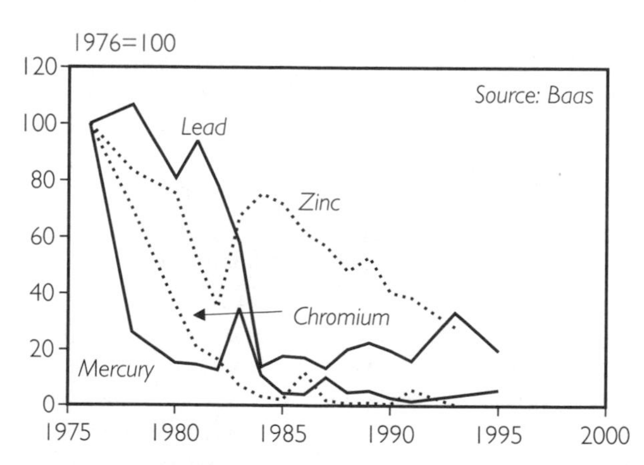

Figure 4–4. Industrial Discharges of Chromium and Zinc, 1976–93, and of Lead and Mercury, 1976–95, into Regional Surface Waters, Netherlands

Shortly after a highly publicized awards ceremony, companies that had regularly ignored regulators started asking how they could improve their grade. Within 18 months, water pollution from the 187 pilot plants fell by 40 percent.[76]

While we clearly need to scale up these and other efforts, an important step in the transition away from toxics—defining what tools should be used—has largely been achieved. This frees up intellectual capital to focus on the more fundamental and challenging task of developing safer materials, products, and processes.

Technological Changes and Opportunities

"We have invested heavily in addressing the effects of the materials in our economy while mostly ignoring the materials themselves," writes Ken Geiser, Director of the Massachusetts Toxics Reduction Institute and author of a new book on materials. In terms of toxicity, Geiser argues that we have barely begun to scratch the surface of opportunities for reduction. Indeed, few sectors of the global economy have been scrutinized in terms of their use of toxic chemicals, let alone subject to actual change. One notable exception is agriculture, where much work has gone into adopting and improving farming methods that are safer for farm workers, consumers, and the environment. But for much of the rest of our economy, opportunities to reduce our use of toxics abound. As an official at the New Jersey Department of Environmental Protection recently stated in an interview on toxics and pollution prevention, not only is "low hanging fruit" going unpicked, some is "rotting on the ground."[77]

In response to human and ecological health concerns, increasing numbers of farmers are abandoning the pesticide treadmill that makes farmers reliant on expensive synthetic chemicals in favor of farming techniques that use pesticides only as a last resort or that avoid them entirely. (See Chapter 3.) Recently, for instance, thousands of rice farmers in China demonstrated that growing multiple varieties of rice in the same paddies could double yields without the use of any synthetic chemicals. In the U.S. Midwest, farmers who produce grain and soybean organically are finding that their net profits equal or surpass those from conventional production, even when they do not charge the premium prices that organic crops generally command.[78]

Many players in the solvents industry have begun to search for—and implement—safer alternatives.

Lucrative global markets—more than $25 billion produced a year in at least 130 countries—combined with growing consumer preferences and labeling have helped make organic food a major influence in world food markets. Currently, between 3 and 5 percent of European food is grown organically. (With 25 percent of the world's pesticides used in household settings and on commercial properties, and with pesticide use in this sector rising, the next step is to apply nonchemical methods of pest control in schools, hospitals, public parks, and private homes and yards.)[79]

The use of pesticides to protect public health is also coming under increasing scrutiny by environmentalists and health professionals. Under the Stockholm Convention, some two dozen tropical countries that need DDT to fight malaria-carrying mosquitoes will be allowed to continue spraying. Indeed, malaria's lethal grip on humanity is the reason DDT is still in use at all: some 950 people become infected every minute by this modern-day plague.[80]

But alternatives are increasingly available here too. Researchers in sub-Saharan Africa have demonstrated that bednets with small amounts of humanmade pyrethroids, which are natural insecticides found in plants, can reduce the transmission of malaria by preventing mosquitoes from biting people who are asleep. Combined with other prevention and treatment strategies, these bednets can prevent half of all childhood deaths from malaria. In addition, they are easily introduced at the local level and relatively cost-effective: $10 for a bednet plus $1 for a year's supply of insecticide. Over the next five years, the Roll Back Malaria program, which involves WHO, the World Bank, and numerous bilateral agencies, is planning a thirtyfold increase in the use of bednets in Africa. Uganda and Tanzania have already reduced taxes on nets to make them more affordable.[81]

By using the least toxic option first, and knowing the ecology of *Anopheles,* the malarial parasite's mosquito host, health officials are beating back this deadly disease in some areas. Although reducing the use of DDT is a primary goal of the POPs treaty, this pesticide will remain in the arsenal of public health protection—and rightly so—until all areas at high risk of malaria have suitable alternatives in place. South Africa's recent experience—a rapid and deadly comeback of malaria following the emergence in 1996 of mosquito resistance to alternative insecticides—has meant the reintroduction of controlled DDT spraying in homes until the outbreak is brought under control.[82]

The same principles at work in organic agriculture and public health campaigns—use the least toxic option first and know your enemy—are equally applicable to the vast range of chemical-intensive processes in our economy. Chlorinated solvents, for example, are "one of the largest and most easily phased out" compounds, according to Joe Thornton, author of a recent book on chlorine. The key phrase is "phase out," since these highly volatile substances are so difficult to contain. Many players in the solvents industry have begun to search for—and implement—safer alternatives. The classic case involves chlorofluorocarbons (CFCs), a group of compounds with a wide range of uses, from aerosol propellants to refrigerants, whose output dropped 87 percent between 1988 and 1997—prompted by the Montreal Protocol that targeted CFCs because they deplete the ozone layer that protects Earth from harmful ultraviolet radiation. Technical ingenuity and innovation on the part of manufacturers played a big role in this international success story.[83]

Because solvents—indeed all chemicals—cost money to use and dispose of properly, phasing them out with safer substitutes makes good economic sense. A 1994 Massachusetts study reported that buying chemicals and disposing of contaminated waste accounted for up to 85 percent of operating costs in companies that regularly used solvents. Moreover, these same companies found that replacing chlorinated solvents with safer alternatives yielded considerable health and environmental benefits as well as economic savings. Most companies in the study reaped enormous benefits by replacing solvents with safer, often water-based alkaline solutions: all but one saved at least 75 percent in net operating costs. The benefits demonstrated in this survey and through the Montreal Protocol have been replicated by numerous multinational companies.[84]

Supplementing these achievements, researchers have made promising advances in "green chemistry." Such efforts have typically focused on finding environmentally benign feedstocks, reagents, catalysts, and chemical products. A variety of traditional industrial materials are now commercially available in bio-based form, and their production is growing steadily. (See Table 4–4.) One company has developed plates, bowls, and other food containers from a mix of potato starch, limestone, and post-consumer recycled fiber. The packaging has been used by several hundred McDonald's restaurants and is being tested in the cafeteria at the U.S. Department of Interior. It is biodegradable and consumes significantly less energy throughout its existence than either polystyrene plastic or paper, which are typically used.[85]

While recent and ongoing research in plant-based industrial materials is gradually gaining a toehold in the market, much of the work remains behind the scenes of commercial markets, off in laboratories. But those involved in such efforts predict that a major breakthrough is closer than it might appear. In the next few years, companies will be building plants that use bio-based materials, predicts Pat Gruber, Vice-President for Technology at Cargill Dow. Her company has invested $300 million to build the world's first facility to produce plastic from corn sugar, known as polylactide polymers, which is an alternative to traditional petroleum-based plastics. Although the processing methods for these and other polymers are still in their infancy, notable technical improvements are expected. Combined with the use of agricultural wastes (rather than high-grade sugars) as the feedstock material and the entrance of several large research companies, plant-based chemical manufacturers and plastics producers could be competitive with high-

Table 4–4. U.S. Industrial Materials Derived from Plant Matter, by Production Volume and Share of Total, 1992 and 1996

Product	Production, 1996	Share of Total 1992	Share of Total 1996
	(million tons per year)	(percent)	(percent)
Wall paints	7.8	3.5	9.0
Specialty paints	2.4	2.0	4.5
Pigments	15 .0	6.0	9.0
Dyes	4.5	6.0	15.0
Inks	3.5	7.0	16.0
Detergents	12.6	11.0	18.0
Surface cleaning agents	3.5	35.0	50.0
Adhesives	5.0	40.0	48.0
Plastics	30.0	1.8	4.3
Plasticizers	0.8	15.0	32.0
Acetic Acid	2.3	17.5	28.0

SOURCE: Kenneth Geiser, *Materials Matter: Toward a Sustainable Materials Policy* (Cambridge, MA: The MIT Press, 2001), p. 262.

volume petroleum-based ones in the next decade or so, if not earlier.[86]

Another promising avenue is the use of plants to absorb and break down toxic metals and pollution, a field known as phytoremediation. University of Florida chemists have found ferns that can accumulate up to 200 times as much arsenic as in highly contaminated soil. In some tests, as much as 2.3 percent of the plant was composed of this toxic metal. Currently, phytoremediation accounts for just 1 percent of the $8-billion environmental remediation market in the United States. But a number of plants, including sunflower, poplar, clover, mustard, and some herbs, can serve as the botanical equivalent of detox centers for polluted soil and water, often working in conjunction with the fungi and bacteria that thrive in the plants' roots and soil.[87]

Although several hundred plant species worldwide have been identified as potential "pollution sponges" for toxic compounds, they do, however, come with a number of cautions: the plants can become so toxic that they must be treated as hazardous waste and kept away from animals, insects, and people; some chemicals may evaporate from the leaves; and although some compounds may break down in plants, this is not true for elements. While they should not be used to justify greater waste generation, these living sponges are already proving useful to contain and concentrate the problem of toxic wastes.[88]

Progress in other cutting edge fields is falling short. To date, advanced and engineered materials that offer significant potential to reduce total materials use have not been adequately tested for toxicity. These include composites and super alloys that are synthesized from byproducts of conventional materials, nanotechnology that requires less materials because equipment is so tiny, and so-called smart materials that change their properties in response to environmental conditions. "For all that is impressive and intriguing about these materials, it is disappointing to consider how little attention has been paid to their effects on human health or the environment....Seldom are even the most obvious health or environmental effects of production or disposal considered," writes Ken Geiser of the Massachusetts Toxics Reduction Institute. In *Materials Matter*, he makes a strong case for materials sciences to integrate the issues of human and environmental health effects as primary design factors along with the traditional concerns for performance, processing efficiency, and cost.[89]

Even before such a fundamental shift can take place in the scientific underpinnings of our economy, consumers can take the lead and demand safer products. This consumer mobilization will not only help spur the transition away from toxic materials in the near term, but also begin to build the political support for lawmakers to make the deeper reforms in our economic and scientific systems that will let us reach far beyond the "low hanging fruit."

Moving Forward

In early 2001, the U.N. Commission on Human Rights declared that living free of pollution is a basic human right. With a number of treaties, programs, and community efforts under way to reduce toxics use and waste, and with the Stockholm Convention expected to be fully ratified as early as 2003, the next decade marks an era of enormous opportunity to give life to this declaration and make the planet a safer and healthier place.[90]

Although toxic chemicals are a unique part of the materials economy, production and consumption of chemicals are just as much a reflection of overconsumption as the volume of material used is. When people think of overconsumption, they typically envision denuded forests, polluted inland and coastal waters, and extinct animals. But the visible stockpiles of chemical substances in our landfills and abandoned industrial sites, as well as those that collect unseen in our bodies, are no less a reflection of global overconsumption of materials. In many ways, it is a more pernicious form of overconsumption. Much of it is undetected and will remain a threat for generations to come, owing to its persistent nature. Moreover, these compounds interfere with normal biological functioning of species in ways we have only begun to identify, let alone fully comprehend.

The key to addressing the challenge of toxics use and wastes rests on a fairly straightforward principle: harness the innovation and technical ingenuity that has characterized the chemicals industry from its beginning and channel these qualities in a new direction that seeks to detoxify our economy. Chemicals and materials researchers will need to make concerted efforts to find nontoxic alternatives. The primary purpose of research should be to find safer substitute materials, products, and processes for those that now contribute to our global toxic burden. Proving the necessity of toxic chemicals should also be foremost in the minds of producers, consumers, and policymakers alike. Only by realigning our uses of chemicals closer to those found in nature will we build an economy that is more accountable to the environment and ourselves.

WORLD SUMMIT PRIORITIES ON CHEMICALS

Short-term

- Phase out leaded gasoline globally.
- Ratify the three major global toxics treaties (Stockholm, Basel, and Rotterdam).
- Secure funding for research on alternative materials and environmentally sound methods of waste disposal.

Long-term

- Adopt a uniform and mandatory system of reporting toxics use and releases.
- Tax commercial and residential pesticide use.
- Eliminate persistent compounds in dissipative uses, such as agricultural pesticide spraying and cleaning agents.
- Minimize the release of mercury, lead, and other toxins as byproducts from the mining of metallic ores and other industrial sources.
- Reduce and eventually phase out coal-based power generation.

Chapter 5

Redirecting International Tourism

Lisa Mastny

Until recently, Kovalam, a small fishing village in India's Kerala state, could not keep up with its rising popularity. Attracted by palm-lined beaches, friendly people, and a relaxed lifestyle, visitors from as far away as Europe began descending on the region in the mid-1960s. Over the next two decades, investors rushed in to meet the demand, building row upon row of new hotels, restaurants, and souvenir shops. But in 1993, the tourist stream—and the revenue it brought—began to slow. By 2000, visitor numbers had dropped by as much as 40 percent.[1]

Tourism experts ruled out economic factors and shifting tourist tastes, and finally attributed the decline to rising visibility of the community's waste management problems. Like many booming destinations in the developing world, Kovalam has no formal plan to deal with the mounting levels of trash generated by tourists. Hotels and other facilities collect recyclable items, such as glass, paper, and metal scraps, for reuse by local industries whenever possible. The less desirable refuse—including human waste, plastic bottles, and other non-biodegradables—simply piles up in towering mounds or is dumped into nearby streams, posing risks of cholera and other disease. Yet according to Jayakumar Chelaton, a local activist, "Nobody bothers about the health issues faced by the locals.... Everybody wants Kovalam beach to be clean so it can get more business."[2]

These concerns are not unique to Kovalam. Increasingly, developing countries are turning to tourism as a way to diversify their economies, stimulate investments, and generate foreign-exchange earnings. Tourism can be a lucrative and less resource-intensive alternative to growing a single cash crop or to traditional industries like mining, oil development, and manufacturing.[3]

Yet tourism is one of the world's least regulated industries, which has serious implications for ecosystems, communities,

An expanded version of this chapter appeared as Worldwatch Paper 159, *Traveling Light: New Paths for International Tourism.*

and cultures around the world. Hotels, tourist transport, and related activities consume huge amounts of energy, water, and other resources and generate pollution and wastes, often in destinations that are unprepared to deal with these impacts. And many communities face cultural disruption and other unwelcome changes that accompany higher visitor numbers. Although fears of terrorism and the safety of air travel have dampened interest in much international travel for the time being, over the long term the demand for tourism is expected to resume its rapid rise.[4]

Many governments, industry groups, and others are promoting "ecotourism"—responsible travel that generates money and jobs while also protecting local environments and cultures. While it does succeed in some circumstances, ecotourism can suffer from many of the same environmental and social pitfalls as conventional tourism, including using resources irresponsibly, creating waste, and endangering ecosystems. In some cases, it is little more than a "green" marketing tool for enterprises hoping to promote an environmentally conscious image.[5]

As tourism's impacts, both good and bad, continue to spread, it is increasingly important to redirect activities onto a more sustainable path. This will require deep sectoral changes that reach far beyond the scope of ecotourism. A broad range of stakeholders—including governments, the tourism industry, international organizations, nongovernmental groups, host communities, and tourists themselves—will need to be involved with sustainability efforts at all levels.

By redirecting tourism, these groups can not only enhance the benefits of tourism, but also help meet many of the goals of *Agenda 21*, the blueprint for sustainability agreed to at the 1992 U.N. Conference on Environment and Development in Rio de Janeiro. These include generating jobs and revenue, protecting the environment, and strengthening cultural diversity. As the World Summit on Sustainable Development approaches in September 2002, many groups are building coalitions on some of the key issues. The challenge is making sure that this activity translates into measurable progress.[6]

A Global Industry

The World Tourism Organization (WTO), an intergovernmental research and support group based in Madrid, defines tourism as the activities of people who travel "outside their usual environment" for no more than a year for leisure, business, and other purposes. Since 1950, the number of international tourist arrivals has increased nearly twenty-eight-fold, reaching 698 million in 2000. (See Figure 5–1.) These numbers are expected to again double by 2020, to 1.6 billion, although all estimates cited in this chapter were made before the September 2001 terrorist attacks. The figures also do not include the millions of people who travel within their own countries—the bulk of the world's tourists, and a figure that would make estimates between 4 and 10 times higher, depending on the location.[7]

Rising disposable incomes, along with the emergence of wide-bodied commercial jets, cheap oil, and low promotional airfares after World War II, have accelerated tourism's growth. And new information technologies like global distribution systems, computer reservation systems, and the Internet enable travel agents as well as individual travelers to check flight availabilities, issue tickets, and make reservations rapidly. Between 1997 and 2000, the number of online bookings of flights and other travel-related services increased fivefold, to

25 million, according to the Travel Industry Association of America.[8]

Despite these numbers, tourism remains restricted to a tiny, more affluent share of the world's population. Nearly 80 percent of international tourists come from Europe and the Americas, while only 15 percent come from East Asia and the Pacific and 5 percent come from Africa, the Middle East, and South Asia combined. Yet even these figures are deceptive: in the United States, a leading source of tourists worldwide, fewer than a fifth of citizens hold valid passports. All told, annual international tourist arrivals represent just 3.5 percent of the world's population. This share is expected to double to 7 percent by 2020 as global prosperity increases and the cost of travel continues to drop.[9]

Nearly two thirds of international tourists travel for vacation, leisure, and recreation as opposed to visiting friends and relatives or health and religious factors. But tourist tastes are gradually changing. According to researcher Auliana Poon, growing displeasure with heavily commercialized, overrun, and polluted destinations is spurring a shift from the highly packaged and standardized mass tourism of the past half-century. In its place, rising numbers of more flexible and independent travelers are pursuing more personalized experiences like culture or nature tourism. A study of U.S. travelers in the early 1990s supported this shift: while 20 percent of respondents were "after the sun," 40 percent sought more "life-enhancing" travel.[10]

These broader trends are reflected in surveys of the most popular destinations worldwide. Although Europe and the Americas continue to attract the most international tourists (the majority from within the regions themselves), the traditional dominance of these destinations is declining. (See Figure 5–2.) Meanwhile, tourism to and within Asia, the Middle East, Africa, and South Asia is growing rapidly. The share of international tourists traveling to East Asia and the Pacific rose from just 1 percent in 1950 to 16 percent in 2000. By 2020, this region is expected to be the most popular destination after Europe. China is expected to unseat France as the world's most visited country, and also to become the fourth largest source of tourists worldwide—behind Germany, Japan, and the United States. Russia and several former Eastern bloc countries also rank among the top destinations of the future.[11]

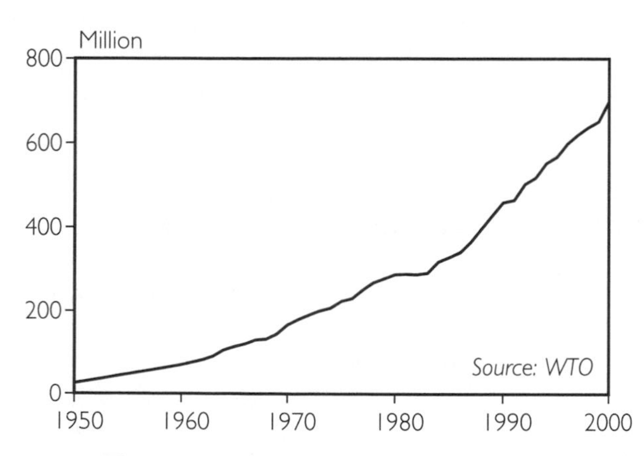

Figure 5–1. International Tourist Arrivals, 1950–2000

As it spreads geographically, tourism is assuming a greater role on the world economic stage, but the complex nature of tourism activities makes measuring this contribution difficult. WTO estimates that between 1975 and 2000, international tourism receipts—the revenue generated from tourist spending abroad on such items as lodging, food, entertainment, tours, and in-country transport—

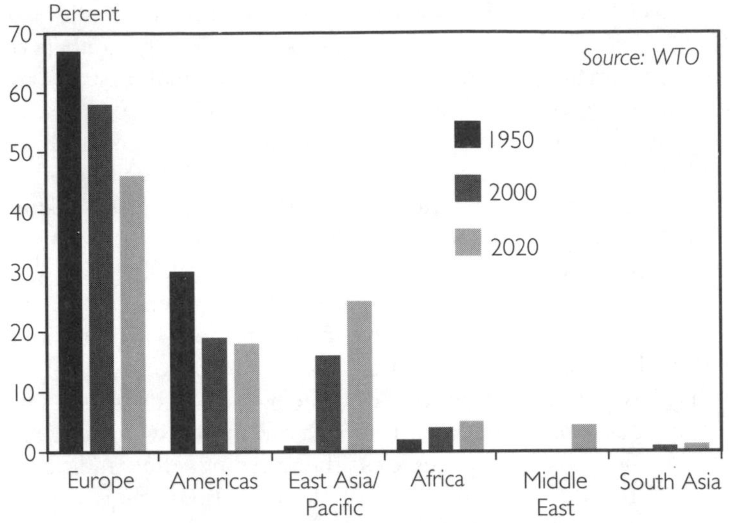

Figure 5–2. Share of International Tourist Arrivals, by Region, 1950 and 2000, with Projections for 2020

grew 35 percent faster than the world economy as a whole, reaching $469 billion in 2000 (in 1999 dollars). Countries in Europe and North America dominate the lists of tourism's top 10 spenders and earners; China is the only developing country on either list. (See Table 5–1.)[12]

Tourism represents a rapidly rising share of world trade. Any tourism service that a visitor buys when traveling abroad is considered an export from the country being visited. In 1999, tourism accounted for more than 40 percent of exports of services and nearly 8 percent of total world exports of goods and services—surpassing trade in such items as food, textiles, and chemicals. And its predominance in trade is widespread: according to WTO, it ranks among the top five export categories for 83 percent of countries and is the leading source of foreign-exchange earnings for at least 38 percent of them.[13]

Another way to measure the economic impact of tourism is to look at its wider effects throughout a country. This means, for instance, calculating not just direct tourist spending, but also the indirect effects of this spending, such as generating income for farmers or creating jobs in construction. Measuring this wider reach, the World Travel & Tourism Council, a London-based industry group, estimates that travel and tourism accounted for some $3.6 trillion of economic activity in 2000—or roughly 11 percent of gross world product, making it the world's largest industry. Direct and indirect tourism activities also supported an estimated 200 million jobs in 2000, representing 8 percent of world employment—1 in every 12 jobs.[14]

Like other sectors in today's global economy, tourism is becoming increasingly centralized. In 1998, the world's 10 leading airlines earned an estimated two thirds of the profits of all airlines that are members of the International Air Transport Association. And in 1999, the top five hotel chains—among them Marriott International, Bass Hotels and Resorts, and Choice Hotels International—managed roughly 14 percent of the world's hotel rooms. Meanwhile, four European tour operators alone handled trips for some 50 million tourists in 2000.[15]

A driving factor behind this rampant centralization is the unregulated nature of the tourism industry compared with other service sectors. It is increasingly easy for international businesses interested in tourism development to enter markets worldwide. This is especially true as more

Table 5–1. Top 10 Spenders and Earners of International Tourism Receipts and Share of Total, 2000

Spenders	Share of Total (percent)	Earners	Share of Total (percent)
United States	14.0	United States	18.0
Germany	10.0	Spain	6.5
United Kingdom	7.7	France	6.3
Japan	6.6	Italy	5.8
France	3.6	United Kingdom	4.1
Italy	3.2	Germany	3.7
Canada	2.6	China	3.4
Netherlands	2.5	Austria	2.4
China	n.a.	Canada	2.3
Belgium/Luxembourg	n.a.	Greece	1.9
Top 10 Total	50.2	Top 10 Total	54.4

SOURCE: See endnote 12.

and more governments privatize national airlines and other state services, reduce domestic subsidies, embrace market reforms, and liberalize trade and investment policies more generally. Many developing countries, in particular, are opening their markets to tourism in an effort to improve their chances on the world economic stage. But whether this actually brings widespread benefits will depend on the extent to which governments and the industry balance the drive for more tourists with the need for more socially and culturally responsible tourism.[16]

A Force for Development?

From Asia to the Caribbean, the developing world has experienced a phenomenal surge in tourism in recent years. One in every 5 international tourists now travels from an industrial country to a developing one, up from only 1 in 13 in the mid-1970s. Rapidly growing destinations include Cambodia, Egypt, Thailand, Turkey, and Viet Nam. In the Caribbean, arrivals to Cuba have risen more than fivefold since 1990. Overall, tourism growth rates in the developing world are expected to exceed 5 percent a year through 2020, outpacing both the world average as well as anticipated growth in industrial countries. (Again, the effect of the recent terrorism events on these projections is not yet clear.)[17]

Across the developing world, governments are pouring money into tourism marketing, infrastructure projects like roads and hotels, and both large and small tourism businesses. To stimulate investments, many countries are offering promotional assistance as well as economic incentives like tax and import duty exemptions, subsidies, and guarantees. By luring tourist dollars, they hope to diversify their economies and attract the foreign exchange needed to reduce heavy debt burdens, pay for imports, strengthen domestic infrastructure, and boost social services like education and health care.[18]

Leading international lenders such as the World Bank and the International Monetary Fund (IMF) are behind many of these efforts. In 2000, the Bank's private-sector arm, the International Finance Corporation, supported some $500 million in tourism-related projects, including hotel rehabilitation and urban revitalization. The IMF, meanwhile, promotes tourism as an important export strategy under its economic structural adjustment policies.[19]

In gross economic terms, these investments are beginning to pay off. In a survey of the world's 100 poorest countries done for the U.K. Department for International Development, researchers found that tourism is "significant"—that is, it accounts for at least 2 percent of the gross domestic product (GDP) or 5 percent of exports—in nearly half of the countries in the lowest income range and almost all in the lower-to-middle income range. The study also found that tourism is significant or growing in all but 1 of the 12 countries that are home to 80 percent of the world's poor. In the world's 49 so-called least developed countries, most of which are in Africa and Asia, tourism is now the second largest source of foreign exchange after oil. In some small island nations in the Caribbean and Pacific, it brings in more than 40 percent of GDP.[20]

The World Trade Organization reports that tourism is the only economic area where developing countries consistently run a trade surplus. And its importance in trade is growing. In 1999, international tourism receipts represented two thirds of services exports in these countries and more than 10 percent of total exports. (In industrial countries, meanwhile, tourism accounted for only about one third of services exports and 7 percent of total exports.)[21]

Despite the potential benefits, however, some countries still invest very little in tourism. Their governments either lack the internal economic capacity or face serious geographic and political obstacles to tourism development. The South Pacific island of Kiribati, for instance, is remote from other tourist centers, while the Solomon Islands and Vanuatu are particularly vulnerable to earthquakes and other natural disasters. The tourism industry of Sierra Leone, after a decade of strong growth, collapsed in the late 1990s as a result of the country's civil war and economic decline. Even Brazil, with its vast land area and high tourism potential, spent only an estimated 2 percent of its public budget on tourism-related activities in 2000, compared with a world average of more than 5 percent.[22]

Yet even in developing countries that do attract growing numbers of tourists, the on-the-ground benefits are not always as significant as the statistics suggest. WTO estimates that as much as 50 percent of the tourism revenue that enters the developing world ultimately "leaks" back out in the form of profits earned by foreign-owned businesses, promotional spending abroad, or payments for imported goods and labor.[23]

Leakage is particularly high in the Caribbean, where 50–70 percent of tourism earnings go toward acquiring imports—from skilled staff to food and consumer goods. Many hotels and other tourism businesses in the developing world rely heavily on foreign inputs either because the host country lacks the luxury goods and services that many tourists demand or because the tourism sector is so poorly linked with other sectors like fishing, agriculture, manufacturing, and transport that it is difficult to procure local supplies.[24]

Today, an estimated 90 percent of the world's tourism enterprises are small businesses, from family-owned restaurants to one-person snorkeling operations. Yet governments are under increasing pressure to grant large-scale investors—including international airlines, hotel chains, and tour operators—easier access to tourism assets. Under a special economic relations treaty with the United States, for example, Thailand is obligated to grant companies owned and operated by U.S. investors the same legal treatment as those owned by Thai nationals. Across the developing world, the increase in foreign investments, mergers, and franchising arrangements threatens to crowd out smaller, local operators who are unable to compete.[25]

Foreign operators dominate the tourism industries of many countries, including Kenya, Tanzania, and Zimbabwe. These businesses typically send their profits back home, leaving little revenue at the destination. Meanwhile, the bulk of a tourist's spending, including the biggest purchases like airline tickets, tour packages, and rental cars, occurs in the home country and never even reaches the destination. According to one estimate, if both the hotel and airline are foreign-owned, as much as 80 percent of a traveler's spending is lost to these businesses. Cruises and other all-inclusive tour packages that cover not only airfare, lodging, and transport but also documentation fees, entertainment, and meals are notorious for funneling away tourism revenues.[26]

The small Central American nation of Belize, which has seen rapid tourism growth in recent years, has witnessed many of these problems. In the early 1990s, expatriates accounted for roughly 65 percent of the membership in the Belize Tourism Industry Association, and an estimated 90 percent of the country's coastal development was in foreign hands. Many Belizeans oppose the rapid growth in luxury resorts and villas yet are unlikely to be able to afford to buy the land back. Meanwhile, the presence of foreign investments, together with the higher purchasing power of tourists, has boosted local inflation, raising property and food prices.[27]

These problems will only intensify as countries implement new international trade and investment policies that give even greater advantage to foreign investors. The General Agreement on Trade in Services (GATS), a 1994 multilateral trade agreement aimed at liberalizing service industries, requires governments to remove subsidies and protections on local enterprises and makes it considerably easier for foreign businesses to establish franchises, transfer staff, and repatriate profits. So far, at least 112 countries are committed to opening up their tourism markets under GATS—more than for any other economic sector—suggesting strong international interest in stimulating tourism investments. A second trade measure, the Agreement on Trade Related Investment Measures (TRIMS), makes it harder for governments to require foreign companies to use local materials and input.[28]

In the world's 49 least developed countries, tourism is the second largest source of foreign exchange after oil.

The employment effects of tourism are also mixed. Roughly 65 percent of the new jobs created by tourism annually are found in the developing world, including positions in restaurants, tour companies, and construction. On small islands like the Bahamas, the Maldives, and Saint Lucia, as many as 45 percent of jobs are tourism-related. Many of these positions go to women: on average, women account for 46 percent of workers in the hotel, catering, and restaurant sectors—a much higher share than in labor markets overall. Women also do much of the informal tourism work, such as running food and craft stalls.[29]

For many people, working in tourism provides a welcome alternative to unemployment. Yet more often than not, foreign or city-based workers hold the more lucrative management positions in tourism, leaving residents with low-wage service jobs—porters, maids, or laborers—that offer little opportunity for skill-building. The International Labour Organisation reports that tourism workers earn 20 percent less on average than workers in other economic sectors. And many of these positions do not meet international labor or other standards: some 13–19 million children under the age of 18 now work in tourism, roughly 2 million of whom have been lured into the booming "sex tourism" industries of Southeast Asia and Latin America, where they risk exposure to AIDS and other sexually transmitted diseases.[30]

Tourism can also divert people from traditional jobs in agriculture and fishing, tightening the local labor supply and increasing dependence on external suppliers. In Grenada, the government is replacing small-scale, organic agricultural plots with large tourism resorts in a push to secure foreign investment, squeezing out local farmers. But if economies become too narrowly dependent on tourism, they are more vulnerable to a collapse resulting from changing tourist tastes or other factors, such as fear of international terrorism. As the aftermath of the September 2001 terrorist attacks illustrated, tourism workers are usually among the first to feel the effects of global insecurity or an economic downturn.[31]

Tourism also has impacts on local cultures. On the one hand, it can heighten respect for minority groups, helping to revive languages, religious traditions, and other practices that might otherwise be lost. Tourist demand for dancing and other arts has reportedly fostered an artistic revival in Bali, Indonesia, while in Peru, rising tourist interest in traditional healing has spurred a resurgence of shamanism. But indigenous communities often end up the "featured attractions" of ventures they have had little input in designing. Industry promoters may reduce entire cultures to brochure snapshots, a depiction of local culture that can ultimately affect community self-perception and behavior. In the Himalayas, for instance, rising tourist interest in Buddhist festivals has led monks to shorten elaborate rituals to satisfy tourist attention spans and has spurred a black market in religious artwork. Meanwhile, local involvement in the events has dropped off.[32]

In general, it is difficult to separate the changes that tourism brings to communities from the wider effects of globalization, westernization, and rising economic prosperity. But tourism can accelerate the influx of western values and material goods into indigenous areas, spurring changes in eating, dress, and other daily activities. Forest tribes in Peru, for example, can now earn more selling traditional cloaks to tourists than trading them for axes or machetes—a shift that has reportedly altered the economic relations of villages. Increased contact with mainstream culture can also lead to replacement of a native tongue with a dominant language, while the promise of tourism and other employment can lure younger members of a community away, threatening its long-term sustainability.[33]

In extreme cases, native communities have been forcefully evicted from their homelands to make room for tourists. In the 1950s, Kenya's colonial government drove the nomadic Masai from their traditional grazing lands to accommodate safari lodges and visitors to the newly created national parks and wildlife sanctuaries. The Masai have since won greater involvement in the management and use of their resources and have been able to meet community needs with money earned leasing their land. Yet they still face the social and cultural repercussions of tourism, including misrepresentation of their crafts and rituals and a rising incidence of prostitution, alcoholism, and drug use. More recently, Burmese authorities gave the 5,200 residents of Pagan only two weeks notice before evicting them in 1990 and turning the ancient pagodas where they lived into a tourist attraction.[34]

Despite the potential negative impacts, many communities still favor increased tourism because they see greater economic and cultural opportunities. Some indigenous groups, such as Panama's Kuna people, hope to maximize the benefits of tourism while fighting some of the unwelcome changes. In 1996, the Kuna ratified a Statute on Tourism that limits the number of hotels yet ensures the collection of tax revenue and the redistribution of benefits among community members.[35]

Environmental Impacts of Tourism

As soaring air travel brings many of Earth's most ecologically fragile destinations within easy reach, concern about tourism's environmental impacts is rising. Travelers from industrial countries often try to replicate their own high consumption lifestyles, increasing the pressures on ecosystems and resources. Yet few developing-country governments have the capacity to protect their attractions adequately from all these new visitors.[36]

Tourism's environmental impacts can begin even before arrival. Studies suggest that as much as 90 percent of a tourist's energy consumption is spent in getting to and from the destination. Increasingly, the passenger jet is overtaking the automobile as the primary means of tourist transport: an estimated 43 percent of international tourists now fly to their destinations, while 42 percent travel by road and 15 percent use either ship or rail. Air travel has been particularly important in the developing world, where in some countries at least 90 percent of tourists arrive by plane.[37]

Unfortunately, air transport is also one of the world's fastest growing sources of emissions of carbon dioxide and other greenhouse gases, responsible for global climate change. The Intergovernmental Panel on Climate Change reports that aircraft emissions contributed roughly 3.5 percent of human-generated greenhouse gases in 1992—and this share is expected to rise steadily as air travel increases. Tourists currently account for about 57 percent of all international air passengers. But they may be responsible for a much higher share of jet fuel use because they tend to travel longer distances.[38]

Once tourists arrive at their destinations, their choices of where to sleep, eat, shop, and be entertained increasingly come at the expense of the environment. Natural and rural landscapes are rapidly being converted to roads, airports, hotels, gift shops, parking lots, and other facilities, leading to deterioration of the scenery, wildlife habitats, and other sites that are the attraction in the first place. The number of hotel rooms worldwide increased by more than 25 percent between 1990 and 1998, to an estimated 15 million. The trend is toward larger hotels, particularly in the newer destinations. At the world-famous Victoria Falls, shared by Zambia and Zimbabwe, a new multimillion-dollar hotel was recently built only a few meters from the water. The Zambezi River there is already polluted with detergents, uncollected garbage, and human waste from existing hotels—the result of poor regional tourism planning.[39]

At coastal destinations in the Caribbean and elsewhere, construction methods like sand mining and dredge-and-fill have destroyed dunes and wetlands, caused groundwater supplies to become brackish, and stirred up nearby waters—choking coral reefs and diminishing fish populations. In Cancun, Mexico, large swaths of mangrove forests, salt marshes, and other wetland areas that harbor wildlife and protect reefs have been cleared and filled to make room for resorts, piers, and marinas. These areas are often shored up with topsoil scraped from inland wetland savanna areas, resulting in the disruption of two valuable ecosystems. Currently, some 21 new resort complexes are being built along Mexico's Yucatan coast—a construction frenzy expected to nearly triple the number of area hotel rooms to 24,000.[40]

In the Cayman Islands, more than 120 hectares of reefs have been lost as a result of cruise ships anchoring in George Town harbor.

The world's hotels and their guests use massive quantities of resources on a daily basis, including energy for heating and cooling rooms, lighting hallways, and cooking meals, as well as water for washing laundry, filling swimming pools, and watering golf courses. This resource use is not only expensive, it can also damage the environment. Tourist facilities are contributing to the drying up of Israel's famed Dead Sea: in the last 50 years, water levels have dropped by an estimated 40 meters, leaving barren, salty mudflats that are hostile to native plants and birds. Environmentalists predict that at current rates of drawdown, the Dead Sea could disappear completely by 2050.[41]

At destinations where fresh water is scarce, overconsumption by tourists and tourism facilities can divert supplies from local residents or farmers, exacerbating shortages and raising utility prices. Tourists in Grenada are said to use seven times as much water as local people, and foreign-owned hotels get preference over residents during droughts. And a popular golf course on an island in Malaysia reportedly uses as much water annually as a local village of 20,000. Meanwhile, in the Philippines, the diversion of water to tourist lodges and restaurants threatens to destroy paddy irrigation at the 3,000-year-old Banaue rice terraces, an important cultural heritage site.[42]

In addition to consuming water, energy, and other resources, tourism creates large quantities of waste. The U.N. Environment Programme (UNEP) estimates that the average tourist produces roughly 1 kilogram (2.2 pounds) of solid waste and litter each day. Hotels, swimming pools, golf courses, marinas, and other facilities also generate a wide variety of harmful residues on a daily basis, among them synthetic chemicals, oil, nutrients, and pathogens. Improperly disposed of, this waste can damage nearby ecosystems, contaminating water sources and harming wildlife.[43]

Many tourist facilities in the developing world possess limited or no sewage treatment facilities, in part because of weak environmental legislation or a lack of money, monitoring equipment, and trained staff. As recently as 1990, none of the 22,000 beachfront hotel rooms in Pattaya, Thailand, were attached to a sewage plant; as of 1996, only 60 percent of that city's sewage was being processed. And a 1994 study for the Caribbean Tourism Organization reported that hotels in that region released some 80–90 percent of their sewage without adequate treatment in coastal waters, near hotels, on beaches, and around coral reefs and mangroves.[44]

Cruise ships are notorious for their waste disposal problems. Worldwide, the number of people taking a cruise nearly doubled between 1990 and 1999, to 9 million passengers annually. The San Francisco–based Bluewater Network reports that on a one-week voyage, a typical cruise ship generates some 3.8 million liters of graywater (water from sinks, showers, and laundry); 795,000 liters of sewage; 95,000 liters of oily bilge water; 8 tons of garbage; 416 liters of photographic chemicals; and 19 liters of dry cleaning waste. Many older vessels have little alternative to dumping much of this waste overboard. According to one estimate, the world's cruise ships discharge some 90,000 tons of raw sewage and garbage into the oceans each day. And untold quantities are dumped illegally: in one highly publicized case, Royal Caribbean Cruises received a record $18 million fine in July 1999 for 21 counts of discharging excess oily bilge and other pollutants into U.S. waters and for attempting to cover up its crime.[45]

These problems will likely worsen as shipbuilders rush to meet the rising demand for cruise vacations. In 2001, at least 53 new vessels were on the order books. Many newer ships resemble "floating cities," boasting more than 2,000 passengers and up to 1,000 crew members. To accommodate these larger vessels, countries often dredge deep-water harbors or modify their coastlines, destroying coastal ecosystems in the process. When ships dock, their massive anchors and chains can break coral heads and devastate underwater habitats: in 1994, one local scientist in the Cayman Islands reported that more than 120 hectares of reefs had been lost as a result of cruise ships anchoring in George Town harbor.[46]

Busloads of cruise passengers, day-trippers, and other visitors are overwhelming fragile cultural and natural sites that are ill equipped to manage rising tourist numbers. Visits to Cambodia's centuries-old Angkor temples more than doubled in 1999 following the government's decision to open the nearby town to international flights—intensifying pressures on the already fragile stone structures. In many of the world's parks, plastic water bottles, soda cans, and gum wrappers are an increasingly common sight.[47]

The presence of tourists in natural areas can affect wildlife behavior and populations. Around the world, whale-watching boats relentlessly pursue whales and dolphins and even encourage petting, influencing the animals' feeding and social activity. Similarly, tourist vehicles that approach cheetahs, lions, and other animals in Africa's safari parks can distract these creatures from breeding or stalking their prey. Safari tourists are also reportedly one of the top markets for illegal elephant ivory, which is banned under international law yet often sold to unsuspecting tourists in the form of souvenir carvings.[48]

At particularly fragile destinations, such as small islands, it can take relatively few visitors to leave a mark. Tourists can unintentionally trample vegetation or disturb nesting seabirds, breeding seals, or other animals, and they can bring invasive plants and animals in with their equipment or luggage. The introduction of these "exotic" species threatens to destroy the unique flora and fauna of Ecuador's Galapagos Islands, where tourism has increased by 66 percent since 1990 and where the local population—attracted by tourism's potential—has doubled in the past 15 years.[49]

In mountain areas, resorts and related infrastructure can disrupt animal migration, divert water from streams, create waste that is difficult to dispose of at high altitudes, and deforest hillsides, triggering landslides. In one Nepalese mountain village, an estimated hectare of virgin rhododendron forest is reportedly cut down each year for fuelwood to support the country's booming trekking industry, causing the erosion of some 30–75 tons of soil annually. And in Tanzania, the number of trekkers on the trails of Mount Kilimanjaro has risen so dramatically that the government had to double the daily climbing fee to $100 per person in September 1999 to slow serious erosion and other environmental harm.[50]

In coastal areas, popular recreational activities such as scuba diving, snorkeling, and sport fishing are damaging coral reefs and other marine resources (though this destruction is minor compared with the impacts of coral bleaching, overfishing, and ocean pollution). UNEP estimates that each year some 300,000 scuba diving trips are advertised to the world's estimated 6 million divers. With their fins and hands, divers and snorkelers have reportedly broken as many as 10 percent of coral colonies at popular Red Sea reefs off Egypt and Israel. A study of self-guided snorkel trails in Australia found similar damage at sites visited by an average of only 15 snorkelers per week. And research off the Caribbean island of Bonaire reveals that heavy diving at many sites has changed the composition of reefs, with more opportunistic, branching corals taking the place of older, large coral colonies. Souvenir shops and restaurants around the world also contribute to the destruction, looting reefs for shells, coral, and seafood to meet tourist demand.[51]

Not surprisingly, the environmental damage caused by tourism can ultimately hurt the industry by destroying the very reefs, beaches, forests, and other attractions that lure visitors in the first place. Already, global warming caused in part by rising aircraft emissions is raising sea levels and damaging coral reefs worldwide, threatening the economies of low-lying tropical countries like the Maldives, where tourism generates more than 85 percent of foreign-exchange receipts. If the environmental damage is significant enough, a destination may begin to lose visitors, as is the case in Kovalam in India and in many popular destinations in the industrial world, including Germany's Black Forest and Italy's Adriatic coast. Environmental deterioration also continues to impede efforts to boost tourism to many cities in the developing world: Cairo's urban sprawl, for instance, often alienates visitors, as does the growing gridlock and pollution in places like Bangkok and Beijing.[52]

Ecotourism—Friend or Foe?

Over the past decade or so, tourism authorities, environmentalists, academics, and others have embraced ecotourism as a way to address some of tourism's negative impacts while simultaneously generating foreign exchange, creating jobs, and stimulating investment. The Vermont-based International Ecotourism Society defines ecotourism as "responsible travel to natural areas that conserves the environment and sustains the well-being of local people." The United Nations has demonstrated its support for the concept by declaring 2002 the International Year of Ecotourism.[53]

Yet whether ecotourism can actually achieve its ambitious goals is increasingly under question. Part of the problem is definitional. Growing numbers of hotels and tour operators now bill themselves as ecotourism outfits, whether they are environmentally responsible or not. One operator in Cusco, Peru, for instance, estimates that less than 10 percent of the local trekking companies really fit the "eco" bill. And many tourists now call any travel that occurs in natural settings ecotourism. The line between genuine ecotourism and nature travel more broadly has become increasingly blurred.[54]

Ecotourism, broadly defined, is one of the fastest growing segments of the tourism industry—though the varying definitions make it difficult to measure. The International Ecotourism Society estimates that this form of travel is growing by 20 percent annually (compared with 7 percent for tourism overall) and generated some $154 billion in receipts in 2000. One 1992 study found that as many as 60 percent of international tourists traveled to experience and enjoy nature, while as many as 40 percent traveled specifically to observe wildlife, such as birds and whales.[55]

Though most of the world's ecotourists come from North America and Europe, many of the top destinations are in the developing world. Popular activities include safaris in Africa, trekking in the Himalayas, hiking in the rainforests of Central and South America, and scuba diving and snorkeling in Southeast Asia and the Caribbean. This demand is expected to continue well into the new century: WTO predicts that the trendiest destinations of the future will be "the tops of the highest mountains, the depths of the oceans, and the ends of the earth."[56]

Rising interest in ecotourism has had many positive benefits. Around the world, governments are setting aside valuable natural areas as national parks or protected areas, sparing them from more environmentally destructive activities like agriculture, logging, or mining. Some of the greatest increases in ecotourism have occurred in places with the highest numbers of protected areas. In 1997, an estimated 60 percent of the nearly 6 million tourists who visited South Africa stopped at a national park or reserve. And nearly half of all respondents in a survey of tourists in Central America cited protected areas as an important factor in choosing their destination.[57]

Once they have established parks and reserves, however, not all governments are willing or able to pay for the upkeep. Worldwide, financial support for these areas is dwindling. Many governments hope to use tourist admission fees and donations to boost park management, strengthen infrastructure, and protect against encroachment. This self-financing mechanism has been more successful in some areas than others. (See Box 5–1.)[58]

Brazil, Chile, Colombia, Kenya, and South Africa have all witnessed an explosion in privately owned nature reserves.

As an alternative, many countries are actively wooing private tourism investments to help protect natural areas. Brazil, Chile, Colombia, Kenya, and South Africa have all witnessed an explosion in the number of privately owned nature reserves, many of which open their lodges and trails to tourists. Two private reserves in Central America—Costa Rica's Monteverde Cloud Forest and Belize's Community Baboon Sanctuary—are well managed and generate sufficient income from tourist fees. In a survey of 32 private reserves in Latin America and sub-Saharan Africa, researcher Jeff Langholz found that more than half were profitable and that their overall profitability had risen 21 percent since 1989. On average, tourism revenues provided more than 67 percent of reserve income.[59]

Some eco-resorts display a high level of environmental commitment, carefully monitoring visitor impacts as well as their own ecological and social footprints. The most basic lodges are fueled by propane, kerosene, solar, or wind energy rather than

BOX 5–1. CAN ECOTOURISM PAY ITS WAY?

As government funding for parks and protected areas dwindles, more and more natural sites in the developing world are relying on tourist dollars to support themselves. The Bonaire Marine Park, for example, began collecting a $10 fee from visiting divers and snorkelers in 1991. Within a year, the park had raised enough money from the program to cover annual operation and maintenance costs.

Yet this self-financing does not work everywhere. In Costa Rica, visitor entry fees provide only about a quarter of the park service's annual budget for management and protection; the rest must be raised from donations. And tourism revenue at Indonesia's Komodo National Park covered only an estimated 7 percent of total park expenditure in the early 1990s.

In some instances, no tourist dollars are reinvested in conservation or park management—going instead to central government coffers or corrupt park authorities. According to one study, not a single cent of the $3.7 million that tourists paid to visit the islands off Mexico's Baja Peninsula in 1993 went directly to the protection or management of these areas.

In other instances, park authorities charge woefully low admission, or else demand no fees at all out of a fear this will deter tourists. Yet studies show that many tourists are willing to pay much more than they do to visit natural areas. Surveys in the United States found that 63 percent of travelers would pay up to $50 toward conservation in the area visited, while 27 percent said they would pay as much as $200. Studies at Komodo National Park suggest that visitors there would be willing to pay as much as 10 times the current entry fee.

The few sites that do charge significantly higher entry fees and apply them to conservation and management are benefiting greatly from this approach, particularly when the system allows for different pay levels for tourists and local people. Ecuador's Galapagos National Park has reportedly recouped nine times its management costs by charging foreign visitors $100. And gorilla viewing—at $250 a day—subsidized all 11 of Uganda's national parks in the late 1990s, providing 70 percent of the revenue of the fledgling park system.

Yet some ecotourism sites may never see enough visitors to support themselves, even with higher entry fees. Studies in the Central African Republic's Dzangha-Sangha protected area suggest that tourist numbers would have to increase nearly eightfold to generate a positive return on investment—a near impossibility—even if entry fees jumped from $16 to $200.

SOURCE: See endnote 58.

electricity or fuelwood, use no indoor plumbing, and generate minimal waste. The Sí Como No resort, in Costa Rica's popular Manuel Antonio National Park, relies on solar energy, uses aerial bridges instead of roads or walkways, puts in native plants to halt erosion, sponsors beach cleanups, and asks guests to reuse sheets and towels, among other environmental actions.[60]

But not all private ecotourism investments are as conservation-oriented. The rising commercial presence of large hotels, restaurants, and other concessions near or inside park boundaries threatens to destroy the natural settings of many destinations. China, for instance, is aggressively transferring control of its important scenic and cultural sites to private development companies, who then profit from their monopolies by charging admission fees and collecting revenues from hotels, restaurants, and gift shops. In some instances, the environment has also benefited: at the scenic mountain site of Huangsan, litter is now virtually nonexistent and forest cover has increased markedly since the 1980s. But two planned hotels and three new cable car runs, as well as increased pedestrian traffic, could ultimately destroy ecosystems in the area.[61]

Indeed, as ecotourism enters the mainstream, it increasingly faces many of the same problems as conventional tourism. Many early ecotourists were motivated by a keen environmental and political awareness and had little choice but to take local transport, stay in locally run accommodations, and eat locally. But today's ecotourists are "less intellectually curious, socially responsible, environmentally concerned, and politically aware" than in the past, according to author Martha Honey. They demand higher-end facilities, many of which are foreign-owned, consume more resources, and produce mounting levels of waste. And because their trips are often only a week or even a day long, they do not always consider the long-term repercussions of their visits or feel the need to follow every rule.[62]

In few places is the risk of "mass" ecotourism more apparent than in Costa Rica, once a little-known tropical destination. It has since become so popular that new airports, beachfront resorts, golf courses, and marinas are being built to accommodate the more than 700,000 tourists who arrive annually, threatening to destroy the lush rainforests and other natural sites that they come to see.[63]

Nevertheless, there are efforts to promote a more "genuine" form of ecotourism that requires less land and resources, generates less waste and pollution, and brings benefits to both local communities and the environment. Initiatives that are either managed by the community or that share a substantial portion of their profits with local residents can be particularly successful at achieving these goals. They can range from low-impact, homegrown efforts like offering an extra room or meal, renting out a small cabana, or showcasing traditional dances, to larger-scale investments like ecolodges or canopy walkways. Although all residents do not necessarily benefit, these initiatives can help to spread tourism's benefits more widely. In Ecuador's Amazon, for example, the Huaorani have set up a community project that distributes nightly tourist fees among all the families and earns residents twice what they could get working for an oil company.[64]

A high level of participation is desirable not only because it can reduce revenue leakage, but because it can heighten local appreciation for wildlife and other natural resources. One Ugandan farmer, talking about the recent boost in gorilla-related tourism at the nearby Budongo Forest Reserve, reportedly remarked of the benefits, "We never thought that vermin like these monkeys could become a source of money. . . now they pay for our schools." When local communities see direct benefits from tourism, they are more likely to slow resource use and to actively protect natural areas. Subsistence farmers participating in Zimbabwe's 23-district CAMPFIRE project recognize they can earn three times more from offering wildlife viewing, sustainable safari hunting, and other tourism-related activities on their land than from resource-intensive cattle ranching. Around the world, many former poachers, hunters, and fishers now guide tourists through nearby jungles or reefs, leaving little time or need for these previous destructive activities.[65]

Alternatively, studies show that when tourism initiatives exclude local people from participating in the management and use of natural areas where they grow food, raise livestock, and gather fuel, they are more likely to resent these efforts and seek to undermine them, ultimately compromising conservation goals. Areas that exclude local participation and use have seen rising incidences of poaching, vandalism, and even armed conflict. One Galapagos fisher reportedly said of government efforts to limit local use of the park's resources: "If the government does not lift the fishing ban we are even willing to burn all the natural areas to finish this tourism craziness."[66]

Areas that exclude local participation and use have seen rising incidences of poaching, vandalism, and even armed conflict.

Many local ecotourism initiatives have benefited from partnerships with outside actors, including government agencies, the private sector, and nongovernmental organizations (NGOs). One Virginia-based nonprofit, the RARE Center for Tropical Conservation, has trained more than 200 nature guides in Costa Rica, Honduras, and Mexico in conversational English and local natural history, boosting individual incomes by 92 percent on average. And some privately owned tour operations support local initiatives by donating a portion of their profits to conservation. Since 1997, New York-based Lindblad Expeditions has given more than $500,000 in client donations from its Galapagos trips to scientific research and environmental preservation efforts in the archipelago.[67]

Nepal's Annapurna Conservation Area Project (ACAP), launched in 1986 with support from the World Wide Fund for Nature, is another example of a successful ecotourism partnership. The project has trained local residents—predominantly subsistence farmers, herders, and traders—in such skills as food preparation and menu costing, safety and security for trekkers, and carpet weaving, allowing them to integrate tourism with their own farming activities and handicrafts. ACAP has helped conserve forests and other resources by setting up micro hydroelectricity plants on streams and installing solar water heaters in the lodges, while residents manage a revolving fund to help pay for latrines and garbage pits. Largely as a result of the project, tourist numbers to the region have jumped from 14,300 in 1980 to more than 63,000 today.[68]

Key players in the international community are also pledging support for ecotourism projects, often in alliance with local or international businesses and NGOs. Since the mid-1980s, the U.S. Agency for International Development has worked with the private sector and conservation groups in more than a dozen countries, including Costa Rica, Jamaica, Madagascar, Sri Lanka, and Thailand—providing funding for new and existing parks, recruiting and training park staff, and helping governments promote regulated investments in lodging, guide services, and other ventures. And since 1991 the Global Environment Facility, sponsored by the World Bank and the United Nations, has channeled more than $1 billion into some 400 biodiversity-related projects in the developing world, many of which have significant ecotourism components.[69]

In the International Year of Ecotourism, however, it is important that any efforts to highlight ecotourism as the solution to tourism's problems be monitored carefully. Although the World Ecotourism Summit scheduled for Quebec in May 2002 aims to be a truly comprehensive effort, allowing all stakeholders to voice their views and to exchange information about ecotourism experiences worldwide, the event is also by its very nature an opportunity for significant tourism marketing and promotion. The heavy involvement of international agencies, governments, and the private sector could distract attention from efforts to develop more low-impact, locally run ecotourism activities, particularly in areas not prepared to handle an onslaught of tourists.[70]

As ecotourism increasingly comes into its own, it is clear that one of the biggest challenges is balancing the potential benefits with the pitfalls. Like other forms of tourism, ecotourism can create its share of social and environmental problems. The degree of impact ultimately depends on the quality of the enterprise, the level of guide training, and the behavior of tourists themselves.

There is also a danger that too much emphasis on ecotourism could distract attention from broader problems. By definition, ecotourism will always remain a niche form of travel, relevant only in the relatively few areas of the world that still possess valuable natural attractions. It can do little to address the very real environmental problems of rampant, mass tourism at more urban destinations, such as downtown Bangkok. As such, it should be viewed as just one possible solution in a range of strategies for more sustainable tourism development.[71]

Toward a Sustainable Tourism Industry

According to the WTO, sustainable tourism should lead to the "management of all resources in such a way that economic, social and aesthetic needs can be fulfilled while maintaining cultural integrity, essential ecological processes, biological diversity and life support systems." Interest in making tourism more sustainable has grown steadily over the past decade, particularly in the wake of the 1992 U.N. conference in Rio. Although tourism was hardly mentioned in that meeting's blueprint for action, *Agenda 21*, countries have since adopted international declarations on a wide range of related topics, including tourism and sustainable development, the social impact of tourism, tourism and biodiversity, and tourism and ethics. In an important milestone, WTO, the World Travel & Tourism Council, and the Earth Council drafted their own *Agenda 21 for the Travel and Tourism Industry* in 1996, outlining key steps for the industry, governments, and others.[72]

Making tourism more sustainable requires careful planning at all levels and the involvement of all stakeholders—including the local communities that will be directly affected by tourism's presence. At its core, however, tourism is a private-sector activity, driven in large part by international hotel chains, tour companies, and other businesses. Sustainability will therefore require systemic change in how this industry operates. But reconciling the industry drive for more tourists with the need for sustainable practices will not necessarily be easy.[73]

Nevertheless, the tourism industry has taken many positive steps to become more environmentally and socially responsible. At least some of this change is a response to growing consumer pressure for more environment-friendly tourism products. A 1997 study by the Travel Industry Association of America reports that some 83 percent of the public supported green travel services, and that people were willing to spend 6 percent more on average for travel services and products provided by environmentally responsible companies. In a similar survey in the United Kingdom, more than half the interviewees said that when planning vacations or business trips, they would find it important to deal with a company that takes environmental issues into account.[74]

Arguably, the bulk of the change in the tourism industry is being driven by financial self-interest rather than genuine environmental concern. Perhaps more than any other industry, tourism depends on a clean environment. Declines in environmental quality can hit industry pocketbooks directly. On the other hand, helping to make destinations more attractive and supporting more environmentally sensitive practices can boost the profits of tourism businesses.[75]

Many of the world's larger tourism companies, from hotels to tour operators, are taking formal steps to restructure their management and operations along environmental lines—including reducing consumption of water, energy, and other resources and improving the management, handling, and disposal of waste. Changes in the hotel industry can be particularly fruitful, not only because these facilities consume large quantities of resources but also because they can have enormous influence over the broader habits and practices of their guests, employees, and suppliers. A simple step such as outfitting rooms with cards that encourage guests to reuse linens and towels when they are staying more than one night can conserve on average 114 liters (30 gallons) of water per room each day, plus energy—at a daily cost savings of at least $1.50 per room.[76]

Spearheading this movement at the global level is the London-based International Hotels Environment Initiative (IHEI), which works with hotels, hotel associations, suppliers, tourist boards, governments, and NGOs to encourage environmentally and socially responsible business practice. Founded in 1992, IHEI now represents some 11,200 hotels in 111 countries, including international chains such as Hilton, Marriott, Radisson SAS, Taj Group, Scandic, and Forte. Many hotels are embracing a wide range of environmental and cost-saving actions, from installing energy-efficient lighting and appliances to purchasing biodegradable housekeeping supplies. (See Table 5–2.)[77]

The cruise industry, too, is making an effort to integrate environmental practices into its activities, though much remains to be done. Some companies are embracing relatively simple initiatives such as recycling plasticware and using recyclable and reusable containers. Others, like Holland America and Princess Cruises, are outfitting newer vessels with on-board water treatment plants, incinerators, or cogeneration incinerators that harness energy from waste burning. And in a significant step, in June 2001 the International Council of Cruise Lines, a powerful industry lobbying group that represents the world's 16 biggest cruise lines, adopted new mandatory waste management standards for its members. Companies risk losing their membership if they fail to abide by the guidelines, which include new rules for the disposal of wastewater, used batteries, and photo processing and dry cleaning chemicals. They also call on members to strengthen compliance with domestic and international environmental laws.[78]

Tour operators and travel agents can play a big part in redirecting tourism because they determine not only where tourists go, but also which services they use. Many tour companies are setting up professional guide accreditation programs and investing in extensive training to ensure that their guides adhere to sound practices. And recently, some 24 of the world's larger tour companies signed on to a new voluntary Tour Operators' Initiative, sponsored by UNEP, UNESCO, and WTO. Members have agreed to integrate sustainability concerns into their operational management and tour designs and to share and implement best practices.[79]

Yet this and many other high-level sustainability efforts fail to reach the bulk of the smaller tour operations, accommodations, and services that many of the world's tourists use. Indeed, a survey in Australia's Gold Coast region found that while energy and water conservation measures were as common in 3- and 4-star hotels as in 5-star ones, they were rarely adopted in 1- or 2-star accommodations. Larger businesses, donors, and lenders can help accelerate the

Table 5–2. Hotel "Greening" Success Stories

Hotel or Hotel Chain	"Greening" Initiative
Hilton International	In recent years, has saved 60 percent on gas costs and 30 percent on both electricity and water costs, while cutting wastes by 25 percent. Vienna Hilton and Vienna Plaza reduced laundry loads by 164,000 kilograms per year, minimizing water and chemical use.
Singapore Marriott and Tang Plaza	Water conservation efforts save some 40,000 cubic meters of water per year—a reduction of nearly 20 percent.
Scandic	Has reduced water use by 20 percent per guest in recent years. Has also pioneered a 97-percent recyclable hotel room and is building or retrofitting 1,500 of these annually.
Sheraton Rittenhouse Square, Philadelphia	Boasts a 93-percent recycled granite floor, organic cotton bedding, night tables made from discarded wooden shipping pallets, naturally dyed recycled carpeting, and nontoxic wallpaper, carpeting, drapes, and cleaning products. The extra 2 percent investment more than paid for itself in the first six months.
Inter-Continental Hotels and Resorts	Hotels must implement a checklist of 134 environmental actions and meet specific energy, waste, and water management targets. Between 1988 and 1995, the chain reduced overall energy costs by 27 percent. In 1995, it saved $3.7 million, reducing sulfur dioxide emissions by 10,670 kilograms, and saved 610,866 cubic meters of water—an average water reduction of nearly 7 percent per hotel, despite higher occupancies.
Forte Brighouse, West Yorkshire, United Kingdom	A transition to energy-efficient lamps reduced energy use by 45 percent, cut maintenance by 85 percent, and lowered carbon emissions by 135 tons. The move paid for itself in savings in less than a year.
Hyatt International	In the United States, energy efficiency measures cut energy use by 15 percent and now save the chain an estimated $15 million annually.
Holiday Inn Crowne Plaza, Schiphol Airport, Netherlands	By offering guests the option of not changing their linens and towels each day, the hotel reduced laundry volume, water, and detergent, as well as costs, by 20 percent.

SOURCE: See endnote 77.

wider adoption of these practices by transferring environmentally sound management tools and technologies such as water-saving and renewable energy systems. Banks and insurance companies could incorporate environmental and social criteria into assessment procedures for loans, investments, and insurance, using green auditing measures to monitor progress.[80]

In addition to structural changes in management and operations, tourism businesses of all sizes and types are embracing a wide range of less formal voluntary initiatives to regulate their impacts, with mixed success. Forty-six of Antarctica's main tour operators, for instance, now belong to the International Association of Antarctic Tour Operators, a voluntary body formed in 1991 that enforces a strict code of conduct for tour operators and their clients. But despite regulations that include landing no more than 100 people per site at a time and making sure visitors do not disturb wildlife, tourists still pick up penguins, approach seals, and drive birds from their nests.[81]

Tourism businesses are also participating in voluntary certification schemes that grant a seal of approval to companies or destinations that demonstrate environmentally or socially sound practice. (See Table 5–3.) Not only do these labels serve as useful marketing tools, but they can spur the tourism industry to develop more environmentally friendly products as well as provide consumers with information about more sustainable travel choices.[82]

Unfortunately, more than 100 competing tourism certification schemes exist worldwide, and there are as yet no international guidelines to help travelers differentiate their value or effectiveness. Though many of these schemes are being developed in partnership with government agencies or NGOs that independently issue or monitor the certification standards, others are based on self-evaluation or paid membership, which may simply allow companies to "buy" their way to a green label. Ultimately, the success of tourism certification will depend on whether it can set a trusted, reliable standard, and on the degree to which the industry and consumers embrace it worldwide.[83]

As the changing rules of the global economy further open markets to tourism development, governments, international institutions, NGOs, and tourists themselves will need to play a more active role in keeping sustainable tourism on track. But this will not be easy. Tourism's rapid growth has been facilitated in large part by an absence of outside interference; like most industries, the tourism industry opposes intervention that it perceives as damaging to competitiveness and profits. Moreover, all signs indicate that instead of tightening regulations, governments are granting ever greater leeway to private actors.[84]

The industry-sponsored *Agenda 21 for the Travel and Tourism Industry*, for instance, places significant emphasis on self-regulation while continuing to uphold the dominant role of open and competitive markets, privatization, and deregulation in spurring tourism's growth. It makes little mention of direct government oversight or international instruments such as tourism taxes. Moreover, while many industry efforts embrace a shift toward environmental sustainability, they are less willing to incorporate social and cultural needs, including addressing labor and employment issues, protecting cultures, and maximizing linkages with local economies and communities.[85]

One way governments can help redirect tourism is by developing regulatory and policy frameworks that support key environmental and social goals without stifling incentives for investment. Planning authorities at the national, regional, and local levels can work to better integrate tourism into overall strategies for sustainable development. Australia's 1992 National Ecotourism Strategy, which recognizes the need for "responsible tourism planning and management to protect the country's natural and cultural heritage," is a good model. Belize and Costa Rica also have

Table 5–3. Selected Tourism Certification Efforts Worldwide

Scheme	Scope	Description
Green Globe 21	Has awarded logos to some 500 companies and destinations in more than 100 countries.	Rewards efforts to incorporate social responsibility and *Agenda 21* principles into business programs. But may confuse tourists by rewarding not only businesses that have achieved certification, but also those that have simply committed to undertake the process.
ECOTEL®	Has certified 23 hotels in Latin America, 7 in the United States and Mexico, 5 in Japan, and 1 in India.	Assigns hotels zero to five globes based on environmental commitment, waste management, energy efficiency, water conservation, environmental education, and community involvement. Hotels must be reinspected every two years, and unannounced inspections can occur at anytime. A project of the industry consulting group HVS International.
European Blue Flag Campaign	Includes more than 2,750 sites in 21 European countries; being adopted in South Africa and the Caribbean.	Awards a yearly ecolabel to beaches and marinas for their high environmental standards and sanitary and safe facilities. Credited with improving the quality and desirability of European coastal sites. Run by the international nonprofit Foundation for Environmental Education.
Certification for for Sustainable Tourism, Costa Rica	Has certified some 54 hotels since 1997.	Gives hotels a ranking of one to five based on environmental and social criteria. Credited with raising environmental awareness among tourism businesses and tourists. But the rating is skewed toward large hotels that may be too big to really be sustainable.
SmartVoyager, Galapagos, Ecuador	Since 1999, has certified 5 of more than 80 ships that operate in the area.	Gives a special seal to tour operators and boats that voluntarily comply with specified benchmarks for boat and dinghy maintenance and operation, dock operations, and management of wastewater and fuels. A joint project of the Rainforest Alliance and a local conservation group.
Green Leaf, Thailand	Had certified 59 hotels as of October 2000.	Awards hotels between one and five "green leaves" based on audits of their environmental policies and other measures. Aims to improve efficiency and raise awareness within the domestic hotel industry.

SOURCE: See endnote 82.

national policies or strategies to promote ecotourism.[86]

Many countries do not yet have such broad plans, however. And those that do typically fail to address social or environmental sustainability. Viet Nam's Tourism Master Plan, for instance, aims to attract large-scale investment primarily through joint ventures between foreign corporations and state enterprises, but it does little to support small-scale entrepreneurs or protect ethnic minorities from exploitation by external operators. In general, because tourism activities cut across a variety of government departments and industry groups, it is often difficult for authorities to coordinate a unified plan of action for addressing the impacts.[87]

In Namibia, local communities can assume legal responsibility for zoning their own agriculture, wildlife, and tourism activities in multiuse areas.

To ensure greater benefits for local communities and the environment, governments will need to balance large-scale investments in hotels, restaurants, and other facilities with smaller-scale initiatives that are actively planned and managed by local communities, such as family-run lodges or informal craft cooperatives. Local participation not only brings residents greater job satisfaction, it gives them greater responsibility for an initiative's outcome and makes them more likely to take a longer-term view toward conserving their local environment and resources. At the same time, smaller-scale tourism growth tends to be slower and more controlled, and can help offset tourism's negative environmental and cultural impacts by allowing more gradual integration of new activities into communities. Many countries, including Belize, Indonesia, Namibia, and Nepal, have begun to incorporate small-scale, community-based initiatives into national tourism efforts.[88]

To help get more-responsible tourism off the ground, governments will need policies and regulations that boost domestic land and resource ownership, facilitate local market access, and sanction exploitative businesses. Tourism agencies and other local government bodies can provide low-cost licensing as well as training in languages, small business development, and marketing, and can offer incentives like tax breaks, special interest rates, or microenterprise loans. They can also encourage externally owned businesses to reinvest their profits at the destination, in order to help support local agriculture and construction, fund area conservation efforts, and train and hire local staff. Strict government regulations can stifle exploitative practices such as sex tourism or child labor. A new Nepalese law, for instance, prohibits children under the age of 14 from working in trekking, rafting, casinos, and other tourism-related jobs—though critics charge that enforcement is weak.[89]

National and regional land use planning that considers the diverse needs of local residents, tourists, and other users, as well as of the environment, is an important element of a sustainable tourism strategy. It gives tourism authorities greater say over whether development occurs in an environmentally or culturally sensitive area or in a controlled manner. A new government plan in Spain's Balearic Islands, for example, oversees the careful zoning of certain areas for facilities like hotels, green areas, sanitary services, and parking. In Denmark, Egypt, France, and Spain, laws forbid developers from building within a defined distance from the coast in order to prevent beach erosion. And at Cuba's Cayo Coco, where hotels must have no more than four stories and be set back from the beach, each new building must go through an extensive government environmental impact assessment before construction is approved.[90]

Another country receiving accolades for its efforts to integrate social and environmental variables into land-use planning is Namibia. Under a bold government plan developed in the early 1990s, local communities can assume legal responsibility for zoning their own agriculture, wildlife, and tourism activities in multiuse areas known as conservancies, and then derive direct financial benefits from these. As of early 2001, 13 communities had registered conservancies, while another 20–24 were under development—bringing large tracts of the country under local tourism management. A national association for community-based tourism, started in 1995, provides advice and training to these communities and helps them to market their lodges and other ventures at international travel fairs and other promotional events.[91]

Elsewhere, governments are mitigating tourism's impacts by restricting the actual number of visitors allowed at a natural area or cultural site—though determining the appropriate level of use is often difficult. The Peruvian government recently decreed that up to 500 people a day can hike to Machu Picchu (down from as many as 1,000), in addition to more than tripling the fee and requiring tourists to trek with a registered company. On a larger scale, the Himalayan kingdom of Bhutan practices an official policy of "high-value, low-volume" tourism and accepted only 7,500 visitors in all of 2000, at a cost of $250 each per day. Elsewhere, natural areas are being roped off completely: visitors to Ecuador's Galapagos Islands are restricted to only 18 sites, while the country's Pasachoa Park closes for a full month each year to allow for environmental restoration.[92]

In addition to regulations, governments are using economic instruments to encourage responsible tourism. These include charging user fees, offering grants that reward "good practice" in tourism, and levying ecotaxes on everything from accommodations to air and marine transport. In 1995, for example, France taxed marine public transport to several protected islands to raise additional funds for their management and protection. By more accurately pricing tourism services, governments can push tourists and the industry to pay a fairer share in maintaining tourism assets.[93]

Yet such levies are often highly controversial because businesses fear they will deter tourists. Local businesses in Spain's Balearic Islands are fighting the regional government's April 2001 decision to charge tourists up to $1.78 extra per night at accommodations, even though the money would pay for improving tourist areas and managing natural spaces against environmental damage. A similar effort by the Indian Ocean island of Seychelles to introduce a $90 ecotax on all foreign visitors fell through in 1998. And a proposed $50 per head passenger tax on Caribbean cruises was reportedly dropped in the early 1990s after threats from U.S. cruise lines. Indeed, rather than levying taxes, many governments instead offer tax holidays, loans, and other incentives to attract tourism investors.[94]

Governments can also take action at the international level by supporting the implementation of environmental treaties that relate to tourism, such as the climate change and biodiversity conventions. They can work to ensure that international trade agreements like GATS and TRIMS do not undermine domestic environmental and labor regulations or compromise broader development goals.

Unfortunately, many governments do not have the capacity to take on a greater oversight or regulatory role. Fiscal and planning instruments are often too weak to influence the direction of tourism investments effectively, while local authorities may have only limited enforcement or other power. Many governments are relying on outside groups for additional support. International lending institutions like the World Bank and the Asian Development Bank (ADB), for instance, have stepped up their funding for sustainable tourism and related infrastructure improvements. In 2001, the ADB approved a $2.2-million loan to improve solid waste infrastructure and management in the Cook Islands—though more could be done to funnel this support to smaller-scale initiatives as well.[95]

Tourists themselves have a growing responsibility to understand the environmental and social impacts of their travel.

Other international institutions are working to create benchmarks for sustainable tourism that will make it easier for governments and businesses to measure progress. WTO has tested nine core indicators to assess the health of tourist destinations and developed a hotel audit program to help owners of smaller hotels become more environmentally responsible. And the International Maritime Organization oversees and enforces international standards on shipping and other maritime activities, including those affecting cruise ships. There is growing concern, however, that the industry preference for voluntary self-regulation could undermine efforts to set more stringent standards for tourism at the global level.[96]

Over the past few decades, nongovernmental actors—including citizen groups, grassroots activists, and tourists themselves—have generated much of the pressure for more sustainable tourism. Notably, it is a citizens' coalition, and not the government or the industry, that is finally taking the initiative to deal with the waste problems at India's Kovalam beach. In February 2001, activists with a local environmental group (Thanal), with support from Greenpeace India, launched Zero Waste Kovalam—a project that aims to convert the village into a zero-waste community by incorporating strategies of reduction, recycling, and reuse into the various waste streams. If the initiative wins industry and government backing, it may be a model for similar efforts across India.[97]

Local communities and international activist groups are having similar success combating unsustainable tourism developments elsewhere, though this remains an uphill battle. In April 2001, these groups played a big role in convincing the Mexican government to revoke permits for five hotel companies to build resorts, golf courses, and other facilities at X'cacel, a 165-hectare stretch of beach south of Cancun that is home to 40 protected species and a key nesting site for endangered Atlantic sea turtles. And the U.K.-based lobbying group Tourism Concern has successfully persuaded many tour operators to stop advertising Myanmar (formerly Burma) as a destination in protest against that country's human rights violations.[98]

Tourists themselves also have a growing responsibility to understand the environmental and social impacts of their travel. Industry groups and NGOs can help promote more sustainable behavior through public awareness campaigns and tourist training. Tourism Concern, for example, has produced five in-flight videos warning tourists about the crime of child sex tourism, and the World Travel & Tourism Council has a video series on tourism's environmental impact aimed at airlines and schools. The relatively low visibility of these initiatives suggests, however, that much remains to be done to boost tourism education.[99]

Before departing for trips, tourists can research whether companies are environmentally and culturally sensitive, hire local staff, or give a portion of their profits to local communities or conservation efforts. The International Ecotourism Society's "Your Travel Choice Makes a Difference" campaign helps travelers select responsible tour operators and guides and encourages them to buy and stay locally. And on its travel Web site, Conservation International selectively advertises tours that benefit local conservation efforts.[100]

Once at their destinations, tourists can seek to stay in lower-impact lodging, follow visitor rules and regulations, buy local food and crafts, and not purchase souvenirs made from endangered animals. They can minimize cultural disruption by learning about local customs, language, or conventions; asking before taking a photograph or entering sacred spaces; supporting local performers or craftspeople; and generally respecting the rights and privacy of others.[101]

Ultimately, sustainable tourism means traveling with an awareness of our larger impact on Earth. This is something that everyone will need to remember—from governments promoting tourism to tourism businesses and tourists themselves. Together, these groups will need to balance the ultimate goal of satisfying tourist demand with key environmental and social objectives, such as reducing resource consumption, eliminating poverty, and preserving cultural and biological diversity.

WORLD SUMMIT PRIORITIES ON INTERNATIONAL TOURISM

For Industry

- Incorporate environmental management principles that minimize both resource use and waste.
- Develop environmental and social "codes of conduct" for staff and clients.
- Adopt environmental and social standards set by international organizations and other certification bodies.
- Engage in efforts that protect and enhance local environments, communities, and cultures.

For Governments

- Create an overall tourism strategy that incorporates key economic, social, and environmental goals.
- Include responsible tourism development in overall land use planning strategies.
- Develop regulations and policies that support smaller-scale, locally run tourism development.
- Implement taxes, entry fees, and other economic tools that reflect the true costs of tourism services.

For International Institutions and NGOs

- Develop environmental and social standards that encourage responsible tourism development.
- Raise government, industry, and public awareness of the impacts of tourism.
- Help travellers select businesses that invest in local communities and that try to minimize environmental and cultural impacts.

Chapter 6

Rethinking Population, Improving Lives

Robert Engelman, Brian Halweil, and Danielle Nierenberg

Sitting in a dark hut in central Mali, a teenager named Djenaba nursed a baby—her second—and said that if she could, she would wait at least three years to have the next one. The truth, she added, was that she would prefer to have few children because "it's too hard [to support a large family]; we don't have any wealth in the village." But she said she knew she was powerless to put either of these wishes into effect, because no health clinic was within walking distance, and even the faraway ones rarely had contraceptives to offer. Survey research suggests Djenaba is not alone, and that nearly two out of every five women who learn they are pregnant wish they had waited at least a couple of years before giving birth again, if at all.[1]

Clearly, if all pregnancies could be the happy outcomes of women and men making earnest commitments to be parents, population growth would slow even more rapidly than it is today. This slowdown is occurring as ideas about childbearing change and as access to contraception improves around the world. Indeed, had average family size not declined from the level in 1960 and had death rates stayed the same, more than 8 billion people would be alive today instead of 6.2 billion. If the decline continues, the growth of world population could conceivably end before the middle of this century. Already, most families in wealthy countries are small enough to reverse population growth eventually, and in a few countries population is actually decreasing.[2]

But in the 48 least-developed countries in the world, population is projected to triple by 2050. And in many more nations the population could double. Three billion people are under the age of 25, with all or most of their reproductive years ahead of them—and without much guidance or help on healthy sexuality and reproduction. There can be no guarantees of a peak in world pop-

---

Coauthor Robert Engelman is Vice President for Research with Population Action International. This chapter is dedicated to John McBride and Kate McBride-Puckett.

ulation this century without major commitments from governments to provide family planning and related services to those who seek them, and such commitments are anything but certain.[3]

There is more to population and the policies that surround it, however, than numbers and distribution of people. Demographers, social scientists, and politicians increasingly see the connections of human numbers to behavior, to relationships, to overall health care, and—especially—to the circumstances and status of women. Evolving from decades of demographic research and field experience, "population" as a concept and a professional discipline now embraces a diversity of efforts to improve the health, livelihoods, and capacities of women at each stage of their lives.

The concept of reproductive health has also evolved to encompass much more than planning and preventing pregnancy; it includes sex education, access to contraceptives, sexually transmitted diseases, infertility, and all matters relating to the reproductive system. The United Nations defines it as "a state of complete physical, mental and social well-being...in all matters related to the reproductive system, and its functions and processes. Reproductive health therefore implies that people are able to have a satisfying and safe sex life and they have the capability to reproduce and the freedom to decide if, when and how often to do so."[4]

Providing education and health services for girls and women can hardly address all needs, however, until boys and men are engaged in efforts to improve unequal gender relations. The population and reproductive health fields have traditionally focused on women, even though men have historically exerted more control over when to have sex and whether to use contraception. Luckily, in many places this is changing. "Increasingly, men—and especially younger men—see the opportunity for egalitarian relationships between men and women as a boon," family expert Perdita Huston has suggested, "a fortunate trend that may allow them to become more involved in family life and less beholden to strict and restrictive gender roles." Any father who spends more time with his child than his own father did with him can appreciate the truth of that statement.[5]

Anyone who seeks to fathom the future interaction between humans and the natural world must consider population change as a dominant force on the human side of that relationship. But any discussion of "population" is increasingly understood to include or at least touch on a host of related issues, including the coexistence of extravagant consumption and degrading poverty and the inability of many governments to meet the basic needs of their people for health care, education, clean water, energy, and shelter.

In considering the links between population and environmental change, a near revolution in thinking has occurred—much of it since the Earth Summit in Rio de Janeiro in 1992. It is increasingly clear that the long-term future of environmental and human health—and, critically, population—is bound up in the rights and capacities of the young, especially young women, to control their own lives and destinies. (See Box 6–1.) What remains unclear is whether political leaders today, still mostly men, will see the potential for positive change that lies in recognizing and responding to the rights and needs of women and children. Societies in rich nations and poor need a new kind of vision to cure the widespread gender myopia that refuses to acknowledge the long-term

BOX 6–1. THE CHANGING FACE OF POPULATION AND WOMEN AT U.N. CONFERENCES

At international conferences throughout the 1990s, from Rio de Janeiro to Vienna and from Cairo to Beijing, women's health and human rights slowly but steadily made their way onto the international agenda. Thanks in large part to the involvement of women themselves, often acting together in nongovernmental organizations (NGOs) and coalitions, women are less likely to be seen as passive recipients of population programs but instead as full participants in a world where all people, including the young, are free to express their sexuality freely, safely, and responsibly.

At the United Nations Conference on Environment and Development (also known as the Earth Summit) in Rio in 1992, women's groups from developing as well as industrial countries lobbied for social change. *Agenda 21*, the plan of action that emerged from the conference, called for women's "full participation" in sustainable development; improvement in women's status, access to education, and income; and attention to the needs of women as well as men for access to reproductive health services, including family planning "education, information and means." This set the stage for the International Conference on Population and Development (ICPD) in Cairo in 1994, where the Programme of Action affirmed that reproductive and sexual health is a basic human right. A year later in Beijing, the Fourth World Conference on Women reaffirmed women's rights and their equal participation in all spheres of society as a prerequisite for human development.

SOURCE: See endnote 6.

implications of current relations between women and men and to see the critical role of gender in human development.[6]

The World by Numbers

Throughout most of human history, parents had on average roughly two children who themselves survived to become parents. We know this not by demographic surveys but by the simple observation that human population grew very slowly until relatively recently. The key word here is survived. Women undoubtedly had many babies, although some women practiced herbal and other means of contraception. But until recently, death rates among infants and children were so high that population growth was episodic and localized rather than consistent and global.[7]

With the advent of better nutrition and basic public health—hand washing, sanitation, immunization, and antibiotics—enough people survived infancy and childhood by the nineteenth and twentieth centuries to boost population growth to unprecedented rates. What had been a billion people around 1800 became 1.6 billion in 1900, 2.5 billion by 1950, and then 6.1 billion by 2000. (See Figure 6–1.) Sometime in the 1960s the global rate of population growth peaked and began to decline—from 2.1 percent a year to just under 1.3 percent today—although the still-growing population base meant that annual additions to human numbers continued increasing until recently. Even today, the planet adds about 77 million people each year, the equivalent of 10 New York Cities.[8]

The direct cause of slowing population growth was that women began having fewer children on average as infant mortality rates declined and as modern means of contraception became available—and

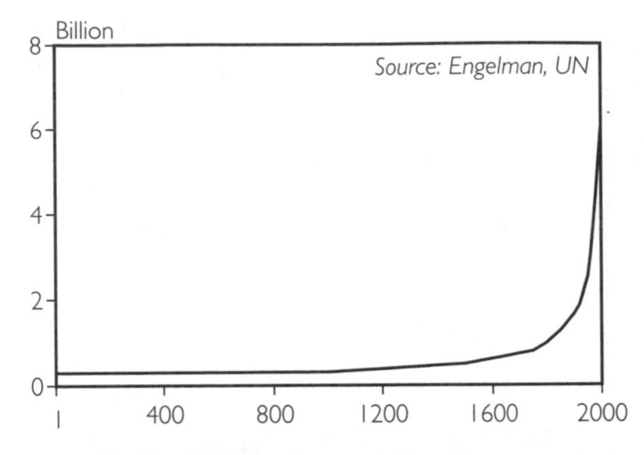

Figure 6–1. World Population Since A.D. 1

increasingly attractive—in most countries. This demographic revolution, however, has developed unevenly around the world. In much of Europe and in Japan, use of birth control rose so rapidly from the 1970s through the 1990s that fertility fell well below the 2.1 average of children per woman needed to replace those who die with those who are born; eventually, continuation of such low fertility will end population growth in these nations. Countries such as Italy, Spain, Armenia, the Ukraine, and Russia now have fertility rates so low that some analysts are concerned about how the nations will adjust to having many fewer working-age people available to support the elderly in their aging populations. Others have countered that such trends are simply the byproduct of the combination of longer life spans and lower birth rates and that changes in tax, social security, and immigration policies can ease the transition to new population sizes and structures.[9]

For most of the world, however, population decline is anything but imminent. Average national fertility rates are at replacement level or higher in more than two thirds of the world's nations. Even with reasonably anticipated declines in fertility, the current population of Nigeria of about 120 million, for example, is expected to grow to between 237 million and 325 million by mid-century. The number of people living on the entire continent of Africa is projected to more than double—from 800 million to between 1.7 billion and 2.3 billion—over the same period. South Central Asia (including India, Pakistan, Bangladesh, and Afghanistan) could more than double its current population of 1.5 billion.[10]

The stark differences between wealthy and poor nations in population trends create the conditions for an increased flow of people across international borders in coming decades. An estimated 150 million people—3 out of every 100 people on the planet—live outside their countries of birth. Between 1985 and 1990, the population of international migrants grew about 50 percent faster than world population as a whole, and given the greater migration of the 1990s and the slowdown in world population growth, it is likely that the gap has grown much wider. In the late 1980s, most migration was from one developing country to another, but in the future the South-to-North axis could dominate migration. (See Box 6–2.)[11]

The United Nations Population Division currently projects that today's world population of 6.2 billion will grow to anywhere from 7.9 billion to 10.9 billion by 2050. Global population by mid-century is projected to be overwhelmingly urban, more tropical, and significantly older than it is today.[12]

Despite this growth, the overwhelming influence on human population today is the fulfillment of parental intentions to have later pregnancies and smaller families. In 1960, women had five children on average worldwide, and more than six in developing countries. By 2000, these numbers had fallen by roughly half, in part because contraceptive usage multiplied sixfold—from 10 percent of couples worldwide in 1960 to 60 percent in 2000. These changes are indicators of a demographic revolution that continues today.[13]

Demographers and population policy analysts increasingly recognize the health and circumstances of women to be among the greatest determinants of how many

BOX 6–2. MIGRATION'S CONTINUING ROLE

In June 2000, 58 illegal Chinese immigrants en route to England through Belgium were packed into a nearly airless truck carrying a shipment of tomatoes. Only four survived the 18-hour journey. The next summer, immigration agents patrolling the U. S.-Mexico border found the bodies of 14 Mexicans dead from dehydration. The effort to cross into the United States kills more than 350 illegal migrants each year. Since migration generally involves great personal risk and expense, given the choice most people would rather stay where they are—close to family, familiar places, and others who speak their language. But the larger the gap between people's current quality of life and that which they believe they can attain in a new land, the more motivation they have to leave.

Among the nations that send the most migrants are China and India. Every year more than 400,000 Chinese leave for other countries and 50,000 Indians migrate to the United States, Australia, the United Kingdom, and Canada. Refugees—migrants forced from their homes by armed conflict or political upheaval—often have little choice but to cross borders. As this chapter was being written, it appeared that more than 1.5 million Afghans—in addition to the 2.5 million already displaced by two decades of conflict—could cross into neighboring countries as a result of U. S. retaliation against terrorists in the region.

In North America and Western Europe, the two regions of the world that receive the most migrants, migration has become a controversial and deeply sensitive topic, all the more so in the wake of the terrorist attacks on the World Trade Center and the Pentagon in 2001. On the one hand, employers and national economies benefit from the generally inexpensive labor that immigrants offer. Societies benefit from cultural diversity unknown to previous generations. On the other hand, migrants make convenient targets for those unhappy about the accelerating pace of change, increasing congestion, or the unevenness of economic growth. In the United States, new fears about terrorism may add to this tension.

Pressures to migrate and opposition to continued immigration are both likely to mount as population density increases and the availability of critical natural resources decreases. Ultimately, each nation must decide how many people to welcome and under what circumstances. Some nations, cities, and communities—especially those without adequate renewable water supplies—may take measures to discourage further in-migration.

Since migration is approaching or even surpassing the number of births as a driver of population growth in many places, nations may ultimately learn to address migration in the context of overall population policy, rather than in response to concerns about the demographic weight of particular ethnic or language groups. The diversity that migration has contributed to the world's nations, especially the wealthy ones, is unlikely to recede for many decades to come.

SOURCE: See endnote 11.

children parents have. When women's education, opportunities, capacity, and status begin to approach those of men, their economic and health conditions improve. Moreover, assuming good access to family planning services, they have fewer children on average, and those they have arrive later in the mothers' lives. An estimated 125 million women worldwide do not want to be pregnant but, like Djenaba in Mali, are not using any type of contraception. Millions more women—survey research has not produced a precise number—would like to avoid pregnancy despite their sexual activity but are using contraception improperly, in many cases because of misinformation about what would be the best method for them. Overall, the U.N. Population Fund estimates, 350 million women worldwide lack any access to family planning services.[14]

A major contributor to later pregnancies and lower fertility is at least six or seven years of schooling. When girls manage to stay in school this long, what they learn about basic health, sexuality, and their own prospects in the world tends to encourage them to marry and become pregnant later in life and to have smaller families. In Egypt, for example, only 5 percent of women who stayed in school past the primary level had children while still in their teens, while over half of women with no schooling became mothers while still teenagers. In high-fertility countries, such as those in Africa, South Asia, and some parts of Latin America, women who have some secondary school experience typically have two, three, or four children fewer in their lifetimes than otherwise similar women who have never been to school.[15]

Educating girls and women also gives them higher hopes for themselves—including raised self-esteem, greater decisionmaking power within the family, more confidence to participate fully in community affairs, and the ability to one day become educated mothers who pass on their knowledge to their own daughters and sons.[16]

Unfortunately, despite some halting progress in international and government commitments to support women's rights, women are still much less likely than men to complete secondary school—or to hold a paying job or sit in a legislature or parliament. (See Table 6–1.) In 1995, an estimated 75 million fewer girls than boys were enrolled in primary and secondary schools, and in all nations women still earn only two thirds to three fourths of what men earn for comparable work.[17]

It is difficult to predict how quickly these less-often-discussed human numbers will change for the better. Until they improve significantly, however, women around the world will be less able to choose to have smaller families.

The Ecology of Population

Whether considering biodiversity or cropland and forests, the number of people on Earth combines with levels of consumption, dominant technologies, and distribution to determine humanity's use of resources. (See Table 6–2.) Consider the potential for population growth to make the planet's finite supply of fresh water inadequate for human needs. Human beings depend on less than one one-hundredth of 1 percent of the world's water; less than a third of this is really usable (much of it falls as rain too far from human settlements or runs to the ocean in floods), and more than half of the usable portion is already being tapped for human purposes.[18]

Hydrologists categorize countries with less than 1,000 cubic meters of renewable water per person a year as water-scarce,

Table 6–1. Gender Disparity in Various Spheres

Sphere	Description of Disparity
Education	Two thirds of the world's 876 million illiterate people are female. In 22 African and 9 Asian nations, school enrollment for girls is less than 80 percent that for boys, and only 52 percent of girls in the least developed nations stay in school after grade 4. In sub-Saharan Africa and South Asia—where access to higher education is difficult for both women and men—only between 2 and 7 women per 1,000 attend high school and college.
Economics	In most regions, women-headed households are much more vulnerable to poverty than male-headed ones. Single-mother households in the United States have 18 percent of American children but one third of the children living in poverty. Throughout most of the world women earn on average two thirds to three fourths as much as men. Women's "invisible" work (such as housekeeping and child care) is rarely included in economic accounting, although it has been valued at about one third of the world's economic production. Women account for 5 percent of the most senior staff of the 500 largest corporations in the United States. At the International Monetary Fund, 11 percent of the economists are women, and women occupy just 15 percent of managerial positions.
Politics	Women's representation continues to increase in all nations, but women are still vastly underrepresented at all levels of government as well as in international institutions. Of 190 heads of state and heads of government, only 10 are female. At the United Nations, women made up only 21 percent of senior management in 1999. While Nordic nations have the highest percentage of women in parliament, with 39 percent of seats in the lower and upper houses held by female representatives, women hold just 15 percent of parliamentary seats in the Americas and a scant 4 percent in Arab states. Only in nine countries is the proportion of women in the national parliament at 30 percent or above. In mid-2001, at least seven countries—Djibouti, Jordan, Kuwait, Palau, Tonga, Tuvalu, and Vanuatu—did not have a single woman in their legislatures.
Civic Freedom	In nations as diverse as Botswana, Chile, Namibia, and Swaziland, married women are under the permanent guardianship of their husbands and have no right to manage property (women's rights for divorce are also widely constrained). Husbands in Bolivia, Guatemala, and Syria can restrict a wife's choice to work outside the home. In some Arab nations, a wife must obtain her husband's consent in order to get a passport.

SOURCE: See endnote 17.

while those with 1,000–1,700 cubic meters are water-stressed. Any inequities in access occur on top of the limitations imposed by basic availability. And the figures say nothing about the quality of the water provided, although as a general rule the scarcer water becomes, the more likely it is to be polluted due to the increasing pressure on each bucketful to serve human needs. (These rules of thumb hold as well for the relationship between population and the availability of other natural resources.)[19]

History shows that few countries have raised living standards successfully while experiencing water scarcity. Sandra Postel of the Global Water Policy Project has found that as water availability drops into the stress and scarcity categories, the importation of food dramatically increases in most countries. More than a quarter of all grain imports, for example, go to water-stressed countries in the Middle East, Asia, and

Table 6–2. Population and Selected Natural Resources

Resource	Description
Fresh water	Today 505 million people live in countries that are water-stressed or water-scarce; by 2025, that figure is expected to be between 2.4 billion and 3.4 billion people (near the equivalent of roughly half of today's world population).
Cropland	In 1960 there was an average of 0.44 hectare for each human being on the planet; today there is less than one quarter of a hectare, a little more than a half-acre suburban lot. By the most conservative of benchmarks of arable land scarcity, nations need at least 0.07 hectare to be self-sufficient in food. Today about 420 million people live with such little cropland; by 2025, that number could top 1 billion.
Forests	Today 1.8 billion people live in 40 countries with less than a tenth of a hectare of forested land for each person—roughly the size of a quarter-acre suburban lot. By 2025, this number could nearly triple, to 4.6 billion. Women and girls in developing countries will walk farther for fuelwood, and there will be less access for all to paper, which remains the currency of most of the world's information.
Biodiversity	In 19 of the world's 25 biodiversity hotspots, population is growing more rapidly than in the world as a whole. On average, population in the hotspots is growing at 1.8 percent each year, more than the global average.

SOURCE: See endnote 18.

Africa. In some cases in sub-Saharan Africa, good farmland may soon be unproductive simply because there is insufficient renewable water to moisten crops and because nonrenewable water sources are drying up.[20]

Between 2.4 billion and 3.4 billion people are projected to be living in countries in water stress or scarcity by 2025, according to calculations by Population Action International, compared with 505 million today. The people of the Middle East, of northern, eastern, and southern Africa, and of southern and western Asia will be especially vulnerable. When water is scarce, the poor tend to suffer—and pay—the most. In urban areas where settlement has outpaced both freshwater availability and the infrastructure needed to distribute water that is safe to drink, the poor pay from 10 to 100 times more for water brought in bottles by trucks than the wealthy pay to get the same or higher-quality water from taps. These pressures on water supplies hamper efforts to reduce the numbers of people who lack access to safe water (currently about 1.1 billion) and sanitation services (2.4 billion).[21]

The sorry state of the world's freshwater supply and distribution services is directly responsible for an estimated 4 million deaths annually, mostly of infants and young children. Entire ways of life are disappearing as water shortages alter landscapes and habitats. Most ominous of all, growing shortages of fresh water are leading to tension along the many rivers—the Nile, the Danube, the Tigris and Euphrates, and the Ganges and Brahmaputra are the chief examples—shared by nations. Once these rivers provided more than enough for all, but under today's economic and demographic conditions, development of water resources by upstream countries reduces levels downstream, which residents of those countries can ill afford to lose. Given the needs of all human beings for water, not to mention those of millions of other species who inhabit land and freshwater bodies, eventually population growth will require reductions in per capita use of water and better conservation practices.[22]

Population is rarely mentioned in debates on a range of other environmental concerns, including climate change. Nonetheless, as world numbers continue to grow, each person has less atmospheric space in which to dispose of carbon dioxide, methane, and other heat-trapping gases. Among the starkest examples of population's impact on greenhouse gas emissions is the United States—the nation with less than 5 percent of world population but 25 percent of all greenhouse-gas emissions. Per capita U.S. emissions of carbon are fairly stable, but over the past decade the emissions total has grown apace with population. The projected carbon emissions of the 114 million people likely to be added to the U.S. population in the next 50 years roughly equal the projected emissions of the 1.2 billion people who could be added to Africa during that period.[23]

As these two examples of environmental linkages suggest, population dynamics cut across all environmental problems, and a host of secondary impacts can themselves affect human health and well-being. As people crowd into popular coastal areas, earthquake-prone urban centers, and floodplains, for instance, the damage to human property and life done by storms, floods, and earthquakes skyrockets. And epidemiologists increasingly see hints of the overarching impact of population growth on the spread of infectious disease, as greater density boosts exposures and shortens transmission distance, making life easier for the organisms that spread infections. One critically important service that undisturbed ecosystems offer, according to Dr. Eric Chivian at Harvard Medical School's Center for Health and the Global Environment, is maintaining equilibria among hosts, vectors, and parasites and between predator and prey. As people open up new swaths of forests and consume the resources there, they are exposed to new infectious agents capable of evolving into vectors of human disease. Indeed, this is one plausible explanation for the emergence of HIV into human populations in the last few decades.[24]

For years economists have debated the relationship between demographic and economic change without reaching any consensus. This is in part because population growth operates in different ways in different countries, and even at different points of time, making it difficult to untangle cause and effect. Some government officials of developing countries are willing to assert that large and growing populations hamper economic development. In the Philippines, for example, economic planning secretary Dante Canlas announced that the country's new administration would act to slow population growth despite the opposition of the Catholic Church in the country. Noting the nation's rapid population growth, Canlas expressed concern that "high fertility in the rural areas is exported into the urban areas and rural poverty gets transformed into urban poverty."[25]

Recent evidence suggests that under some conditions, falling fertility and slower population growth can powerfully boost some economies. A number of countries in East and Southeast Asia, for example, invested strongly in health—including mother and child health care and family planning services—in the 1970s, specifically hoping that smaller families would produce economic and developmental dividends. These governments also committed themselves to education and to helping growing industries that promised to be major employers. The strategies worked. Having fewer children meant that parents could invest more in their schooling and health. And studies indicate that as average family size declines, savings increase, and household savings are among the major sources of internal investment in developing countries. Harvard economists recently calculated that between 1965 and 1990, the slowing of population growth accounted for as much as one third of the rapid growth in per capita income in East Asian countries like South Korea and Taiwan.[26]

A stable or gradually declining population can be seen as a helpful side benefit of efforts that improve people's lives directly.

Rapid growth can put an enormous strain on governments and other institutions. From schools and hospitals to low-cost housing and waterworks, growing numbers of people generate a larger demand for public services—a demand that inefficient or heavy-handed governments often cannot meet. The rapid expansion of school-age populations, for instance, puts tremendous pressure on nations to train more teachers and build more schools. This is especially worrisome because many of these nations already lag in meeting educational needs. In sub-Saharan Africa—where only 56 percent of people are literate and secondary education reaches only 4–5 percent of the population—the number of school-age children is projected to expand by over 30 percent in the next three decades. Without additional investments in education, today's average student-teacher ratio of 39 to 1 in sub-Saharan Africa will balloon to 54 to 1 by 2040.[27]

Many of these demands converge in the mushrooming urban centers of the developing world, which are projected to be home within a few decades to virtually all future population growth. Many of these cities have doubled in population just over the past 12–15 years. One analysis found that young children in the largest cities of Latin America, North Africa, and Asia were less likely than children in smaller cities to have received health care or schooling and were more likely to be suffering from diarrhea because of a lack of clean drinking water, safe food, and sanitation. And the most rapidly growing cities in Latin America and Africa suffered from the highest levels of infant and child mortality. Long-term population growth rates in excess of 5 percent a year raised the odds of infant mortality by 24 percent in North Africa and Asia, by 28 percent in Latin America and the Caribbean, and by 42 percent in tropical Africa. In Mali, where Djenaba is struggling to raise her children and where both birth and death rates are among the highest in Africa, moving to the rapidly growing capital city of Bamako may no longer represent the improvement in life chances it once did.[28]

Bulges in young age groups may precipitate social upheaval or international aggression. Researchers at York University have argued that most of the major wars and conflicts of the past few centuries have been precipitated by nations in which young men predominated. A large cohort of young men does not make aggression inevitable, but it can provide the tinder that despotic leaders can spark for bellicose ambitions when grievances are acute. Along similar lines, other researchers have argued that population-related scarcities of natural resources can also provide fuel for conflict, especially when the needs of dense and rapidly growing populations strain weak institutions.[29]

Healthy Reproduction, Healthy Families

In 1994, representatives from international institutions, national governments, and NGOs gathered in Cairo at the International Conference on Population and Development. They sketched a vision of a world in which an end to population growth is one of many outcomes of policies and programs that put individuals, especially women and young people, in control of their own productive and reproductive lives. This was a breakthrough event, bringing to policymakers and the public an intellectual revolution that had been brewing for years within the population and women's health movements. The consensus among governments paved the way for a new people-centered—and ultimately much more effective—way to craft human development and population policies.[30]

Through the lens of Cairo, a stable or gradually declining population can be seen as a helpful side benefit of efforts that improve people's lives directly. That is, greater access to health care and education not only yields personal and community benefits, it also has the effect of reducing the size of families and raising the average age of first pregnancy. When participants from virtually all countries gathered in Cairo, they agreed to adopt precisely this strategy for addressing population change—framing population as an issue of people, especially their capacities and their rights, more than numbers. Such thinking went a long way toward reconciling tensions among ecologists, demographers, and feminists regarding the causes and consequences of high fertility rates and population growth.

The capacity to plan, prevent, and postpone pregnancy is essential to reproductive health, reducing maternal and child deaths and setting the stage for women and men to manage their own sexuality and reproduction. There is much more to this aspect of health, however, than family planning alone. According to Jodi Jacobson of the Center for Health and Gender Equity, in order to address unwanted fertility, HIV/AIDS, and the whole range of women's reproductive needs and concerns, health care systems need to be sensitive to the realities women face on a daily basis. Recent programs in India and South Africa are addressing that challenge by asking difficult—but much needed—questions: Can women negotiate contraceptive use with a partner? And if not, how can services be tailored that allow them to protect themselves in secret?[31]

Young people in all regions of the world also face a variety of challenges related to reproductive health, whether or not they are sexually active at the moment. At the ICPD in 1994, and even more so at the conference's five-year review in New York in 1999, people in their teens and early twenties expressed their desire to be recognized and included in population and reproductive health policies and to be agents of change for implementing those initiatives.[32]

"Wait until you're older" is hardly helpful advice for the millions of adolescents already having sex or preparing to enter into intimate relationships. Research in several countries has demonstrated that access to sound information and guidance on sexuality and reproduction helps young people postpone sexual activity and avoid infection and pregnancy when they do become sexually active. The young need adult guidance and support, as well as access to safe and effective contraception and reproductive health services, in order to protect themselves from violence, unplanned pregnancies, and infection from HIV/AIDS and other sexually transmitted diseases. They also need the self-confidence to say no to unwanted sex or to insist that their sexual partners use contraception.[33]

While the mere presence of contraceptive options is hardly sufficient to change women's lives and world population trends, without that access even the most highly motivated women and couples are unlikely to be sexually active for long without a pregnancy. Lack of access to services, lack of knowledge, and opposition of family members are among the most commonly cited reasons for not using contraception. Prohibitively high costs—in some sub-Saharan African nations, condoms and birth control cost 20 percent of the average income—also keep many women from taking action to prevent pregnancy. The correlation is straightforward: where contraceptive use in the world is high, families are smaller. (See Figure 6–2.) In Angola, Chad, and Afghanistan, for example, fewer than one in 20 couples uses contraception, and family size is close to seven children per woman. In Italy, in contrast, contraceptive prevalence exceeds 90 percent and average fertility stands at 1.2 children per woman, close to the lowest fertility level in the world.[34]

If contraception were simply a means of slowing population growth, it is unlikely that most of the world's sexually active couples would be using it. The capacity to experience sex and sexuality without fear of becoming a parent is among the most liberating aspects of contemporary life—especially for women. By one analysis, the influx of women into U.S. medical, law, and other professional graduate schools in the 1970s was in large part a product of widespread availability and popularity of the oral contraceptive pill starting late in the 1960s. In developing countries, women often express their gratitude to family planning for new opportunities to earn an income, pursue an education, or participate more actively in civic life.[35]

Family planning also directly improves health, especially for mothers but also for their infants and children. In developing countries, children are significantly more likely to die before their fifth birthday if they are born fewer than two years after their next older sibling, whereas a gap of four years or more between births raises infant and child survival chances above the average. Mothers themselves are more likely to survive childbearing if they use family planning to have fewer children, as it gives their bodies time to recover between each birth.[36]

In the past 40 years, most developing countries have launched programs to subsidize or otherwise make more widely available sterilization, condoms, pills, injectable

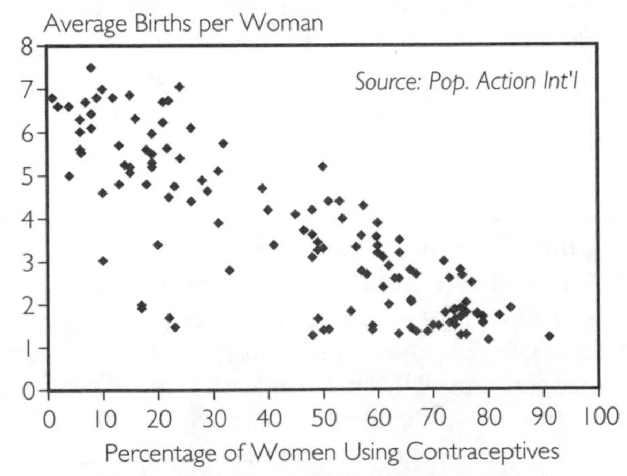

Figure 6–2. Cross-Country Analysis of Contraceptive Use and Childbearing

contraceptives, intrauterine devices, and other methods of avoiding pregnancy. Quantifying the role of different factors in social change is difficult, but by one analysis, the international family planning movement can take credit for just under half the decline in birth rates since 1960, with cultural and social change accounting for the rest. One interesting glimpse of the impact of government-sponsored family planning programs on national fertility comes from a comparison of six nations with strong programs begun before 1980—Bangladesh, Ghana, India, Mexico, South Korea, and Zimbabwe. In each case, the use of contraception rose fairly continuously, with attendant declines in average family size that have helped slow the growth of world population appreciably. For example, after the Zimbabwean government launched its program in 1968, contraceptive use jumped from just 5 percent of the population in 1975 to 50 percent by 1993.[37]

The international community can help to close gaps where government provision of family planning and reproductive health services is constrained by tight budgets, debt, entrenched bureaucracies, or narrow political conflicts. Many industrial countries have contributed funds and technical expertise to such programs. Nongovernmental sources also shoulder a heavy load. Private U.S. foundation expenditures may now rival official U.S. overseas assistance for family planning. In Bangladesh, one quarter of reproductive health services comes from nongovernmental groups. In Colombia, an affiliate of the International Planned Parenthood Federation called Profamilia provides more than 60 percent of family planning services.[38]

Still, the gap between the need for contraception and its availability in developing countries is particularly worrisome, because supplies for reproductive health appear to be entering a period of scarcity. Two waves are reaching shore simultaneously and reinforcing each other. First, the largest generation of young people in human history—1.7 billion people aged 10–24, projected to approach 1.8 billion by 2015—is now reaching reproductive age. The number of women already aged 15–49 is at an all-time high at 1.55 billion and could increase to 1.82 billion by 2015. At the same time, greater proportions of young women and men want to delay childbearing and to have at most two or three children.[39]

Today, 525 million women use contraception, and that number is projected to reach 742 million by 2015. In Rwanda, Guatemala, and other developing nations, however, surveys among men have found that between one quarter and two thirds are not using any form of contraception with their partners, even though they do not want any more children. Presumably as this gap between intentions and practice shrinks, demand for contraceptives will rise even faster. For the foreseeable future, it is unlikely that this growth in demand can be satisfied without increased assistance to the developing countries where the growth is most dramatic.[40]

In some countries, the contraceptive shortage has already arrived. In July 2001, Indonesia revealed that its stock of contraceptives needed for 8.3 million low-income couples would run out by the end of the year. Few nations or agencies have developed strategies for meeting the rising demand for contraception, and the gap between demand and supply could simply widen over time. Worldwide, between rising numbers of young people and growing proportions wanting to plan their families, total demand for contraception is expected to grow by 40 percent between 2000 and 2015. The U.N. Population Fund has estimated that global spending on contraceptive supplies will need to more than double, from $810 million annually in 2000 to $1.8 billion in 2015. The cost of making these contraceptives accessible through quality services will also more than double, from $4 billion in 2000 to $9 billion in 2015.[41]

The impact of these two waves—more young people, with higher proportions wanting to plan pregnancies—combines with yet another wave: the soaring need for male and female condoms to prevent HIV and other sexually transmitted infections. More than any other single threat, HIV has the potential to turn population growth around for the worst of reasons: increases in death rates.

It appears that AIDS-related deaths could become a major factor in a reversal of population growth rates in at least one country—South Africa—which is something few demographers would have predicted 20 years ago. South Africa combines a 20-percent HIV infection rate for adults aged 15–49 with a fertility rate that is low for Africa, at fewer than three children per woman. Moreover, HIV/AIDS deaths are concentrated in the prime of life, among those who have the most to offer their societies. And women and young people in many countries are disproportionately affected. In South Africa, for example, a recent study indicated that death rates are higher among women in their twenties than among those in their sixties. By hollowing out the core of a nation, HIV/AIDS could cause economic and social havoc unprecedented in the modern world.[42]

The approach most likely to slow the further spread of the infection is the one agreed to in Cairo in 1994: a holistic effort to maximize the prospects of every human being to enjoy sexual expression and intentional reproduction in good health for themselves and their children. But intervention is not likely to be fully successful in combating HIV/AIDS while prevailing attitudes of sexual and gender relations make women so vulnerable to sexual predation in many societies. "Prevention strategies," says Noeleen Hayzer, Executive Director of UNIFEM, "must be designed with full recognition of the social factors that leave most women, particularly young women and girls, unable to negotiate safer sex or to refuse unwanted sex."[43]

The Politics of Population

After the Earth Summit, the Cairo conference, and the Beijing conference on women, the community of nations knows why and how to slow world population growth. And this work is moving forward. The global fertility rate has fallen almost by half in just 40 years. Yet the promise of reproductive health for all and equality for women remains unfulfilled. As a result, so does the vision of a world moving swiftly toward a population peak based on intended childbearing.[44]

At the ICPD in Cairo in 1994, governments agreed to spend $17 billion a year (in 1993 dollars) by 2000 to achieve universal access to basic reproductive health services for all by 2015. This was to include $10.2 billion for family planning services, $5 billion for maternal health and care at delivery, and $1.3 billion for prevention of HIV/AIDS and other sexually transmitted diseases. Since Cairo, the emerging deadliness of the HIV/AIDS pandemic has framed it almost as a separate health issue in international dialogue, with agreement that much more will need to be spent than the ICPD envisioned. But so far there is no consensus on just how much money will be needed in the effort to contain HIV/AIDS, what it will buy, and who will pay for it.[45]

Of the original Cairo sum for family planning and other reproductive health needs, wealthy nations pledged to cover one third of the cost, with the developing world agreeing to pay the remainder. In 1998, the most recent year with comprehensive data, wealthy nations contributed less than 40 percent of their Cairo commitment. By contrast, developing nations have been spending close to 70 percent of their committed levels. (This proportion is somewhat distorted, however, by high spending in China, India, and Indonesia, with much lower spending in sub-Saharan Africa.)[46]

The U.S. contribution to Cairo spending levels has been the most disappointing. The nation with the world's largest economy should be spending, according to calculations by Population Action International, $1.9 billion annually on family planning and related health programs in developing countries. The current U.S. contribution, however, is $500 million for reproductive health programs, as appropriated for fiscal year 2001, including $450 million for family planning and ancillary services and $50 million specifically for maternal health. Abortion-related restrictions—the "global gag rule" reinstated by the Bush administration—complicate the allocation of these funds. Consistent with greater attention to the HIV/AIDS pandemic, the U.S. government appropriated $320 million to combat the disease, but so far there is no roadmap for how the money will be spent or whether it will be integrated with any other aspects of reproductive health or with needed changes in gender relations.[47]

For the vision of Cairo to be realized, the ICPD Programme of Action needs to be fully funded to provide reproductive health services, including contraception, maternal care, and sexually transmitted disease prevention, for all who seek them. Ideally, more generous spending than the Cairo conference foresaw would be forthcoming to improve and fully integrate the entire range of reproductive health services, including HIV/AIDS prevention and basic treatment as well as access to safe abortion. At a minimum, honoring the Cairo spending goal could well be more effective than any other single effort in improving the lives of women and bringing population growth to an early peak based on intentional and healthy childbearing.

Historically, the world's major religions have erected some of the most formidable barriers to increased availability of family planning services and reproductive health care in general. Some Catholic, Islamic, and other religious leaders continue to preach abstinence as the only effective and moral means of controlling births. Nonetheless, from Iran to Italy, nations in which religion plays a major role have made great progress in widening access to family planning and reproductive health care and improving the status of women.[48]

Many religious leaders are coming to realize that there is no inherent conflict between family planning and religion, and that in fact lack of reproductive rights represents a grave social injustice. In Iran, Islamic clerics have even issued *fatwas*, or religious edicts, approving family planning methods—from oral contraceptives and condoms to sterilization. This approval, along with the integration of family planning services with primary health care, the provision of free contraceptives, and the strengthening of men's role in reproductive health, resulted in the total fertility rate in Iran dropping from 5.6 children in 1985 to 2.8 children in 2000—among the most precipitous declines in family size in the modern demographic transition.[49]

The influence of religious leaders tends to occur at the level of policymakers—undermining agreements on population and reproductive health, for example, and discouraging government health programs that include effective access to a range of contraceptives. At the household level, in contrast, women and men make the choices that affect their daily lives. In the privacy of their bedrooms, many see contraception not as a sin or a sign of lack of faith, but as an important part of loving, committed relationships.

Where religion continues to hamper efforts to give people greater control over their reproductive lives, the world's religious leaders may need to reconcile their actions with their humanitarian ideals. For instance, Bishop Kevin Dowling recently risked his career when he introduced a proposal at the Southern African Catholic Bishops conference in support of condom use as part of the wider effort to stop the spread of HIV in his region—home to the highest HIV infection rates in the world. Although the proposal was rejected, and the Church remains aggressively opposed to condom use, Bishop Dowling's efforts give some sense of the leadership that will be needed if religions are to work with others in the fight against HIV/AIDS and other public health problems related to reproduction.[50]

The gap between the opinions of church leaders and church members on reproductive issues mirrors a wider chasm between elected officials and their constituencies. According to a recent Gallup poll, for example, over 75 percent of Mexicans believe in a woman's right to choose abortion. Yet Mexico's politicians oppose reforms allowing women and couples greater access to safe abortion procedures. Conservative U.S. politicians, too, would like to see *Roe v. Wade*, the 1973 Supreme Court decision legalizing abortion in the United States, overturned. And they continue to stymie efforts to fund international family planning programs, even though opinion polls show that the vast majority of Americans support both a woman's right to control her own fertility and U.S. efforts in this area overseas.[51]

From some political and religious organizations, yet another misconception clouds discussion and muzzles debate—the idea that providing choices about pregnancy and childbearing is synonymous with the promotion of abortion. In the United States, a consistent effort by a small number of groups and politicians to promote this point of view has politicized what was once a bipartisan effort to guarantee worldwide access to contraception, and it has created a web of restrictions on U.S. spending to support international family planning. Ironically, demographic research confirms what logic tells us: wider provision of good family planning services reduces the numbers of abortions that would otherwise occur. When researchers looked at two similar areas of rural Bangladesh, one with good family planning services and the other without, they found that abortion rates had increased over the past two decades in the one with poor family planning services but had held steady at low rates in the area with good services.[52]

Just as important as spending levels are the political attitudes that shape and expand population policies and reproductive health programs around the world. In the spirit of Cairo, countries in Africa, Asia, and Latin America are rethinking population policies and programs and looking to the Programme of Action for guidance on new directions related to overall health and development. Progress is uneven, of course. The governor of the Indian state of Andhra Pradesh, for example, publicly urges the parents of large families to "immediately go" for state-sponsored sterilizations. And China's central government resists the key principle of reproductive freedom of choice by continuing to insist that most Chinese couples limit their families to a single child. Nonetheless, the government has at least acknowledged the importance of the principles agreed to at Cairo. And India's federal government is abandoning its decades-long history of targets and quotas for family planning and its reliance on sterilization rather than the contraceptives that are more appropriate for tens of millions of couples.[53]

The overall movement among national governments in developing countries is clearly away from bureaucratic population "control" and toward supporting the choices of couples and individuals to have children, when desired, in good health.

Correcting Gender Myopia

In the long view of where population policy is heading, the most daunting issues include not only religious obstacles or public division over abortion rights, but also the social and psychological shifts that will occur as women approach equal status with men. The more we learn about the interconnections between population growth, fertility, timing of pregnancy, and reproductive health, the more we see their links to ingrained attitudes about the relative roles and power between females and males.

As long as girls and women are envisioned as less able than boys and men to navigate human experience and decide for themselves how to live, population policy will always be imperfect. When girls go to secondary school free of fear of violence and sexual coercion and when women approach economic, social, and political parity with men, they have fewer children and give birth later on average than their mothers did—and, assuming good access to health and family planning services, fertility almost invariably reaches replacement level or lower. That slows the growth of population.

Yet this centrality of women to population's future also introduces discomfort, implying that interest in slowing population growth can turn women into instruments for some "larger" purpose, or into commodities to be counted and valued for the results of their reproductive decisions and actions. Those who work to slow the growth of population and those who work for women's parity with men sometimes are the same people, aiming at many of the same interim objectives: access to comprehensive and integrated reproductive health care, ending the gender gap in education and in economic opportunities, eliminating violence against women. The fact is that certain changes are essential for women themselves—simply from a perspective of fairness and equal rights for all humans—while simultaneously contributing to broader improvements in population trends and in human and environmental welfare.

The pervasiveness of violence against women around the world—verbal, physical, sexual, or economic—stands as the strongest indictment against current relations between the sexes. (See Box 6–3.) As many as half of all women have experienced domestic violence, according to the World Health Organization. Abuse from an intimate partner is the most common form, and this occurs in all countries—transcending economic, cultural, and religious boundaries. This picture of abuse is a conservative one at best: shame, fear, lack of legal rights, and gender inequality inside

BOX 6–3. VULNERABLE BY GENDER

Abuse shadows women from birth. Sex-selective abortions, female infanticide, and neglect of female children are common in India, China, and other nations. By some estimates, today's world population should include more than 60 million girls whose absence can only by explained by the fact that their own parents wanted them only if they were boys. Many girls who survive early childhood experience other abuses, including enforced malnutrition, incest, female genital cutting, denial of medical care, early marriage, prostitution, and forced labor. An estimated 130 million women and girls worldwide have undergone a ritualized cutting of their genitals; another 2 million girls a year still experience this ancient tradition, which can lead to a lifetime of painful urination, menstruation, and sexual intercourse, and which adds to the risk of death in childbirth.

Girls and women are more likely than boys and men to be sold into slavery, and trafficking in women condemns thousands annually to lives of essentially forced prostitution. In 2000, as many as 5,000 young girls died at the hands of their parents or another relative for shaming their families under prevailing social mores—they were suspected of having had sex or sometimes simply of socializing with the opposite sex. In some cases, the "dishonor" was that they had been raped.

It is tragically no surprise that women in some societies are much more likely than men to take their own lives. Eighty percent of all suicides in Turkey are women, and similarly high rates are found in other repressive societies, such as China, Afghanistan, and Iran. In all these cases, the link to population change is complex but significant: societies that treat women as property, or cause their disappearance because they are not male, or drive them to take their own lives are unlikely to support the conditions needed for planned families and the delay of pregnancy and childbirth.

SOURCE: See endnote 54.

and outside the household keep many women from reporting their attackers.[54]

Many men consider sex an unconditional right, and fear of reprisal can prevent girls and women from discussing contraception or their sexual rights with partners. The United Nations reports that women in Kenya and Zimbabwe hide their birth control pills for fear that their husbands might discover that "they no longer control their wives' fertility." Young girls married off to older men are neither emotionally nor physically prepared for their first sexual experience, which can set them up for years of having no say in when they have sex. In several African countries, most HIV-infected teenagers are female, reflecting the power of older men—sometimes including teachers—and the relative inability of girls to negotiate whether and under what conditions they have sex.[55]

Used as a weapon, sexual violence in all its forms—coerced sex, rape, incest—inhibits women's ability to control their own reproductive health. Ending this violence will be first and foremost its own reward. The supplemental benefit for positive demographic change comes from the simple fact that women can scarcely be free to decide for themselves when and with whom to become parents if they cannot even control the security of their persons.

Gender-related violence, however, is simply the most direct form of discrimination against women. Economies and societies generally undervalue women's work, from the household to the farm, the factory, and the office. Women typically work longer hours than men—nurturing children, caring for elders, maintaining homes, farming, and hauling wood and water home from distant sources. This labor is largely invisible to economists and policymakers, but by some estimates it amounts to a third of the world's economic production.[56]

When women's contributions do emerge from this obscurity, opportunities sometimes open up for broader social development as well as slower population growth. Making sure that girls and young women are in school, for example, can sometimes be even more effective than improved sanitation, employment, or a higher income in helping children survive. The nations in sub-Saharan Africa with the highest levels of female schooling—Botswana, Kenya, and Zimbabwe—are also the nations with the lowest levels of child mortality, despite higher levels of poverty than some of their neighbors. A study from the International Food Policy Research Institute (IFPRI) found that improvements in women's education were responsible for 43 percent of the reduction in child malnutrition in the developing world in the last 25 years.[57]

These benefits across generations appear to result from women's tendency to devote higher proportions of their personal income than men do to the needs of their children. A study in Brazil found that additional income in the hands of mothers was 20 times more likely to improve child survival than the same income earned by fathers. In general, as David Dollar and Roberta Gatti of the World Bank state, "Societies that have a preference for not investing in girls pay a price in terms of slower growth and reduced income." In countries where fewer than three fourths as many females as males are enrolled in primary and secondary school, for example, per capita income is roughly one fourth lower than in other countries. There can be little doubt that increases in income that have their roots in the education of girls and women also help build societies in which women on average have fewer children and give birth later in their lives.[58]

With the emergence of strong women's NGOs in the decade since Cairo, it seems likely that full political participation by women in national politics may become the last and most important frontier in achieving the gender equity needed for truly sustainable societies. Women remain underrepresented at all levels of government in almost all countries. There has been progress, but it has been slow, with women's share of seats in lower chambers of parliaments growing from 3 percent in 1945 to 14 percent in 2001. Typically, women's leadership is parceled out in less powerful sectors of government, such as health and education, with much smaller numbers of women holding key economic, political, and executive positions. Higher rates of illiteracy, poverty, and other social and economic handicaps conspire against political participation by women. Although sexual and reproductive rights occupy minimal space in debates over democracy, notes Marta Lamas of Mexico, once gained they allow women to achieve self-determination, and are thus intimately linked to the meaning of modern citizenship.[59]

Evidence from Sweden, South Africa, India, and other nations shows that when more women hold political office, issues important to women and their families rise in priority and are acted on by those in power. Over the past decade, the Swedish government—where women currently hold almost 43 percent of the seats in Parliament and 82 percent of the cabinet ministries—has passed expansive equal opportunity and child care leave acts. And in South Africa, which established a quota for women candidates to parliament in 2000, women hold 119 of the 399 seats in the National Assembly and 8 of the 29 cabinet positions. These female politicians have played a key role in lobbying for the Choice of Termination of Pregnancy Act and the Domestic Violence Act and in establishing governmental institutions that promote gender equality.[60]

When women gain rights to land or other resources, they also gain power that reaches well beyond forests or watersheds.

Despite such progress and the evidence of its benefits, gender myopia continues to cloud the vision needed by development agencies, international lenders, and governments. From agriculture to trade liberalization to health care reforms, policy decisions affect women in quite distinct ways. If their specific concerns are not made part of the policy process, the results can be disastrous. But seeing things through a gender lens requires a very different course for development—one that includes women and other marginalized groups—in planning and decisionmaking. Rachel Kyte, a senior specialist at the International Finance Corporation, argues that even now, a full decade after Rio, "it's very difficult to talk about the rights of women when the development industry remains truly patriarchal." Gender, by this view, is still not a central issue in development, perhaps in part because it so fundamentally challenges men's power.[61]

Gender myopia can be especially damaging in natural resource policy—for example, when development agencies offer technical and agricultural assistance mostly to men in areas where women are the ones toting the fuelwood and water and tilling the soil. In the past decade, the international development community has made strides in focusing its efforts on women's stewardship of natural resources. "Since rights to natural resources are so heavily biased against women," reasons Agnes Quisumbing of IFPRI, "equalizing these rights will lead to more efficient and equitable resource use." Indeed, when government officials or community leaders fail to recognize the different ways women use natural resources—in the spaces between male-managed cash crops, for example—the resources are easily destroyed.[62]

When women gain rights to land or other resources, they also gain power that reaches well beyond forests or watersheds. By commanding a concrete resource, notes Indian economist Bina Agarwal, women can take more control in existing relations by improving their self-sufficiency, reducing their dependence on men, and boosting their bargaining position within the marriage, including their ability to negotiate contraceptive use with their husbands. All these produce benefits that ripple out into the broader community.[63]

The strong role women play in environmental stewardship points to the opportunity for integrating reproductive health and family planning components into conservation programs. In the 1970s, some western NGOs concerned with improving rural environments and reducing poverty in the Philippines and Nepal began to offer improved access to family planning services. As interest in family planning expanded, other organizations partnered with national and regional family planning organizations to respond to women's requests for help with avoiding pregnancy. These initiatives demonstrated that incorporating improved access to contraception and other reproductive health services can increase women's participation in natural resource conservation or functional literacy programs and vice versa—a real-life demonstration that health and family planning cannot be separated from other aspects of people's lives.[64]

More recently, in Madagascar's Spiny Forest Ecoregion—home to the greatest concentration of baobab trees in the world—the World Wildlife Fund (WWF) produced maps showing that where female literacy levels were the lowest, both population growth rates and deforestation were the highest. Based on this, WWF fieldworkers and local stakeholders formed a partnership with Madagascar's regional public health organization to deliver literacy programs, reproductive health information, and family planning services to communities with both the highest population growth and the greatest levels of biodiversity.[65]

As the connections between conservation and population projects become clearer, the environmental community and environment ministers can become an important new constituency for discussions of reproductive health and women's rights. Investments to slow the rate of population growth will significantly reinforce efforts to address many environmental challenges, and considerably lower the price of such efforts.[66]

The river of thought on human rights and development runs inexorably toward the emancipation of women everywhere and equality between men and women. But eddies and rivulets carry the water backwards every day—as when pregnant girls are expelled from school, or when the genitals of young women are cut in a ritual destruction of their capacity for sexual pleasure.

Unfortunately, it is likely that even today people in Djole, the central Malian village in which teenaged Djenaba was interviewed in the 1990s, have no easy access to the health services that would allow a new mother to wait a few years before being pregnant again. But there are positive signs that such isolation cannot endure much longer. More NGOs than ever consist of women advocating for women's rights, empowerment, and well-being. The gap between the numbers of boys and girls in schools is beginning to close. Governments increasingly acknowledge the principles that were affirmed in 1994 in Cairo—that the capacity to plan a family is a basic right and that population trends should flow from the free decisions of women and couples.[67]

As the growing concerns about population aging and decline in some countries illustrate, it is increasingly possible that world population growth will end within the next 50 years. By the end of this century, there may be few countries whose populations are still growing. For the sake of the environment and healthy human relations, we should encourage this historic process, resisting the urge to try to roll back population aging in some countries by stoking continued population growth. We can work, as well, to make sure that the inevitable end to that growth is driven by intended reductions in births, not by increases in deaths.

If we succeed, history will note that world population growth ended not because governments commanded it to do so, but because the free decisions of women and men made that end inevitable. And the population peak will arrive as one momentous ripple from an equally momentous drop of a stone in a pond—the stone by which women at last gain their full rights, choices, and standing as equal members of the human family.

WORLD SUMMIT PRIORITIES ON POPULATION AND GENDER EQUITY

- Meet the goals of the 1994 International Conference on Population and Development, including funding universal access to reproductive health care and closing the gender gap in education.
- Aggressively respond to the global HIV/AIDS pandemic, stressing prevention of further infections as well as treatment of those already infected.
- Change laws and work for social change to ensure that women enjoy equal protection and equal rights.
- Increase female participation in all levels of politics.
- Correct gender myopia in all levels of private and public planning, including international lending, natural resource policy, and globalization.
- Guarantee equal access to economic opportunities for women and men.
- Enact and enforce strong laws to protect women from all gender-based violence.
- Involve men in reproductive health services and discussions, and educate them about the importance of gender equity.
- Ensure that young people have better access to reproductive health care choices and to education on sexuality and the changing roles of men and women.

Chapter 7

Breaking the Link Between Resources and Repression

Michael Renner

The United Nations Children's Fund has described Angola as "the worst place in the world to be a child." Almost 30 percent of children die before they reach the age of six. Nearly half of all Angolan children are underweight, two thirds of Angolans scrape by on less than a dollar a day, and 42 percent of adults are illiterate. Food shortages, unsafe drinking water, and a pervasive lack of sanitation and health services have combined to limit life expectancy to just 47 years—short even by the standards of sub-Saharan Africa. The 2001 Human Development Index of the U.N. Development Programme (UNDP), a broad measure of social and economic progress, ranked Angola 160th out of 174 nations.[1]

Endowed with ample diamond and oil deposits and other natural resources, Angola should not be on the bottom rungs of the world's social ladder. But more than a quarter-century of brutal "civil" war has imploded the economy, displaced close to 4 million people—one out of three Angolans—and left about a million people dependent on foreign food aid. While the bulk of the population lives in misery and terror, the leaders of both the government and the rebel UNITA forces have devoted most of the money they gained selling Angola's resources to buying weapons and lining their own pockets. The ideological differences that first sparked the war now reside in the dustbin of history, but resource-driven greed and corruption have proved to be powerful fuel for its continuation. Instead of a promise, diamond and oil wealth has turned out to be a curse.[2]

Though a somewhat extreme case, Angola is merely one of numerous places in the developing world where abundant natural resources help fuel conflicts. (See Table 7–1.) Altogether, about a quarter of the 49 wars and armed conflicts waged during 2000 had a strong resource dimension—in the sense that legal or illegal resource exploitation helped trigger or exacerbate violent conflict or financed its continuation.[3]

As of late 2001, conflict is not on the agenda of the World Summit on Sustainable

Table 7–1. Selected Examples of Resource Conflicts

Location and Resources	Observation
Colombia – oil	Since 1992, a "war tax" (more than $1 per barrel) has been levied on foreign oil firms to finance the army's defense of oil installations against rebel attack. Occidental Petroleum also made direct payments to the army. Guerrilla groups have generated some $140 million in extortion money from oil firms. Oil has become Colombia's largest export earner, but most people see few benefits, and indigenous groups like the U'wa fear growing encroachment by the oil industry. Protests against oil projects have brought military repression.
Sudan – oil	Civil war restarted in 1983 (the government reneged on a peace pact after oil was discovered in 1980), leading to more than 2 million deaths, 1 million refugees, and 4.5 million people displaced. Oil exports, started in 1999, now escalate the conflict: oil revenues pay for arms imports and helped triple military expenditures; oil industry roads and airstrips are used by the army. To depopulate oil-producing and potentially oil-rich areas in southern Sudan, government forces are bombing villages, destroying harvests, and looting livestock, and they are encouraging intertribal warfare by supplying arms to some factions. Opposition forces have targeted oil installations.
Chad and Cameroon – oil	Suppression of a revolt in Chad's Doba region (where oil production is to start in 2003) led to hundreds of deaths. In 2000, the government of Chad bought weapons with part of $25 million in "bonuses" paid by ExxonMobil, Chevron, and Petronas. Construction of a pipeline to Cameroon's coast threatens the land of the Baka Pygmies and may bring poaching and unregulated logging to Atlantic rainforest areas.
Afghanistan – emeralds, lapis lazuli, opium, heroin	Opium trafficking helped finance the anti-Soviet struggle and then civil war among Mujahideen factions. It has been a crucial source of revenue for the Taliban regime in the ongoing civil war since the mid-1990s, earning it up to $50 million a year. Opium production surged from 10 tons in the late 1970s to 1,200 tons in 1989 and then 4,600 tons in 1999. Under international pressure, the Taliban banned poppy cultivation in July 2000, but scrapped this ban following U.S. attacks in October 2001. A 25-percent tax has also been levied on timber shipments to Pakistan. The opposition Northern Alliance has relied mostly on earning up to $60 million annually from the sale of emeralds and lapis lazuli (an azure-blue semiprecious stone).
Cambodia – sapphires, rubies, timber	Following the end of Chinese aid in 1989, Khmer Rouge rebels resorted to resource looting to finance their operations. Mining and logging licenses granted to Thai companies in Khmer Rouge territory earned the group as much as $120–240 million a year in the early to mid-1990s. Gem depletion and Thai restrictions on the timber trade caused a sharp income drop after 1995, severely weakening the Khmer Rouge. The Cambodian government was making some $100 million a year in the mid-1990s from secret, illicit deals that gave Vietnamese loggers access to timber concessions. But extensive deforestation cut earnings to $20 million.

Note: Examples discussed in some detail in this chapter (Angola, Sierra Leone, Democratic Republic of Congo, Nigeria, Indonesia, and Papua New Guinea) are not included here.
SOURCE: See endnote 3.

Development in Johannesburg. But it is unquestionably an issue of prime importance. Basic human security—the absence of violent conflict—is a precondition for establishing a sustainable society. And many contemporary resource-related conflicts are being fought in areas of great environmental value. The Democratic Republic of Congo (formerly Zaire), Indonesia, Papua New Guinea, and Colombia, for example, together account for 10 percent of the world's remaining intact forests. These and other countries in which resource conflicts are raging are also home to some of the world's biodiversity hotspots. Yet they suffered from the world's highest net loss of forest area in the 1990s, due to illegal logging and a host of other factors.[4]

The Relationship Between Resources and Conflict

There is growing awareness of the close links among illegal resource extraction, arms trafficking, violent conflict, human rights violations, humanitarian disaster, and environmental destruction. Expert panels established by the United Nations have investigated cases in Angola, Sierra Leone, and the Democratic Republic of Congo. Civil society groups have launched a campaign against "conflict diamonds" from those countries and have shed light on other conflict resources as well. Company and industry practices are coming under greater scrutiny. Media reports have helped carry these concerns from activist and specialist circles to a broader audience. All of this also comes against the background of an intensifying debate over the unchecked proliferation of small arms—the weapons of choice in resource-based conflicts.

In some places, the pillaging of oil, minerals, metals, gemstones, or timber allows wars to continue that were triggered by other factors—initially driven by grievances or ideological struggles and bankrolled by the superpowers or other external supporters. Elsewhere, nature's bounty attracts groups that may claim they are driven by an unresolved grievance such as political oppression or the denial of minority rights, but are actually criminal entrepreneurs trying to get rich through illegal resource extraction. They initiate violence not to overthrow a government, but to gain and maintain control over lucrative resources, typically one of the few sources of wealth in poorer societies. They are greatly aided by the fact that many countries are weakened by poor or repressive governance, crumbling public services, the lack of economic opportunities, and the presence of deep social divides.

There is another dimension to the relationship between resources and conflict. It concerns the repercussions from resource extraction itself. In many developing countries, the economic benefits of mining and logging operations accrue to a small business or government elite and to foreign investors. But in case after case, an array of burdens—ranging from the expropriation of land, disruption of traditional ways of life, environmental devastation, and social maladies—are shouldered by the local population. Typically, these communities are neither informed nor consulted about resource extraction projects. This has led to violent conflict in places like Nigeria's Niger Delta, Bougainville in Papua New Guinea, and a variety of provinces in Indonesia. Rather than full-fledged war, these conflicts usually involve smaller-scale skirmishes, roadblocks, acts of sabotage, and major human rights violations by state security forces and rebel groups.

This chapter is concerned with natural

resources extracted through mining and logging activities, but other resource-related conflicts are also taking place around the world. Many local and regional disputes revolve around equitable access to arable land and water, although these resources are not easily traded and therefore do not lend themselves to financing hostilities. Due to space constraints, the chapter also does not address conflicts arising out of the depletion of resources and degradation of natural systems.

The examples discussed here are all "civil" conflicts in the sense that the violence takes place within a given country, although there are important global connections through the world market for illegal resources and the supply of arms, and through spillovers into neighboring countries. The years ahead will also likely witness the threat of growing resource wars across borders. A recent study by Professor Michael Klare of Hampshire College underscores that as demand for fuels, minerals, water, and other primary commodities continues to rise at environmentally unsustainable rates, disputes over ownership are multiplying, and emerging scarcities are increasing the likelihood that industrial powers will intervene to secure "their" supplies of raw materials.[5]

Anatomy of Resource Conflicts

In contrast to the cold war era, today's conflicts are less about ideologies and more about the struggle to control or loot resources—less about taking over the reins of state and more about capturing locations that are rich in minerals, timber, and other valuable commodities or controlling points through which they pass on the way to markets. Although some of today's conflicts have their roots in long-standing grievances, changed circumstances have altered the dynamics of these conflicts and provided them with a powerful momentum of their own: a vicious cycle in which the spoils of resource exploitation fund war, and war provides continued access to these resources.

The end of cold war rivalry meant that much of the support previously extended by the two superpowers to their Third World allies—whether governments or rebels—has fallen by the wayside. While external patrons (either governments or nationals living outside the country) have not vanished altogether, warring factions are increasingly relying on a variety of criminal means, including extortion, pillage, hostage-taking, monopolistic control of trade, drug trafficking, exploitation of coerced labor, and commandeering of humanitarian aid within their borders.[6]

But possibly the most important revenue source is the illicit extraction and trading of natural resources. Paul Collier, director of the Development Research Group at the World Bank, suggests that greed and the availability of "lootable" natural resource wealth are key factors. What is required is the presence of primary commodities that can be captured or taxed by armed groups. While Collier overstates the importance of greed and downplays other factors such as grievance, his work underscores that "some countries are much more prone to conflict than others simply because they offer more inviting economic prospects for rebellion."[7]

Most resource-based conflicts are unlike traditional wars. Pitched battles between opposing sides are generally avoided because the objective is maintaining conditions conducive to resource looting. In fact, some of the different armed factions are known to have engaged in simulated attacks against opposing groups, sold arms and supplies to them, or collaborated in other ways. The common objective is to facilitate looting and to perpetuate conditions that permit activities that otherwise would be plainly understood as criminal.[8]

The bulk of the violence is directed against civilians. Since establishing undisputed control over resources is a key objective, armed groups seek to intimidate the local population into submission or use terror to drive people away. "Hence the importance of extreme and conspicuous atrocity," observes Mary Kaldor of the University of Sussex, including directly expelling people, rendering an area uninhabitable by the indiscriminate spread of landmines, shelling houses and hospitals, chopping off people's limbs, imposing long sieges and blockades to induce famine, and applying systematic sexual violence. Unlike ideologically based rebel movements, those pursuing resource wealth do not compete for "the hearts and minds" of the local population. Young boys are often turned into child soldiers, and girls into sex slaves for the fighters. Many fighters are forced to commit atrocities, often against their own relatives, in order to traumatize them and to spread a sense of complicity that will prevent them from being accepted back into their communities later.[9]

Actions that are often described as chaos, collapse, and senseless violence in media reports actually flow from a certain logic, albeit a perverted one. David Keen, a lecturer at the London School of Economics, argues that violence serves an economic function, maintaining a conflict economy that benefits certain groups—government officials, warlords, combatants, arms smugglers, and unscrupulous traders and businesspeople. Those who benefit from this violent "mode of accumulation" derive profit, power, and status, even as it spells impoverishment, broken lives, and death for society at large. Groups living off a lucrative resource have a vested interest in prolonging conflict. They are likely to find this to be a more attractive choice than settling conflict because it allows them to maintain their privileged position and bestows a quasi-legitimacy on their actions.[10]

But the lure of easy wealth through illegal resource exploitation also encourages the proliferation of competing groups of ruthless and well-armed people, often deepening the lawlessness and prolonging violence. The implications for those in the international community who seek to end such conflicts are disturbing. It may be possible to arrange cease-fires or even peace accords, but these tend to be respected only as long as they suit the interests of the armed groups. Achieving a true peace will require long-term and substantial involvement by the international community.[11]

Why are some countries susceptible to resource-based conflicts? While the availability and "lootability" of natural resources is a key factor, this alone does not explain the conflict. Many countries with a rich resource endowment have not fallen prey to violence. A number of factors—political, social, economic, and military—result in weak states and vulnerable economies.

Ample resource endowment can actually have negative economic consequences, as countries grow overly dependent on these resources and allocate little capital and labor to other sectors—agriculture, manufacturing, and services. The result is a failure to diversify the economy and to stimulate innovative energies and the development of human skills. And the volatility of world commodity market prices can trigger distorting boom-and-bust cycles.

Some researchers argue that societies

whose main income is derived from resource royalties instead of value added are prone to develop a culture with widespread corruption, a growing gap between rich and poor, and state institutions that do not function properly and fail to serve the public. Resource extraction industries tend to have enclave characteristics—that is, they have few linkages to the rest of the local economy, particularly if the resources are exported before any processing takes place. Frequently, enclaves are also physically separated, as mineral deposits or timber resources are often found in remote areas. The benefits to the larger economy are therefore quite limited.[12]

The resources over which so much blood is being shed have consumers in the richest countries as their destinations.

Another factor is the extremely poor governance of many countries, in some cases leading to what William Reno of Northwestern University has called the "shadow state": a situation where corruption and patronage are rife, public goods and services are being withheld from most people, and state institutions (like the civil service, universities, the central bank) are weakened to thwart potential challengers to the ruler, while a parallel network outside these formal institutions is created for the benefit of leaders and their cronies. State revenues are diverted to generate huge illicit fortunes for rulers and payments to key regime supporters. (Zaire's dictator Mobutu, for instance, was thought to have amassed illegal wealth worth an estimated $4–6 billion, more than the country's annual economic output.)[13]

Rulers of shadow states often foster and manipulate conflicts among different communities, factions, and ethnic groups as a means to maintain control. However, ruling in such a fashion intensifies frictions within society. In such conditions, discontented and aggrieved groups turn increasingly to protest and perhaps violence, rivals rise to challenge the discredited leadership, and ruthless political or criminal entrepreneurs who sense an opportunity for pillaging resources use violence to achieve their objective.[14]

Many developing countries, particularly in sub-Saharan Africa, face a situation where government forces are in decay and private security formations, including paramilitary units, citizens' self-defense groups, corporate-sponsored forces, foreign mercenaries, and criminal gangs, are on the rise. In fact, it is becoming more difficult to make clear-cut distinctions between legitimate and illegitimate, between public and private, security forces.

This is happening for a number of reasons. Without cold war–motivated sponsorship and under increasing pressure from western donors for belt-tightening, many governments can no longer maintain large armies. Soldiers go unpaid or underpaid and often turn to other sources of funding, including looting and extortion. Some military commanders become de facto local warlords. Such fragmentation is even more likely where rulers have deliberately created rival security forces that keep each other in check, preventing a serious challenge to central control.[15]

During the 1990s, a number of private mercenary firms rose to prominence. Companies like Executive Outcomes (now defunct), Sandline International, Defense Systems Ltd., and Ghurka Security Guards attracted military personnel from western industrial and former Warsaw Pact armies who lost their jobs at the end of the cold war, as well as veterans of apartheid-era South Africa. They offer a range of "services" that include training and consulting, guarding facilities, procuring or brokering weapons, and running combat operations—and corporations and governments the world over have contracted with them. Several beleaguered governments, including those of Angola, Sierra Leone, and Papua New Guinea, turned to them to help fight rebel groups, paying them with revenues derived from natural resources or, in some cases, granting them (or affiliated companies) concessions to diamonds and other resources.[16]

Multinational oil and mining corporations often rely on private security forces to guard their operations and facilities. And in some cases, such as Occidental Petroleum in Colombia, Shell in Nigeria, Talisman Energy in Sudan, and ExxonMobil and Freeport-McMoRan in Indonesia, they have subsidized or helped train and arm government security forces or have made equipment and facilities available. These units have been involved in severe human rights violations.[17]

The massive proliferation of small arms and light weapons plays a key role in all of this. Resource-based conflicts are primarily carried out with such weapons because they are cheap, widely available, easy to conceal and smuggle, and easy to use and maintain. Using them does not require complex logistical arrangements. A rough estimate of global small arms production suggests that about 6 million pistols, revolvers, rifles, submachine guns, and machine guns were manufactured each year during the past two decades; all in all, it is thought that at least 550 million firearms exist worldwide. For 2000 alone, it is estimated that at least 15 billion rounds of ammunition were produced. The picture that emerges, despite uncertainty surrounding the data, is one of a world exceedingly well equipped with these tools of terror and death.[18]

Because many activities along the resource-conflict spectrum are illicit in nature and involve actors of questionable legitimacy, grey and black market transfers carry special significance. The trafficking of arms is closely linked to illegal trade of raw materials such as minerals, timber, and diamonds. The routes on which arms and commodities travel in opposite directions are often the same. Revenues from selling off commodities finance the purchase of arms, ammunition, military equipment, uniforms, and other items; sometimes weapons are directly bartered for natural resources, drugs, animal products, and other commodities.[19]

Resource-based conflicts in places like Kono, Aceh, and Bentai seem far removed from the shopping malls of the western world. But the resources over which so much blood is being shed have consumers in the richest countries as their destination, no matter how complex and circuitous the networks of delivery are. For consumers, this connection is easiest to grasp in the case of diamonds, a highly visible and prominently marketed product. For materials like petroleum, timber, copper, and coltan, the connection is harder to make because they undergo extensive processing before they find their way into complex final products. (See Box 7–1.) But a portion of the western world's cell phones, mahogany furniture, and gold chains bears the invisible imprint of violence.[20]

In the final analysis, it is the strong demand for commodities and the consumer products made from them that makes illegal resource exploitation so lucrative. The enormous expansion of global trade and the growth of associated trading and financial networks have made access to key markets relatively easy for warring groups. They have had little difficulty in establishing international smuggling networks, given either a lack of awareness and scrutiny or a degree of complicity among international traders, manufacturers, and financiers, as well as lax controls in consuming nations.[21]

Some major international companies have in effect helped perpetuate resource-based conflict in several ways:

- by purchasing "hot" commodities from combatants, as De Beers did until recently when it bought conflict diamonds;
- by providing revenues to governments that are at war, as the oil companies Chevron and Elf have done in Angola;
- by facilitating the shipment of illicit raw materials, such as Sabena flying coltan derived from occupied Congo to Europe; and
- by operating in countries with repressive or illegitimate governments, as Exxon-Mobil and Freeport-McMoRan in Indonesia, Shell in Nigeria, Talisman Energy and others in Sudan, and Occidental in Colombia do.

BOX 7–1. THE COLTAN CONNECTION

Few owners of mobile phones realize that their technical gadgets may link them to one of the deadliest of contemporary wars—the conflict in the Democratic Republic of Congo. Coltan, short for columbite-tantalite, is one of the raw materials that warring factions have battled over. With the appearance of gritty black mud, coltan is an ore containing tantalum, a highly heat-resistant material. Tantalum is crucial for the manufacturing of capacitors, tiny components that regulate the flow of current on circuit boards and help make the modern world go round. As one journalist put it, "for the high-tech industry, tantalum is magic dust." More than half the global supply is used by the electronics industry for products like cell phones, laptops, and pagers, but there are also important applications in the aerospace, defense, chemical, pharmaceutical, medical, and automotive industries.

World tantalum supply runs to about 3,000–3,500 tons a year. Perhaps three quarters of this comes from legitimate mining operations in Australia, Canada, and Brazil. But Congo, with the world's fourth-largest coltan reserves, is also an important supplier. Rwandan troops and their rebel allies, the Rally for Congolese Democracy (RCD), took control of 1,000–1,500 tons of coltan stocks in 1998–99. They also drove Congolese farmers off their coltan-rich land and had Rwandan prisoners dig for coltan in exchange for reduced sentences.

The high-tech industry's soaring demand for tantalum triggered a temporary global supply shortage in 2000. Prices surged from less than $20 per pound in 1998 to more than $200, making the coltan business extremely lucrative for the warring parties and individual miners. In late 2000, the RCD rebels said they produced 100–200 tons of coltan a month, yielding the group a larger windfall than its diamond mining activities. Then in 2001, prices crashed in response to slumping cell phone sales, putting a damper on the gold rush–like conditions in illegal mining camps in eastern Congo. Still, coltan deposits retain their lure—the promise of a better life in a country where most incomes are measured in mere cents per day.

SOURCE: See endnote 20.

These corporate practices do not necessarily all constitute wrongdoing. Oil companies in Angola are contracting with a recognized government, for example. But their presence plays an enabling role in situations where the majority of the population suffers from violence and deprivation.[22]

How Conflicts Are Financed by Natural Resource Pillage

A campaign against "blood diamonds" launched by civil society groups has highlighted the fact that several violent conflicts in developing countries are funded by the sale of glittering stones that advertisers work hard to associate with the idea of love and personal commitment. Diamonds and other commodities have been of particular concern in three conflicts discussed in some detail here: Sierra Leone, the Democratic Republic of Congo, and Angola.

It is difficult to know the share of resources derived from war zones. For diamonds, industry giant De Beers estimates that in 1999 blood diamonds accounted for about 4 percent of the world's rough diamond production of $6.8 billion. But other estimates go as high as 10–20 percent. Besides conflict diamonds, there is also a substantial quantity of illicit diamonds—ones that have been mined illegally or stolen, but not derived from conflict areas. Because both types rely on gray and black markets, it is extremely difficult to distinguish between them. A U.N. group of experts estimated that about 20 percent of the global trade in rough diamonds is illicit in nature.[23]

Diamonds have played a central role in the conflict that devastated Sierra Leone during the 1990s. Ibrahim Kamara, Sierra Leone's U.N. ambassador, said in July 2000: "We have always maintained that the conflict is not about ideology, tribal or regional difference....The root of the conflict is and remains diamonds, diamonds and diamonds."[24]

Even prior to the 1990s, corruption, cronyism, and illegal mining had squandered the country's diamond riches, to the point where few government services were functioning, and educational and economic opportunities were few and far between. Sierra Leone became a "model" shadow state. Pressure from international lenders for financial austerity and retrenchment in the government workforce only succeeded in worsening the situation. The International Rescue Committee has reported that one third of all babies in the diamond-rich Kenema District die before age one. UNDP placed Sierra Leone dead last on its Human Development Index.[25]

Throughout the 1990s, Sierra Leone suffered from rebellion, banditry, coups and coup attempts, and seesawing battle fortunes. (See Table 7–2.) In March 1991, the Revolutionary United Front (RUF) invaded Sierra Leone from Liberia, with strong backing from then warlord and now president Charles Taylor, and seized control of the Kono diamond fields. The ranks of the RUF contained disaffected young men from slum areas, illicit diamond miners, Liberian and Burkinabe mercenaries, and others who welcomed the opportunity for pillage and violence. But many others were kidnapped and forced to commit atrocities, including a large number of children. Though the RUF professed to act on unresolved grievances, its principal aim was to gain control over the country's mineral wealth. The rebellion was characterized by banditry and brutality. It claimed more than 75,000 lives, turned a half-million Sierra Leoneans into refugees, and displaced half of the country's 4.5 million people.[26]

Faced with the RUF rebellion, the gov-

Table 7–2. Key Events in Sierra Leone's Civil War

Year	Events
1991–95	RUF invades Sierra Leone, triggering a civil war involving government troops, civil defense militia, several private mercenary firms, and Nigerian forces. RUF controls diamond-rich areas. Army splits into factions. The private military firm Executive Outcomes (paid in cash and with diamond profits) pushes the RUF back.
1996	Elections in March won by Ahmad Tejan Kabbah; new government signs peace agreement with RUF in November calling for disarmament, demobilization, and withdrawal of foreign forces.
1997	In May, Kabbah overthrown by Armed Forces Revolutionary Council (AFRC—a disgruntled faction of the army). AFRC invites the RUF to join its regime in June; systematic murder, torture, rape, and looting follow. In October, the United Nations imposes an arms embargo.
March 1998	Nigerian troops and Sandline, a private mercenary firm, drive the AFRC/RUF from the capital of Freetown; Kabbah reinstated. By June, the United Nations narrows arms embargo to nongovernmental forces.
Late 1998	RUF regains control of diamond areas and attacks Freetown; massive human rights violations.
1999	Lomé peace accord signed (amnesty and cabinet positions for RUF and AFRC leaders); Nigeria begins troop withdrawal; U.N. Security Council establishes UNAMSIL peacekeeping force, but it is "peacekeeping on the cheap" (slow arrival of underequipped, poorly trained troops).
May 2000	RUF takes several hundred UNAMSIL troops hostage, full-scale fighting resumes; British troops intervene; RUF leader Foday Sankoh captured.
July 2000	U.N. embargo imposed on any diamonds that do not have a government certificate.
November 2000	Cease-fire agreement signed between the government and the RUF; cease-fire largely observed but situation remains volatile.
March 2001	U.N. Security Council threatens sanctions against Liberia unless it demonstrates that it is not supporting the RUF.
July 2001	U.N. Security Council approves plans for a war crimes tribunal, but proposed budget is cut in half. Government, the United Nations, and RUF agree on diamond mining ban in Kono district.
Fall 2001	RUF continues to mine diamonds in violation of ban, using forced labor; 15,000 U.N. peacekeepers enforce cease-fire, but policing the mining ban is not part of their mandate. Some 16,000 RUF and militia combatants disarmed, but lack of funding hinders reintegration of fighters, and RUF retains weapons and its military structure.

SOURCE: See endnote 26.

ernment expanded its armed forces from 3,000 to 14,000. This undisciplined, ineffective, ragtag army brought together ill-trained soldiers, militiamen from neighboring Liberia, urban toughs, and street children involved in petty theft. Mary Kaldor of the University of Sussex comments about the latter that "they were given an AK47 and a chance to engage in theft on a larger scale." Government soldiers often supplemented their meager payments through looting and illegal mining.[27]

Rebel forces and parts of the government army actually collaborated at times. Government soldiers by day sometimes became rebels by night. This cooperation between supposed adversaries culminated in May 1997 when disgruntled government soldiers staged a coup against a government that had been elected just several months earlier, and invited the RUF to join the new junta. The prospect of peace was seen as an unacceptable threat to their system of criminal exploitation.[28]

Sierra Leone is a comparatively small diamond producer, but a large share of its gemstones are of very high quality and therefore sought after. The RUF purchased arms and sustained itself through its control of the diamond fields, but diamond wealth has been a constant source of internal friction. At first, RUF fighters did the mining, but later the group relied more on forced labor, including that of children. The group's annual income has been estimated at $25–125 million, though some estimates are considerably higher.[29]

RUF diamonds enter the world market disguised as Liberian, Guinean, and Gambian diamonds. An investigative U.N. panel reported in December 2000 that it had "found unequivocal and overwhelming evidence that Liberia has been actively supporting the RUF at all levels, in providing training, weapons and related matériel, logistical support, a staging ground for attacks and a safe haven for retreat and recuperation, and for public relations activities." Under Charles Taylor, Liberia has become a major center for diamond smuggling, arms and drug trafficking, and money laundering.[30]

Taylor has grabbed exclusive control over Liberia's natural resources and is using the profits from timber exports to support the RUF in Sierra Leone. As international sanctions succeed in clamping down on the trade in conflict diamonds, the importance of timber revenues is rising. Taylor receives extra-budgetary payments from a small number of logging companies that get special privileges in return and are involved in arms smuggling. One is Exotic Tropical Timber Enterprise run by Ukrainian arms and diamond dealer Leonid Minin, who was arrested in Italy in July 2001 for gunrunning. But the key player in the illicit timber trade appears to be the Oriental Timber Co. (OTC). Controlling some 43 percent of Liberia's forests, the company has been implicated in smuggling weapons to the RUF along its timber roads.[31]

Liberia still has a considerable amount of its original rainforest cover and a rich array of plant and animal species, including forest elephants and the endangered Pygmy hippopotamus. But the scale of the timber trade now is such that its forests are likely to be denuded in little more than a decade; according to current plans, the pace will actually intensify further. OTC has not only engaged in rapacious clear-cutting methods, it has also bulldozed through homes and entire villages with little warning and no compensation. Forest management and replanting efforts are virtually absent.[32]

The U.N. expert panel also found conclusive evidence that Burkina Faso is a key conduit in facilitating small arms shipments to Liberia and the RUF. In addition, arms have been transferred through Senegal, Gambia, and Guinea. And Côte d'Ivoire has directly assisted the RUF. Weapons originated primarily in Libya, Ukraine, Slovakia, and Bulgaria, and sometimes were shipped with the help of western air cargo companies.[33]

Resource pillage has also been a key factor in the conflict in the Democratic Republic of Congo, a war that has killed some 1.7 million people and displaced another 1.8 million. In August 1998, Ugandan and Rwandan troops invaded, assisting rebel groups seeking to overthrow the government of Laurent Kabila. Angola, Zimbabwe, and Namibia dispatched troops in support of Kabila. According to one estimate, more than 100,000 foreign troops have been involved in Congo. Several of the intervening forces wanted to thwart their own rebel groups operating from Congolese soil. Rwanda, in particular, was concerned that remnants of the Hutu Interahamwe militias that had carried out a campaign of genocide against the Tutsi in Rwanda in 1994 were using Congo as a staging ground for their hit-and-run attacks.[34]

But in addition to political and military factors, the opportunity to plunder the enormous resource wealth of Congo, in the context of lawlessness and a weak central authority, came to be a powerful incentive for continued conflict. Congo is extremely rich in minerals, gemstones, and agricultural raw materials such as diamonds, gold, coltan, niobium, cassiterite, copper, cobalt, zinc, manganese, timber, coffee, tea, and palm oil. In addition, the country's wildlife, including okapis, gorillas, and elephants (for their tusks) attract poachers. The losers have been the vast majority of Congo's population and the natural environment. (See Box 7–2.)[35]

At first, the invading forces and their allies resorted to outright plunder of stockpiled raw materials; later they organized a variety of methods to extract additional resources. Individual soldiers work for their own or their commanders' benefit, while local Congolese have been put to work by Rwandan and Ugandan forces. Local miners were made to relinquish some of their finds (either by force or through extortion rackets). Foreign nationals, including Rwandan prisoners, have worked for the Rwandan army's or the commanders' benefit. Companies of questionable reputation were given concessions, and child labor was used in gold and diamond mining. Occupying forces and their rebel allies have also forced coffee growers and palm oil producers to sell their commodities at depressed prices.[36]

The conflict has enabled Rwanda and Uganda to become major exporters of raw materials that they do not possess at all or have only in limited quantities. Looted resources have become a major source of their foreign exchange. Uganda, for instance, is re-exporting gold, diamonds, cassiterite, coltan, coffee, tea, timber, elephant tusks, and medicinal barks. It now exports 10 times more gold ore than it did five years ago. Resource pillage has allowed both countries to finance their military presence. Rwanda has even set up an extra-budgetary system for this purpose. In Uganda's case, the individual enrichment of top military commanders and businessmen—including Salim Saleh, who is the brother of President Museveni, and James Kazini, the former chief of staff of the Ugandan army—appears to be the main driving force.[37]

Congo's government itself has used its natural resources as payment in kind to buy weapons. For instance, a Chinese company was brought into a mining joint venture as

BOX 7–2. ENVIRONMENTAL IMPACTS OF RESOURCE CONFLICT IN CONGO

An expert panel of the United Nations on illegal resource exploitation in Congo found that okapis, gorillas, and elephants have dwindled to small populations in several national parks. First, in 1994, came the influx of more than a million refugees from Rwanda, leading to widespread deforestation and wildlife poaching. In 1998, they were joined by Rwandan and Ugandan troops and their Congolese allies, rebels against the Kabila regime. The promise of rich coltan deposits and other resources also lured some 10,000 miners into Kahuzi-Biega National Park and the Okapi Wildlife Reserve; both of these are UNESCO World Heritage sites, but severe environmental degradation has landed them on the organization's list of sites in danger. Poaching of elephant tusks, in violation of an international treaty, left only 2 out of 350 elephant families in Kahuzi-Biega in 2000. Likewise, the number of eastern lowland gorillas has been driven so low that they are threatened by extinction.

Miners strip off the bark of eko trees to fashion troughs in which they flush out coltan from ore-bearing mud; thousands of trees have been destroyed, undermining the livelihoods of the local indigenous people, the Mbuti, who use the eko trees for gathering honey.

Logging companies connected to the rebels have engaged in rapacious clear-cutting operations. DARA-Forest Co., for example, had been denied a logging license in early 1998, but obtained a concession in Orientale Province in 2000 from RCD-ML, a rebel group allied with Uganda. It subsequently carried out logging "without consideration of any of the minimum acceptable rules of timber harvesting for sustainable forest management," according to the U.N. expert panel. Satellite images show deforestation taking place at an alarming rate. Although DARA-Forest failed to satisfy Forest Stewardship Council procedures and evaded international requirements for timber certification, the U.N. panel found that companies from Belgium, Denmark, Switzerland, China, Japan, Kenya, and the United States nevertheless imported the company's timber via Uganda.

SOURCE: See endnote 35.

part of a deal to secure Chinese military equipment. And although the motivation of countries intervening in support of the government seems primarily political and strategic, Congo has nevertheless used resource wealth as an incentive for its allies to stay involved. The government has granted several concessions, including offshore oil wells to Angola, diamond and cobalt to Zimbabwe, and a share of a diamond mine to Namibia. Ridgepoint, a Zimbabwean firm whose officials include Zimbabwe's justice minister and a nephew of President Mugabe, became a partner of Congolese state-owned Gécamines in a copper-mining venture. Zimbabwe also received timber as barter payment for its military assistance.[38]

Responsibility for the conflict lies with not only regional leaders but also more distant countries, international donors, and private companies that have wittingly or unwittingly facilitated the exploitation of Congolese resources by shipping and buying illegally obtained commodities. A U.N. expert panel named Mombasa (in Kenya), Dar es Salaam (in Tanzania), and Douala (in Cameroon) as the main ports used. It also listed 34 companies based in Western Europe, Canada, Malaysia, India, Pakistan, and Russia as importers of illicit Congolese minerals. Finally, the panel criticized the World Bank for turning a blind eye to the illegal exploitation of resources; Bank staff either failed to question or even actively defended Uganda's suddenly increased raw materials exports.[39]

Even if international efforts succeed in establishing a lasting cease-fire and a withdrawal of foreign troops, a number of illegal networks headed by military officers and unscrupulous political and business leaders continue to control vast areas of Congo and operate them as their personal fiefdoms. They have a vested interest in the continuation of conflict as a cover for their plundering activities.[40]

Angola's involvement in the Congo war is but the most recent episode in its own history of interminable conflict. Since its independence struggle against Portugal from 1961 to 1975, Angola has been at war except for short interludes. At first, it was superpower support (and Cuban and South African intervention) that sustained fighting between the MPLA government and UNITA rebels. But when the outside powers phased out their assistance in the late 1980s, both sides turned to the country's ample natural resources. Three cease-fires and peacemaking efforts failed, primarily because UNITA reneged on its commitments and returned to war.[41]

Angola's oil and diamond wealth (and to a lesser extent its gold, coffee, timber, and wildlife) has fueled arms purchases, but also served to enrich a small elite on both sides. Angola is the world's fifth-largest producer of nonindustrial diamonds, and the second-largest oil producer in sub-Saharan Africa. While the offshore oil wells have remained in government hands, control over the diamond mines has shifted back and forth. (Despite the fierce war, there have also been allegations that senior members on the government side have sold military supplies to UNITA and sold diamonds on behalf of the rebels.) Both sides have succeeded in mortgaging the country's natural bounty in pursuit of a crippling conflict, severely clouding prospects of future generations.[42]

UNITA derived an estimated $3.7 billion from diamond sales in 1992–98. In the early 1990s, UNITA controlled about 90 percent of Angola's diamond exports, but after some defeats its share declined to about two thirds in 1996–97. Its revenues have now further declined due to additional territorial losses, depletion of some deposits, and the (limited) impact of U.N. sanctions. It is believed that UNITA currently sells about $80–150 million worth of diamonds a year, down from as much as $600 million a decade ago. Diamond dollars purchase weapons, fuel, and food for troops, but they have also been used to curry favor with the leaders of Burkina Faso, Togo, and the former Zaire. And a considerable portion of the income has apparently been siphoned off by corruption.[43]

UNITA has some of its own people involved in diamond digging, but much of the mining is carried out by what in effect are bonded laborers deprived of even basic rights and working under dangerous conditions. The rebel group is also imposing a "tax" payable either in diamonds or in cash from diggers working in territory under its control, and receives "commissions" from diamond buyers operating in its realm.[44]

Until 1999, when De Beers decided to stop buying Angolan diamonds, UNITA had little difficulty selling its gemstones. For several years, De Beers pursued a no-questions-asked diamond-purchasing policy, being more interested in maintaining its market control than in the suffering that "blood diamonds" perpetuate. In 1996 and

1997, Angolan diamonds are thought to have accounted for about one fifth of De Beers' business.[45]

Its diverse smuggling routes have apparently enabled UNITA to circumvent a 1998 U.N. embargo on its diamonds. Burkina Faso, Zaire (until the fall of the Mobutu dictatorship), and Rwanda (since 1998) have served as safe havens for illicit transactions. UNITA has been able to smuggle diamonds through Côte d'Ivoire, Morocco, the Central African Republic, Namibia, South Africa, and Zambia, with or without the knowledge of the governments of these countries. The Zambian Ministry of Mines, for instance, provided false Certificates of Origin. The origin of UNITA gemstones was further disguised by having them polished, most likely in Israel and Ukraine.[46]

UNITA was similarly able to evade a U.N. arms embargo by relying on a variety of arms brokers and delivery routes and securing the complicity of several governments that provided false end-user certificates for weapons. Mobutu's Zaire, Burkina Faso, and Togo (since 1996) were major conduits for arms from Eastern Europe; Zaire and the Republic of Congo were also used to store UNITA weapons; after 1998, Rwanda allowed UNITA to hold meetings with arms brokers in its capital, Kigali. Weapons—mostly small arms, but also including major items such as tanks and artillery—came primarily from Bulgaria and other East European countries.[47]

What diamonds are to UNITA, oil is to the Angolan government. At $2–3 billion per year, oil revenue accounts for about 90 percent of Angolan exports and a similar share of the government's budget. Oil money buys arms and keeps the war going: almost three times as much is allocated to the war than to social programs. Meanwhile, a small elite surrounding President Eduardo dos Santos and his top generals rakes in considerable profits through corrupt oil and weapons contracts, control over the allocation of scarce foreign-exchange and import licenses, and other opaque financial deals. For these individuals, the war is lucrative.[48]

Angola's oil and diamond wealth has fueled arms purchases, but also enriched a small elite.

Many of the world's largest oil firms, including Chevron, Elf Aquitaine, BP, and ExxonMobil, operate in Angola and are planning major investments to expand their presence. Global Witness, a British nongovernmental organization (NGO), charges that the oil companies are complicit in perpetuating the war because they provide the necessary revenues. Much of the close to $900 million in "signature bonuses" that these companies were required to pay in order to secure exploration and production rights in ultra-deep offshore blocks in the late 1990s was apparently used to buy arms. The consortia of companies that were awarded two of these blocks, led by Elf and Exxon, include firms that have been involved in arms dealing.[49]

Since the mid-1980s, the Angolan government has resorted to securing loans from international banks by mortgaging future oil production. Much of the money from these high-interest loans has financed military spending. A substantial portion of oil revenues flows directly into a foreign bank account for debt servicing instead of being available for badly needed social expenditures. With a significant portion of revenues being outside the formal state budget—more than two thirds in 1997—financial accountability is nearly absent, and opportunities for corruption are rife.[50]

How Resource Extraction Triggers Conflict

In many instances, resource extraction is itself the source of conflict. Around the world, the operations of oil, mining, and logging companies are causing severe tensions with local populations, often indigenous peoples. From Colombia, Ecuador, and Peru to Nigeria, Cameroon, Indonesia, and Papua New Guinea, the same scenario is unfolding.

Typically, these operations confiscate land from local people without proper compensation. They cause an array of environmental problems by poisoning drinking water, destroying arable land, and disrupting hunting and fishing grounds. And they introduce social disruptions because they bring a heavy influx of construction workers, miners, and loggers. Buildings and roads that are etched into previously inaccessible areas may bring boomtown conditions and attract additional outsiders. While the burdens and disruptions are all too real, the economic benefits from resource extraction mostly accrue to outsiders: the central government, multinational corporations, and assorted foreign investors. But when the affected communities resist, they are often met with severe government repression.

Several places in Indonesia are the site of some of the most intense resource-triggered struggles. At the northern tip of Sumatra, Aceh has seen increasing violence. Aceh is home to Arun, Indonesia's largest gas field and the site of a huge liquefied natural gas (LNG) plant. Operated by ExxonMobil and owned by the state company Pertamina, Arun generates 30 percent of the country's oil and gas export income, or about $1.2 billion a year. The facility gave rise to local resentments in a number of ways. Construction in the late 1970s displaced several villages and hundreds of families. Gas leaks and chemical spills caused health and environmental problems, devastating local communities depending on agriculture and fish farming. Revenues from the LNG facility fed rampant corruption but proved of little benefit to the local population, one third of whom live below the official poverty line.[51]

Aceh is also rich in timber, minerals, and fertile land. But these resources, too, were exploited by cronies of the Suharto dictatorship. Land traditionally owned by indigenous people was expropriated; deforestation resulting from excessive logging has caused landslides and flooding and has destroyed homes and rice paddies. "Transmigrants" from Java that came to Aceh under Suharto to set up timber, pulp, and wood-processing industries have also been a source of intense resentment for the Acehnese.[52]

The Aceh Freedom Movement known as GAM (Gerakan Aceh Merdeka) began in 1976, but its first uprising was easily crushed by the military. A second rebellion in the late 1980s met with arrests, torture, and rape; it is estimated that more than a thousand civilians were killed by the military. Aceh was put under martial law from 1989 to 1998. But the fall of the Suharto regime allowed exiled GAM guerrillas to return and popular support for independence to rise. Today, GAM is well equipped and financed. Renewed violence has killed more than 5,000 people, mostly civilians.[53]

GAM guerrillas have long targeted military installations and Javanese migrants, but ExxonMobil has now become a prime target. Intensifying attacks forced the company to suspend operations from March to July 2001, costing the government an estimated $100 million in lost revenue per month. Military commanders responded with a counterinsurgency operation that resulted in numerous executions and disappearances and that led thousands of Acehnese to flee their homes.[54]

ExxonMobil has sought to portray itself as an innocent bystander of the violence, but nongovernmental groups have charged the company with a "complicity of silence" in the face of severe military abuses. Several mass graves with more than 5,000 bodies have been discovered. ExxonMobil paid the military to provide security for its operations, and reports allege that the company provided equipment to dig the mass graves and allowed its facilities to be used by the military for torture and other activities. The International Labor Rights Fund filed a lawsuit against the company on behalf of 11 Acehnese villagers, suing for complicity in murder, torture, kidnapping, and sexual abuse by Indonesian soldiers.[55]

Some 5,000 kilometers to the east, in Indonesia's Irian Jaya province (also known as West Papua), resource wealth has contributed to a conflict that began even earlier. After the area was forcibly incorporated into Indonesia in 1961, a rebel movement known as OPM (Organisasi Papua Merdeka, the Papuan Freedom Organization) arose in the mid-1960s and advocated the establishment of a separate state. But OPM did not gain wider support until the 1970s, when it harnessed grievances against a large-scale mining operation.[56]

U.S.-based Freeport-McMoRan Copper & Gold Inc. is operating the Grasberg mine—the world's largest open-pit gold mine, which is roughly as large as the state of Vermont. Profits from the operation have been the single biggest source of tax revenue for Indonesia. Land owned by the indigenous peoples, the Amungme and Kamoro, including a mountain sacred to them, was taken over without their consent by a 1967 agreement between Freeport and the Suharto regime. Not only have many villages been displaced, but mine wastes have been dumped on downstream tribal lands. In 1998, for example, some 200,000 tons of ore were dumped into the Ajkwa river system. These mine "tailings" have turned 230 square kilometers of the river delta into a lifeless wasteland.[57]

Land owned by the Amungme and Kamoro, including a mountain sacred to them, was taken over by a 1967 agreement between Freeport and the Suharto regime.

From the beginning, the local tribes opposed Freeport's presence, but this opposition was not linked to OPM's armed separatism until 1977. Indonesian security forces retaliated by bombing and burning villages. But land rights conflicts, compensation demands, human rights violations, and environmental damage kept triggering violent and nonviolent protests. As in Aceh, the migration of Javanese migrants into West Papua added fuel to the conflict. Freeport has maintained close ties with the armed forces. The company relies heavily on military protection in return for providing transportation, accommodation, and funding to the troops. Financial reports for the company show that it has made more than $9 million available to the military since the mid-1990s.[58]

Since 1998, pro-independence sentiments have heightened due to two opposite factors: the greater political freedom of the post-Suharto era and the increasing military repression of separatist movements. The movement has grown to become a broad, civilian-based Papuan independence movement. But Jakarta dispatched thousands of additional troops after the Papuan Congress declared independence in June 2000. Civilians were attacked, peaceful protests banned, key Papuan leaders arrested, and access by journalists and human rights observers severely restricted. Papuan militants in turn have attacked military forces and non-Papuan migrants. Although the violence is currently less intense than in Aceh, the death toll since 1961 may be as high as 100,000.[59]

Under special autonomy packages being discussed, both West Papua and Aceh are supposed to receive a larger share of the revenues derived from resources—80 percent of the income from mining and forestry industries, 30 percent from natural gas, and 15 percent from oil. But this may not satisfy the rebels, and these provinces are too valuable for Jakarta to grant full independence.[60]

Aceh and West Papua are currently the most visible and thorniest conflict spots. But conflicts between resource extractors and local populations are on the rise across much of Indonesia. The impacts on indigenous populations are as severe as those on the natural environment. (See Box 7–3.)[61]

In Bougainville (an island that is part of Papua New Guinea), similar issues led to a 12-year conflict. The world's largest open-pit copper mine, owned by London-based mining giant Rio Tinto, started operating at Panguna in 1972. But the loss of land and other impacts severely affected the subsistence agriculture and the hunting and gathering activities of the area's inhabitants. Mine tailings and other pollutants damaged about one fifth the total land area, decimated harvests of food crops and cash crops like cocoa and bananas, contaminated rivers, and depleted fish stocks. The mine also led to major social disruptions, including rising crime.[62]

The Panguna mine produced up to $500

BOX 7–3. DEFORESTATION AND CONFLICT IN BORNEO

The island of Borneo is the scene of three decades of conflict between indigenous peoples (the Dayak) and loggers and rubber and palm oil plantation businesses. The forests of Borneo are among the largest remaining tropical forests, but commercial logging has been eating into these areas at a mighty pace since the 1960s, cutting down 12 percent of Kalimantan's forest cover in the 1980s alone (Kalimantan is the portion of Borneo that belongs to Indonesia). The enormous wealth that a small but politically well connected elite (military officers and businessmen close to the Suharto dictatorship) has derived stands in stark contrast to the mortal threat that logging presents to the Dayak, whose livelihood—food, shelter, clothing, and medicine—is intimately connected to healthy forests. Unsustainable logging has resulted in soil degradation, silted streams, diminished wildlife and biodiversity, and unprecedented floods and droughts.

The government-subsidized "transmigration" of unemployed people from Java and Madura islands to Kalimantan provided cheap labor for clearing forests and converting the land to commercial rubber and palm oil plantations. But the Dayak saw the migrants as the agent of their growing marginalization; clashes between the two groups grew in frequency and violence. By early 1997, a low-level insurgency drew Indonesian troops. Military repression succeeded in imposing only a temporary calm, with fighting erupting again and again. Thousands have been killed, and tens of thousands displaced.

SOURCE: See endnote 61.

million worth of copper, gold, and silver per year. But Bougainvilleans received next to nothing, and their concerns were ignored. In 1988, they began a sabotage campaign that quickly developed into guerrilla war. The mine fell to the rebels and was closed in May 1989. Because income from Panguna was critical—Papua New Guinea lost 40 percent of its foreign-exchange earnings and a large portion of government income—widespread social and political unrest followed.[63]

Bougainville declared independence in May 1990, but was not recognized internationally. Unable to recapture the mine and defeat the rebels, the government saw itself compelled to agree to peace negotiations. Still, a new attempt to invade the island was organized in 1996. Prime Minister Julius Chan offered $36 million in World Bank funds to the private military firm Sandline International, but senior army officers—incensed that their own budget was cut—forced Chan to resign and cancel the Sandline contract. A cease-fire was signed and a small peacekeeping force was deployed in 1998. In June 2001, the government and the rebels agreed in principle that Bougainville would gain some autonomy and an eventual referendum on independence. A resolution to the conflict is now in sight, after hundreds, and possibly thousands, of deaths.[64]

In Nigeria, one of the world's leading petroleum producers, oil development has enriched a tiny minority and foreign oil companies, but it has translated into environmental devastation, health problems, and impoverishment for the inhabitants of the oil-producing areas that have traditionally lived from fishing, agriculture, and palm oil production. The Niger Delta, where oil production is taking place, forms Africa's largest wetlands area, harboring extensive mangrove forests and providing habitat for a number of unique plant and animal species. Poor industry practices such as constant flaring of natural gas, along with frequent oil spills from antiquated pipelines and leaks from toxic waste pits, have exacted a heavy toll on soil, vegetation, water, air, and human health. Local communities complain of respiratory problems, skin rashes, tumors, gastrointestinal problems, and cancers. They have seen a drastic decline in the fish catch and agricultural yields.[65]

Throughout the 1990s, local communities staged protests, often directed against multinational oil companies in Nigeria—primarily Royal Dutch/Shell as the largest producer, but also Chevron, Mobil, France's Elf, and Italy's Agip. The Ogoni are one of the Niger Delta communities that gained world attention for their cause. The Movement for the Survival of the Ogoni People (MOSOP) organized mass protests that succeeded in shutting down Shell operations in Ogoni territory in 1993. The military dictatorship—which got 80 percent of its revenues from oil—responded with a campaign of violence and intimidation, and instigated various ethnic groups in the delta to attack each other. Some 2,000 Ogoni were killed and 80,000 uprooted; MOSOP leaders were detained or forced to flee, and in October 1995, the regime executed Ken Saro-Wiwa, MOSOP's well-known spokesman, and eight other leaders.[66]

Aided by weak enforcement policies and oppressive government, the oil companies have failed to abide by Nigeria's environmental laws and have largely evaded paying compensation to delta communities for any damages. And although corporate representatives deny knowledge of the government's repressive tactics, they frequently summon the notoriously abusive security forces to intervene against unarmed protesters. Chevron helicopters were reportedly used in a 1998 assault against protesters. Elf and Agip are alleged to have instigated deadly attacks against, respectively, female protesters and a village that refused to let oil drilling go forward. Shell helped finance and arm a local paramilitary force in Ogoniland. Exposed to increasingly unfavorable world opinion, Shell undertook a major review of its activities and attitudes toward Niger Delta communities, but as a 1999 Human Rights Watch report comments, the company's actual performance on the ground will be the judge of whether this amounts to more than changed rhetoric.[67]

The death of military dictator Sani Abacha in June 1998 allowed a transition to an elected government in 1999. According to Human Rights Watch, this brought a "significant relaxation in the unprecedented repression...inflicted on the Nigerian people." A Human Rights Commission is investigating cases going as far back as 1965, and more than 10,000 petitions have been brought before it. Although western media attention has faded, protests and occupations of oil facilities surged after Abacha's death. The government withdrew the feared Internal Security Task Force from Ogoniland, but human rights abuses against those attempting to raise grievances in the oil-producing areas continue nevertheless. In this sense, at least, conditions in the delta have changed little.[68]

While democratization efforts in Nigeria, Indonesia, and elsewhere give greater hope that these conflicts can be resolved, far more needs to happen to bolster the human and development rights of affected communities. Greater awareness and scrutiny are also needed in major consuming countries if the link between resources and repression is to be broken.

Sanctions, Certification Systems, and Economic Diversification

Resource-related conflicts have been raging in large part because of a business-as-usual approach by governments and corporations. But prodded by NGOs, the situation is beginning to change.

Confronted with unending conflicts in Sierra Leone, Angola, and Congo that threaten to spiral out of control, the U.N. Security Council has increasingly examined the role of resources in perpetuating these wars. It imposed a number of embargoes on the illicit diamond trade and on the purchases of arms, equipment, and fuel paid for with diamond money. (See Table 7–3.) These efforts are only a beginning. Observers from NGOs and expert U.N. panels have called for similar measures that would cover additional types of resources. But governments have blocked action in some cases; for instance, France and China, the two leading importers of timber from Liberia, have opposed U.N. sanctions against Liberian timber exports.[69]

It has also become painfully obvious that existing sanctions are being violated by unscrupulous commodities producers, traders, bankers, and governments. There is an urgent need to step up international efforts to monitor compliance with sanctions and to improve the capacity to enforce embargoes and investigate violations so that traffickers can no longer operate with impunity.[70]

Growing energy is being directed at efforts to make it more difficult for resources gained through conflict to be sold on world markets. By far the most attention has gone to the diamond industry. The governments of Sierra Leone, Angola, and the

Table 7–3. Resource Conflicts and United Nations Sanctions

Date	U.N. Security Council Action
November 1992	Arms embargo against Liberia.
September 1993	Embargo on deliveries of arms, military equipment, and fuel to Angola's UNITA rebels after their rejection of the 1992 election results.
August 1997	Additional sanctions against UNITA: freezing of bank accounts; prohibiting foreign travel by senior UNITA personnel; closing of UNITA offices abroad.
October 1997	Embargo on arms and oil supplies to Sierra Leone; travel ban on members of military junta (oil embargo terminated in March 1998).
June 1998	Arms embargo and travel ban on anti-government forces in Sierra Leone.
June 1998	Embargo on direct and indirect import of Angolan diamonds that have not been approved under an Angolan government certificate-of-origin regime.
May 1999	Panel established to investigate violations of sanctions against UNITA.
July 2000	Embargo on direct and indirect import of rough diamonds from Sierra Leone; following establishment of a new monitoring regime, embargo was narrowed to nonofficial exports in October 2000.
March 2001	Demand that Liberia cease financial and military support for RUF, and cease imports of Sierra Leonean rough diamonds without an official certificate of origin; embargo on arms deliveries to Liberia and travel ban against its political and military leaders; threat of embargo against Liberian diamond exports unless Liberia can show that it is not supporting RUF.

SOURCE: See endnote 69.

Democratic Republic of Congo are backing schemes under which only diamonds with proper documentation are considered legal. All gems are to be accompanied by certificates of origin, whose digital "fingerprint" is shared with authorities in importing countries. While polished diamonds cannot be traced to their origin, a recent technological breakthrough allows some high-tech sleuthing to pinpoint the source of rough stones by comparing trace amounts of impurities in the diamonds.[71]

But a certificate-of-origin system can be undermined by poor enforcement and circumvented by intricate international smuggling networks. A U.N. report in October 2001 found that $1 million worth of diamonds are still smuggled out of Angola every day. Lax government controls in the major diamond trading and cutting centers (Belgium, Switzerland, the United Kingdom, Israel, and others) and the opaque, unaccountable nature of the diamond industry are also major obstacles in the struggle to root out conflict diamonds. A March 2000 U.N. investigative report on how sanctions against UNITA were circumvented concluded that Belgian authorities "failed to establish an effective import identification regime" or to effectively "monitor the activities of suspect brokers, dealers and traders." The Belgian and British governments have now expressed determination to crack down on conflict diamonds. Efforts are also continuing in the United States, the largest importer of diamonds, to ban imports of illegally mined diamonds. Although the industry initially threw its support behind a bill with weak standards and loopholes, it now supports more stringent legislation introduced in both chambers of the U.S. Congress.[72]

So far, western governments have been all too ready to turn a blind eye in order to protect the interests of their own corporations.

In recognition of the ease with which country-by-country diamond certification schemes can be evaded, support has been growing for establishing a standardized global certification scheme. Since May 2000, representatives from some 38 nations, the diamond industry, and a number of NGOs have conducted negotiations (referred to as the Kimberley Process) to develop an international system. The Kimberley controls were expected to be finalized and presented to the U.N. General Assembly by December 2001. But NGOs have complained of backtracking and stalling maneuvers by some governments, and they were worried that instead of a binding and credible system, a voluntary one might emerge.[73]

Similar measures may be needed for other conflict resources. A certification system for timber, for example, could build on existing efforts by the Forest Stewardship Council (FSC) to ascertain whether lumber is being produced in a sustainable manner. The FSC effort, initiated in 1993, entails independent audits to verify compliance with a series of requirements. Of particular interest is its "chain-of-custody" certification, which seeks to trace the lumber or furniture on consumer store shelves all the way back to the forest where the trees were felled. A comparable system of accounting could determine whether timber had been produced in conflict situations.[74]

It is clear that a number of businesses—oil and mining companies, trading firms, airlines and shipping companies, manufacturers, and banks—carry some responsibility for the events that have triggered campaigns against blood diamonds and other conflict resources. This responsibility ranges from an active role (in which companies are directly and knowingly involved in illicit resource exploitation), to a silent complicity (in which firms do business with repressive regimes or rebel groups because of lucrative contracts), to a passive "enabling" role (in which few questions are asked about the origin of raw materials or about money being laundered).

International embargoes and U.N. reports have begun to create greater transparency. NGO campaigns have tugged at the cloak of complicity through investigative reports and by "naming and shaming" specific corporations, in an effort to compel them to do business more ethically or to terminate their operations in certain locations. Such campaigns have been most potent in the case of companies that sell highly visible consumer products or whose corporate logos and slogans are familiar to millions.[75]

At the end of the 1990s, the diamond industry was hit by a wave of bad publicity and faced the threat of consumer boycotts. De Beers, the industry's monopolist, was sufficiently embarrassed by London-based NGO Global Witness, which revealed that the company had knowingly purchased diamonds from Angola's UNITA rebels, that it adopted a more responsible policy and urged the rest of the industry to follow suit. Similarly, when the role of coltan in the Congo war become more widely known, consumer electronics companies scrambled to avoid the kind of negative publicity that the diamond industry had endured. Companies like Ericsson, Nokia, Motorola, Compaq, and Intel suddenly scrutinized their supply chains and put pressure on mineral processing firms to stop purchasing illegally mined coltan from Congo. The Belgian airline Sabena stopped its coltan shipments to Europe.[76]

There is growing awareness that natural resources will continue to fuel deadly conflicts as long as consumer societies import and use materials irrespective of where they originate and under what conditions they were produced. Support is growing for the idea that companies need to adopt more ethical ways of doing business. Shareholder activism and campaigns for ethical investing can help achieve these goals. But it is clear that activities to date are only a beginning. Governments and international organizations will need to work hard to create greater transparency in the dealings of financial and other companies. So far, western governments have been all too ready to turn a blind eye in order to protect the interests of their own corporations.[77]

Another priority area for action concerns the massive proliferation of small arms. As awareness of the impact of small arms in resource-related conflicts and other settings has grown, national governments, regional organizations, and the United Nations have become more active in seeking ways to check the spread of these weapons, particularly illegal transfers. Especially noteworthy is a moratorium on the trade and manufacture of such weapons in West Africa, which was signed in October 1998 and renewed for another three years in 2001. Since West Africa is awash in small arms, a U.N.-assisted effort is also being made to collect weapons already in circulation.[78]

It has become clear that the small arms plague can be tackled successfully only with broad international cooperation. A U.N. conference on small arms was held in July 2001, with the expectation of launching efforts to conclude international agreements on marking and tracing weapons, regulating arms brokers, and establishing stricter export criteria. The opposition of a few governments, most notably the United States, nearly derailed the conference, however. The Bush administration opposed a number of measures, including restrictions on civilian ownership of such weapons, prohibitions against sales to nongovernmental entities such as rebel forces, and any limitations on the legal trade. Although the outcome was a low-common-denominator action program, it nevertheless provides a basis for stepped-up efforts to pursue post-conflict small arms disarmament, to destroy surplus and illegal arms, to demobilize soldiers and reintegrate them into civil society, and, most important, to improve transparency and greater knowledge about transfers.[79]

Experience to date also provides a strong case for improving peacekeeping capabilities. The conflicts in Angola and Sierra Leone have attracted two of the largest U.N. peacekeeping efforts, and the Security Council has considered the feasibility of a large presence in Congo. But U.N. efforts confront a number of severe handicaps. The first concerns the warring parties. They may agree to cease-fires or even peace agreements as an expedient move that allows them to maneuver for advantage, only to return to violence at an opportune moment.

There are also systemic weaknesses in U.N. peacekeeping. Since there is no standing peacekeeping force, the United Nations relies on national governments to make personnel and equipment available. Typically, it takes several months for a mission to reach its authorized deployment strength. The numbers of peacekeepers are often inadequate to the task, and many of them are ill equipped and poorly trained. National contingents frequently do not work together well and sometimes fail to adhere to the mission's mandate.[80]

Fixing the deficiencies inherent in the current approach to peacekeeping would not only help brighten the chances of success in ending ongoing resource-based conflicts, it would probably also constitute something of a deterrent to future resource looters. An effective peacekeeping system that deploys well-trained and well-equipped troops in a timely fashion and that is able to protect victims instead of adopting a false neutrality would make a significant difference. An effective system would provide capacities to wrest control of resource-rich areas that are being illegally exploited, intercept smuggling routes, enforce peace agreements, and facilitate disarmament and demobilization of combatants. To establish such a system, governments must be prepared to invest adequate money, effort, and political support.

The policies discussed here are largely concerned with reacting to resource-based conflict rather than preventing it. Prevention is not an easy task, and there is no silver bullet. Promoting democratization, justice, and greater respect for human rights are key tasks, along with efforts to reduce the impunity with which some governments and rebel groups engage in extreme violence. Another challenge is to facilitate the diversification of the economy away from a strong dependence on primary commodities to a broader mix of activities.

The quest for sustainable development that is the focus of the Johannesburg conference is of crucial importance in this context. Investing in human development, improving health and education services, and providing adequate jobs and opportunities for social and economic advancement will go a long way toward reducing the risk that a country's natural resource endowment will become its undoing. This is an investment that needs to be made not only by every government but also by the World Bank and other multilateral development agencies that have generously funded oil, mining, and logging projects. It must also be a priority for the rich nations that have for so long benefited from cheap raw material supplies while turning a blind eye to the destruction at their source.

WORLD SUMMIT PRIORITIES ON CONFLICT

- Develop strong global certification systems for diamonds, timber, and other resources to improve ability to ascertain origins of commodities and to screen out those produced and traded illicitly in conflict areas.
- Secure better compliance with U.N. sanctions against illicit resource trafficking by improving the capacity of the United Nations, regional and international organizations, and governments to monitor and enforce embargoes.
- Increase the transparency and accountability of oil, mining, and logging corporations in areas of conflict, of trading and shipping companies, and of banks and other financial institutions. Develop strong codes of conduct for corporations and brokers.
- Reduce the availability of small arms by establishing stricter national export criteria, regulating arms brokers, marking and tracing weapons, and improving collection of surplus arms.
- Promote democratization and greater respect for human rights, particularly the rights of indigenous and minority groups.
- Support diversification of economies away from a heavy dependence on a handful of primary commodities.
- Increase consumer awareness of the connections between resource exploitation and conflict.

Chapter 8

Reshaping Global Governance

Hilary French

In late July 2001, tens of thousands of protesters gathered in the streets of the ancient port city of Genoa, Italy, while the Group of Eight major economic powers held its annual summit meeting. These demonstrations were the latest in what has become a steady stream of massive public protests related to globalization—a much-used though ill-defined term that covers the broad range of societal transformations that have accompanied the rapid growth in international trade and investment in recent years, as well as the virtual shrinking of the planet due to computers, cell phones, and other accoutrements of the information age. Less than two months later, the world watched in horror as hijacked airplanes crashed into the World Trade Center and the Pentagon, causing some 5,000 deaths and seemingly reordering the world's preoccupations and priorities in the course of a few hours. Suddenly, globalization protests were off the front pages, and the world's war on terrorism dominated headlines.[1]

In the days following the terrorist attacks, scores of meetings and events were cancelled, including the annual meetings of the World Bank and the International Monetary Fund (IMF) in Washington in late September, as well as the public demonstrations that had been expected to accompany them. Although the terrorists had struck at the heart of the global economy by targeting the World Trade Center, the leaders of anti-globalization protests moved quickly to distance themselves from the terrorist attacks and to express their sympathy for the victims.[2]

At the same time, the horror of the events of September 11th has caused people everywhere to contemplate the root causes of the disaster. Not all of the terrorists who hijacked the airplanes were themselves impoverished. But the growing gap between the rich and the poor in many regions and worldwide and the persistence of extreme poverty among more than a billion people have undoubtedly helped to create a climate that is ripe for fundamentalism and extremism. As Klaus Töpfer,

Executive Director of the U.N. Environment Programme (UNEP), recently put it: "When people are denied access to clean water, soil, and air to meet their basic human needs, we see the rise of poverty, ill-health and a sense of hopelessness. Desperate people can resort to desperate solutions."[3]

Although the globalization trends of recent years enriched economic elites and added to the ranks of the global middle class in some countries, they also bypassed billions of destitute people and in some cases directly undermined the welfare of marginalized people by destroying the ecological and social fabric that has formed the backbone of traditional, subsistence-based societies. Reorienting our current globalization path—one of the primary goals of massive public protests of recent years—may thus turn out to be a key pillar in any successful long-term strategy against terrorism.[4]

The term globalization was not in widespread use in June 1992 at the U.N. Conference on Environment and Development, better known as the Rio Earth Summit. But in retrospect, that meeting can be seen as a process that was at least partially aimed at reshaping the global economy to make it less environmentally harmful and more socially equitable—the essence of the concept of sustainable development and something that has more recently been among the demands of demonstrators in the streets.

The list of formal results from the Rio conference was substantial, including major new international treaties on climate change and the loss of biological diversity as well as *Agenda 21*, a lengthy action plan for achieving sustainable development that covers an exhaustive set of issues—from agriculture and chemicals to poverty and institutional reform. But formal agreements were only one part of the Rio story. Perhaps even more significant was the international mobilization it brought about, as tens of thousands of people from around the world convened to express their concern for the fate of the planet, including heads of state, indigenous peoples, local officials, business representatives, environmental activists, and journalists.[5]

With the world now preparing for the World Summit on Sustainable Development in Johannesburg in September 2002, this is an appropriate time to assess the Earth Summit's legacy. (See also Chapter 1.) Although international environmental negotiations have mostly plodded along at a snail-like pace in the decade since Rio, the world at large has been changing rapidly. Within a few years of the Earth Summit, the underlying forces of globalization were sweeping the world at breakneck speed.

As the Rio conference wound down, the Uruguay Round of world trade negotiations was gathering force, paving the way for the creation of the World Trade Organization (WTO) at the beginning of 1995. The final text of the Uruguay Round agreement was over 26,000 pages long (mainly detailed tariff and services schedules) and covered an enormous array of issues, including agriculture, intellectual property rights, investment, and services. In comparison, the 273-page *Agenda 21* reads like a brief call to action. The Uruguay Round negotiators made little effort to incorporate the Rio commitments into their deliberations. Indeed, many WTO provisions contradict the spirit and in some cases arguably even the letter of the Rio accords.[6]

But the events of September 11, 2001, put a monkey wrench into what had seemed to be an almost inexorable march toward a globalized world. International travel and tourism plummeted in the wake of the terrorist attacks, and the global economy was in a dangerously precarious state. Even before September 11th, public unease about globalization had been growing fast, as evidenced by the strength of the anti-globalization protest movement. At the root of this rising public concern are some basic questions: What rules govern today's increasingly global economy? Who sets them? And whose interests do they serve?[7]

The growing power of global economic institutions such as the WTO juxtaposed against the relative weakness of international institutions charged with environmental protection and social welfare is leading to a persistent imbalance in today's emerging structures of global governance. And as globalization is pushing decisionmaking up to the international level on more and more issues, many people around the world worry that democracy and accountability are being lost in the process.

Despite these dilemmas, collaborative action at the international level is essential if we are to successfully address the debilitating environmental and social trends that are undermining prospects for a livable and secure world. The Johannesburg Summit offers us an opportunity to create new, more transparent global governance structures that can protect the ecological integrity of the planet while improving the quality of life of the more than 6 billion people who currently inhabit it.[8]

Reinvigorating International Environmental Governance

In March 1999, the World Trade Organization convened a high-level symposium to examine the connections between trade and environmental policymaking. When Director-General Renato Ruggiero spoke, the most notable remarks he made focused not on international trade rules but on the need to strengthen international environmental governance by creating a World Environment Organization to be the "institutional and legal counterpart" of the WTO. Some nine months later, as anti-globalization protestors dramatically took to the streets of Seattle, Washington, during a WTO ministerial meeting, *Washington Post* editorial writers came to a similar conclusion, arguing that "Trade these days is so entwined with social issues that selective internationalism is decreasingly possible. The health of the WTO may turn out to require something like a world environment organization."[9]

It is ironic that some of the staunchest advocates of building stronger international environmental governance structures have emerged from the community of trade policy experts rather than environmental ones. And some skepticism is warranted, as arguments in favor of creating a World Environment Organization are often used to deflect attention from the need to overhaul WTO rules. But though the message may have come from unusual quarters, it is nonetheless fundamentally on target: in this age of globalization, there is a crying need for some environmental rules of the road for the global economy, and it is environmental institutions rather than economic ones that are best equipped to provide them.

Determining how to make international environmental governance work better requires understanding the nature of the current system. The number of environmental treaties has soared over the last few decades. UNEP estimates that there are now over 500 international treaties and other agreements related to the environment, more than 300 of which have been agreed to since the first U.N. conference on the environment was held in Stockholm in 1972, and 41 of which UNEP considers "core environmental conventions." But international environmental governance has to some degree become a victim of its own success. As the number of treaties has climbed, problems of duplication, fragmentation, and lack of coordination have arisen that are undermining the efficacy of the system as a whole.[10]

Each environmental treaty creates its own mini-institutional machinery, including annual meetings of the treaty members (called Conferences of the Parties) as well as small offices called secretariats that are charged with overseeing treaty implementation. The secretariats and the various meetings of treaty members are scattered around the world, causing international environmental diplomacy to at times resemble a moving circus. The growing number of environmental treaties has caused the sheer number of international meetings and negotiating sessions to climb, straining the ability of environmental diplomats, nongovernmental organizations (NGOs), and other interested parties to keep pace. This proliferation of international meetings poses a particular challenge for developing countries, who generally have only a few diplomats available to cover the sprawling international environmental agenda.[11]

One result of fragmentation in the current system of international environmental governance is that the provisions of different environmental conventions sometimes act at cross-purposes. The negotiations that led to the Montreal Protocol on ozone depletion, for instance, paid little heed to the complex interconnections between ozone depletion and climate change. One of the perverse results of this was the development of hydrofluorocarbons (HFCs) as common substitutes for ozone-depleting chlorofluorocarbons, despite the fact that HFCs are potent greenhouse gases. More recently, provisions aimed at encouraging the establishment of tree plantations to absorb carbon dioxide were included in the Kyoto Protocol on climate change with little regard for the impact of uniform stands on biological diversity, despite the fact that most of the countries that are members of the Kyoto Protocol are also parties to the Convention on Biological Diversity (CBD.)[12]

In this age of globalization, there is a crying need for some environmental rules of the road for the global economy.

More fundamental than problems of overlap and coordination are weaknesses in the individual agreements themselves. Most environmental treaties contain few specific targets and timetables, and provisions for monitoring and enforcement are generally weak to nonexistent. Nonetheless, negotiators have made substantial headway since the Earth Summit in fleshing out the two major conventions that were concluded there, those on climate change (see Chapter 2) and biological diversity. They have also reached agreement on four other international treaties that grew out of Rio—on combating desertification, managing migratory fish stocks, controlling trade in hazardous chemicals and pesticides, and phasing out persistent organic pollutants. (See Table 8–1 and Chapter 4.)[13]

Perhaps most notably, in late 1997 nations forged the Kyoto Protocol under the rubric of the climate change convention of 1992, creating binding limits on carbon emissions for the first time. But the post-Kyoto years have been marred by continuing disagreements among the signatories to the protocol, particularly between the United States and the European Union (EU), about how and even whether its provisions should be implemented. Despite the contin-

Table 8–1. The Rio Conventions—A Progress Report

Convention on Biological Diversity, 1992

Status: 168 signatories, 182 parties; in force since 1993

Accomplishments:
- Provides broad guidelines for the conservation of biodiversity at the national level and requires participating countries to formulate national biodiversity strategies
- Recognizes national sovereignty over biological resources and affirms the principle of prior informed consent (PIC) before resources may be transferred out of a country
- Stipulates that biodiversity use must be sustainable and resulting benefits must be equitably shared between source country and receiving country
- Global Environment Facility (GEF) funding has channeled $1.02 billion into biodiversity projects in 120 developing countries

Challenges:
- Biodiversity is difficult to measure and data are hard to collect
- Of 182 parties, only about 70 countries have submitted National Strategies
- Most resources have gone into creating national reports, yet still only 54 countries had met the May 2001 deadline for submitting them
- Biosafety Protocol of 2000 allows governments to choose whether to allow imports of products containing genetically modified organisms, but it has so far only been ratified by 7 of the 50 states required for it to enter into force

UN Framework Convention on Climate Change, 1992

Status: 165 signatories, 186 parties; in force since 1994

Accomplishments:
- Annex I countries (24 industrial nations, the European Union, and 14 countries with economies in transition) agree to adopt policies to stabilize greenhouse gas emissions at 1990 levels by 2000
- Annex II countries (24 industrial countries and the European Union) agree to provide financial resources for technology transfer
- Non-Annex I parties (developing countries) are eligible for GEF funding to meet national reporting requirements
- GEF has funneled $884 million into climate change projects and leveraged an additional $4.9 billion from recipient governments and other organizations
- 1997 Kyoto Protocol requires Annex I countries to reduce overall emissions by 5.2 percent from 1990 levels by 2012

Challenges:
- Protocol remains contentious, with recent negotiations leaving many dissatisfied; 40 states have ratified the protocol, but to come into force, it requires ratification by at least 55 nations, including Annex I parties representing 55 percent of greenhouse gas emissions
- Emissions continue to rise in industrial countries, and the United States, the world's largest emitter of greenhouse gases, has declined to participate in the Kyoto Protocol

Convention to Combat Desertification, 1994

Status: 115 signatories, 176 parties; in force since 1996

Accomplishments:
- Flexible structure creates a network of four regions: Africa, Asia, Latin America and Caribbean, and Northern Mediterranean; each has the power to design and implement a plan tailored to local needs
- Increasing number of national, subregional, and regional action plans have been submitted, and implementation has begun in some areas
- Approximately 175 reports have been filed from donors as well as countries afflicted by desertification

Table 8–1. *(continued)*

Challenges:
- Projects are not eligible for GEF funding and treaty stipulates that most funds are to come from the countries themselves, leaving many projects without stable financing
- Commitments on the part of both developing and industrial countries are vague, leaving ample room for inaction

UN Agreement Relating to the Conservation and Management of Straddling Fish Stocks and Highly Migratory Fish Stocks, 1995

Status: 59 signatories, 29 parties; not in force (requires 1 more ratification)

Accomplishments:
- Advocates a cooperative, precautionary approach to the management and conservation of relevant fish stocks
- Requires coastal states and those fishing in international waters to adopt national measures to restore stocks to levels capable of producing maximum sustainable yields
- Encourages regional planning and information exchange, recognizes the needs of developing states and subsistence fishers, and contains provisions on pollution control, related ecosystems, and domestic monitoring and compliance
- Includes provisions allowing parties to board and inspect vessels of other parties on the high seas in order to verify compliance

Challenges:
- Only 12 of the top 20 fishing nations have signed and just 4 have ratified, weakening it when it does enter into force

Rotterdam Convention on the Prior Informed Consent Procedure for Certain Hazardous Chemicals and Pesticides in International Trade, 1998

Status: 73 signatories, 16 parties; not in force (requires 34 more ratifications)

Accomplishments:
- Exporting states must receive explicit permission from importing state before shipments of 27 types of restricted substances may take place
- Details safety and labeling requirements for the handling of these substances
- States refusing shipments containing a chemical must halt domestic production of the substance, avoiding conflict with trade rules

Challenges:
- Not yet in force, but builds on existing voluntary PIC procedures, which many states continue to honor until the treaty becomes binding
- Developing countries often lack the infrastructure and capacity for implementation
- Excludes many categories of substances, such as pharmaceuticals, narcotics, radioactive materials, and food products

Stockholm Treaty on Persistent Organic Pollutants, 2000

Status: 105 signatories, 2 parties; not in force (requires 48 more ratifications)

Accomplishments:
- Regulates the production and use of 12 persistent, toxic substances; the 9 Annex A chemicals are slated for elimination, while Annex B lists chemicals such as DDT that are subject to restricted use
- Mandates the identification and elimination of stockpiles, products, and wastes containing persistent organic pollutants

Challenges:
- Numerous exemptions exist, including for articles manufactured or in use before the convention enters into force and a conditional renewable 10-year exemption for hexachlorobenzene and DDT
- Funding provisions are vague, delegating authority to GEF while acknowledging that the existing GEF mandate and resources limit its ability to serve this function

SOURCE: See endnote 13.

uing defiant opposition of the U.S. government to the terms of Kyoto Protocol, 178 other nations jointly agreed at a negotiating session in Bonn in July 2001 to a range of key provisions that filled in all-important missing details of the pact. (As this book went to press, governments were meeting once again on climate change in Marrakesh, Morocco, with the outcome not yet clear.)[14]

Most environmental treaties lack clear criteria for monitoring and measuring effectiveness.

The moment the deal was struck in Bonn was a euphoric one for the negotiators and NGO activists who had labored so long and hard to breathe life into the Kyoto Protocol—and for good reason, given the collapse of the negotiations eight months earlier and the U.S. government's intransigence in the intervening period. But it is sobering to note that the commitments contained in the protocol represent just a small first step down what is sure to be a long and challenging road. Under the terms of the Kyoto agreement, 38 industrial countries agreed to collectively reduce their annual greenhouse gas emissions to 6–8 percent below 1990 levels by 2012. Scientists estimate that emissions cuts on the order of 60–80 percent will ultimately be required to achieve the convention's overarching objective of stabilizing greenhouse gas concentrations in the atmosphere at a level that will prevent dangerous interference in the climate system.[15]

Although the Kyoto Protocol is far from perfect, the priority now is for countries to press ahead with ratifying it, with the goal of bringing the agreement into force by the time of the Johannesburg Summit. Experience with other environmental treaties, particularly the Montreal Protocol on ozone depletion, suggests that the agreement will then be strengthened over time as technologies advance, as scientific understanding of the problem deepens, and as public support for action grows. If the rest of the world forges ahead with putting the Kyoto Protocol into practice as the United States looks on from the sidelines, many in the U.S. business community are likely to conclude that they are at a disadvantage because they do not have a seat at the table where key decisions are being made about the future of the world's energy system. An industry about-face combined with growing public pressure might then pave the way for the United States to join in the accord. (See Chapter 2.)[16]

Like the climate change treaty, the Convention on Biological Diversity has also had a somewhat checkered history since Rio. The most tangible outcome of the convention so far has been the Cartagena Protocol on Biosafety. This agreement aims to regulate international trade in genetically modified agricultural commodities by putting in place a system known as prior informed consent, in which importing countries must be informed of and explicitly grant their approval before shipments of genetically modified commodities can proceed. Negotiations on the protocol finished successfully in January 2000, but so far it has been ratified by only 7 of the 50 countries that must do so in order for it to come into force.[17]

In addition to bringing the biosafety protocol into operation, it is also important that countries move ahead with other efforts to implement and strengthen the CBD. Governments have been slow to develop the national-level strategies and action plans for biodiversity preservation that are called for under the convention, and the treaty itself suffers from a lack of clear targets, timetables, and ways to measure progress and trends. Further initiatives are needed to protect endangered ecosystems, in part by developing specific provisions and initiatives related to major causes of ecological disruption, such as the introduction of invasive species.[18]

The CBD's lack of performance indicators highlights a broader set of problems with many environmental treaties: most of them lack clear criteria for monitoring and measuring effectiveness. Environmental treaties are also generally characterized by nonbinding and voluntary dispute resolution procedures for cases where countries are suspected of violating a treaty's rules, in striking contrast to the WTO's system of binding rulings that are ultimately enforceable by trade sanctions.[19]

The divergences in specificity and enforceability between environmental treaties and WTO rules become a particular problem in cases where the two bodies of international law contradict one another. Several environmental treaties, including the Montreal Protocol, the Convention on International Trade in Endangered Species of Wild Fauna and Flora, and the recently agreed biosafety protocol contain provisions that arguably are at odds with WTO rules. These inconsistencies stem from different philosophical underpinnings: environmental treaties often aim to limit certain ecologically harmful forms of commerce, such as trade in endangered species and hazardous wastes, whereas the WTO is in the business of tearing down obstacles to the flow of goods across international borders.[20]

One of the more pronounced contradictions is the divergence between the intellectual property rights stipulations of the WTO and those of the Convention on Biological Diversity. The WTO requires countries to put in place strict systems for recognizing the intellectual property rights of plant breeders and biotechnology companies. The CBD, on the other hand, affirms that any economic benefits of commercializing seeds, pharmaceuticals, and other products based on indigenous knowledge gained over thousands of years should be shared with the farmers and communities that developed them in the first place.[21]

Although no country has thus far lodged a formal WTO challenge against the provisions of a multilateral environmental agreement, arguments about WTO consistency often arise during environmental treaty negotiations. These tensions were much in evidence during the negotiations on the biosafety protocol, which endorses the need to sometimes take precautionary steps to prevent the possibility of irreversible harm, even in the face of scientific uncertainty. The United States has resisted incorporating this "precautionary principle" into the biosafety protocol and other international agreements, preferring the WTO's insistence that food safety policies and a range of other standards related to human, animal, or plant health be based on scientific evidence. Although there is broad agreement that science should inform the regulatory process, conflicts can arise in cases where a clear scientific consensus does not yet exist about the extent of the harm posed by a specific threat or substance, a common predicament in environmental policymaking.[22]

One way to respond to the power imbalance between the more enforceable rules of the WTO and comparatively weak environmental treaties would be to give the latter sanctioning powers similar to WTO's. A few environmentally related treaties are beginning to do just that. The Law of the Sea convention, for example, created an International Tribunal as one of several possible vehicles for resolving disputes about implementation and compliance; it is empowered to impose fines and other penalties in cases where a company or country is found to be violating the terms of the accord.[23]

But in many cases, noncompliance with environmental treaties stems more from a lack of ability owing to shortages of funds and weak administrative capacities than from deliberate ill will. As Calestous Juma, who formerly ran the CBD's secretariat, puts it: "The real task is deciding how to get national governments to comply fully with environmental laws. If governments promote greater compliance with domestic environmental laws, they will find it easier to reflect this in international agreements. What is perceived as deficient global environmental regulations is really an indication of poor domestic housekeeping." In other words, making international environmental governance more effective will require making governance in general work better, as it is primarily at the national and local levels that environmental treaties are translated into on-the-ground results.[24]

Making international environmental governance more effective will require making governance in general work better.

Grants provided by the Global Environment Facility are among the main tools available for promoting national-level implementation of treaties within developing countries. The GEF was created on a pilot basis in 1991, and emerged as a major institutional player at Rio and in subsequent years. As a joint initiative of the World Bank, UNEP, and the U.N. Development Programme, the GEF's mandate is to finance the additional costs that developing countries incur in responding to global environmental problems, particularly those covered by major international treaties. Since it was created, the GEF has focused primarily on four areas—climate change, ozone depletion, the loss of biological diversity, and the degradation of international waterways. In May 2001, GEF members decided to also finance projects that help to implement the Stockholm Convention on Persistent Organic Pollutants. (See Chapter 4.)[25]

Projects financed by the GEF have, among other things, helped Ethiopian farmers preserve genetic diversity in local agriculture, have encouraged a partnership between an NGO, a local government, and a cement plant to preserve the Dana Nature Reserve in Jordan, and have helped thousands of households, health clinics, and schools in some 20 countries to install solar power systems. Over the last decade, the GEF has committed $3.4 billion in grants to over 650 projects in 150 countries, an average of some $300 million per year. But raising even this relatively small sum from donor governments has proved to be a continuing challenge.[26]

Like the GEF, other environmental institutions have also suffered from scarce funding. Budgets of the secretariats charged with administering critical environmental treaties such as the Montreal Protocol and the biological diversity convention are generally in the range of $1–10 million, and UNEP has struggled to maintain its annual budget of roughly $100 million. In comparison, the U.S. Environmental Protection Agency had a budget of $7.8 billion in 2000, while the U.S. military budget was over $300 billion, and global military expenditures added up to more than $750 billion.[27]

As the Johannesburg Summit approaches, many observers are questioning the adequacy of our current structures of global governance for environmental protection and sustainable development. UNEP has been particularly active in raising this issue. In a laudable exercise in introspection, it has launched a broad-ranging review of the current complex system of environmental treaties and institutions and convened a series of high-level meetings of government officials as well as consultations with academic experts, NGOs, and other interested parties aimed at producing recommendations for reform that can be acted on in Johannesburg.[28]

A range of proposals is receiving serious consideration. One idea with wide support is to cluster related environmental conventions together either physically or virtually, in order to facilitate coordinated action. Among the possible groupings are treaties related to atmospheric issues, biodiversity protection, chemicals and hazardous wastes, and the control of marine pollution. A second focus of attention has been the need to provide UNEP with a more secure funding base, perhaps by shifting from voluntary government pledges to an automatic "assessed" contribution. In the background of these deliberations are questions about the relationships among the major international agencies with important roles in environmental protection and sustainable development—from UNEP and the GEF to the U.N. Commission on Sustainable Development (CSD), the U.N. Development Programme, and the World Bank.[29]

At the time of this writing, a consensus has not yet emerged on any single step that will revolutionize the current system of global environmental governance. But there is widespread agreement that the world community needs to put more political muscle behind the task of creating international agreements and institutions that are up to the task of reversing ecological decline.

Striking a Global Fair Deal

The Earth Summit attempted to bridge the interests of countries of the North and the South in forging a sustainable development path through what is often called the Rio bargain. The essence of this deal was that industrial and developing countries would agree to implement the range of environmental provisions contained in *Agenda 21* and other Rio documents, but that industrial countries would provide substantial financial resources to help others accomplish this. This financing was to come from a range of sources, including increased foreign aid, debt relief, and improved market access for developing-country exports. Besides merely generating resources for implementing the commitments in *Agenda 21*, governments deemed these steps to be important for combating poverty and improving living standards in the developing world—one of the main goals of the conference.[30]

Ten years later, frustration is running high in many quarters over a perceived failure of industrial countries to uphold their end of this bargain. At the same time, the strength of the anti-globalization protest movement in recent years has focused public attention on the importance of addressing persistent inequities between the North and the South. As the Johannesburg Summit approaches, many observers are hoping that it will provide a platform for reinvigorating and updating the Rio bargain to form a Johannesburg Global Deal.[31]

Agenda 21 put a price tag on its own implementation in developing countries of over $600 billion annually, $475 billion of which was expected to be generated from domestic resources and $125 billion of which was to come as foreign aid. The aid sum was widely viewed as unrealistic at the time, as it amounted to twice the overall level of spending on foreign aid then. But northern governments nonetheless agreed to strive to meet it, in part by reaffirming the commitments of many donor countries to contribute 0.7 percent of their gross national product (GNP) annually to development assistance.[32]

But in the decade since Rio, aid spending has declined substantially rather than increased. According to Organisation for Economic Co-operation and Development (OECD) figures, official development assistance amounted to $53 billion in 2000, down from $69 billion in 1992 (in 2000 dollars). (See Figure 8–1). Aid spending as a share of GNP also declined, from 0.33 percent in 1992 to 0.22 percent in 2000. But spending levels vary greatly by individual donor country. In relative terms, Denmark leads the list, contributing over 1 percent of its GNP in aid, with the Netherlands, Sweden, and Norway following close behind. (See Table 8–2.) The United States ranks as the least generous donor by this measure, spending just 0.1 percent of its national income. In absolute terms, however, the United States is the world's second largest donor, following Japan.[33]

The overall decline in aid spending since Rio has meant that, as described earlier, key international environmental programs and agencies such as the GEF and UNEP have been starved for funds. The shortage of funds for worthy initiatives such as these is troubling. At the same time, it must be acknowledged that the overall decline in aid spending over the last decade comes at a time when questions are being raised from many quarters about the record and role of official development assistance. In his recent memoir, Maurice Strong, who promoted the Rio bargain as Secretary-General of the Earth Summit, concedes that "the era of foreign aid as we came to know it in the last half of the twentieth century is . . . coming to an end. . . . There is a tiredness and frustration on the part of both donors and recipients—the donors because they see so much money being 'wasted,' and the recipients because they see it surrounded by so many restrictions and limitations and because they understand as well as anyone that a culture of dependence will never be a long-term solution."[34]

The roles of the World Bank and the International Monetary Fund have come under particular scrutiny in recent years, with the anti-globalization protesters leveling strong critiques from the streets. The World Bank has long been lambasted by environmentalists for its support of large projects such as dams and power plants, which often leave enormous environmental destruction in their wake. These kinds of projects have declined in importance as a share of the Bank's standard public-sector portfolio. But they continue to be financed through loan guarantees and through partnerships with the Internation-

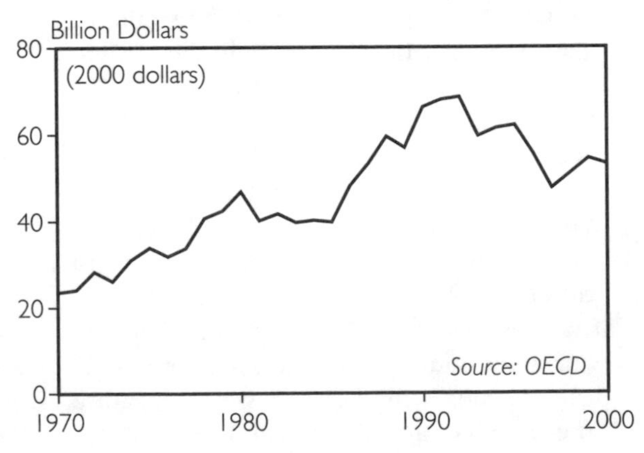

Figure 8–1. Official Development Assistance, 1970–2000

Table 8–2. Development Assistance Contributions, Top 15 Countries and Total, 1992 and 2000

Country	1992 Total (million 2000 dollars)	1992 Share of GNP (percent)	2000 Total (million 2000 dollars)	2000 Share of GNP (percent)
Denmark	1,583	1.02	1,664	1.06
Netherlands	3,132	0.86	3,075	0.82
Sweden	2,798	1.03	1,813	0.81
Norway	1,448	1.16	1,264	0.80
Belgium	984	0.39	812	0.36
Switzerland	1,296	0.46	888	0.34
France	9,407	0.63	4,221	0.33
United Kingdom	3,659	0.31	4,458	0.31
Japan	12,685	0.30	13,062	0.27
Germany	8,613	0.39	5,034	0.27
Australia	1,107	0.35	995	0.27
Canada	2,861	0.46	1,722	0.25
Spain	1,727	0.26	1,321	0.24
Italy	4,689	0.34	1,368	0.13
United States	13,319	0.20	9,581	0.10
All Countries	68,808	0.33	53,058	0.22

SOURCE: See endnote 33.

al Finance Corporation and the Multilateral Investment Guarantee Agency, two Bank-affiliated institutions that underwrite private-sector investment.[35]

The World Bank and the IMF have also been heavily criticized in recent years for the impact of their economic policy advice. Structural adjustment loans, in which recipient countries agree to implement specified policies in exchange for access to large infusions of cash, have come under particular fire. The policies commonly include slashing government budgets, opening up to trade and foreign investment, and privatizing government-owned enterprises. Critics maintain that these conditions have often exacerbated poverty and environmental destruction. In Nicaragua, for example, the Montreal-based Social Justice Committee charges that the IMF is making an already bad situation worse by demanding that the government "slash spending, pull money out of circulation, and privatize public utilities" at a time when the country is suffering from flooding, drought, and collapsing coffee prices. And a recent report by the World Wildlife Fund and the Center for International Forestry Research concludes that export promotion policies imposed by the World Bank and the IMF in Indonesia following the financial crisis of the late 1990s led to rapid expansion of the country's pulp and paper industry at the expense of the health of its forests.[36]

The last few years have also brought growing understanding that World Bank and IMF lending is inextricably linked with the persistent problem of Third World indebtedness, as these institutions are mainly in the business of making loans rather than grants. Despite pledges made in *Agenda 21* to reduce indebtedness and an energetic campaign for debt cancellation by NGOs, the total debt burden in developing and former Eastern bloc countries has climbed 34 percent since the Earth Summit, reaching $2.5 trillion in 2000. Some 17 percent of this total is owed to the World Bank, the IMF, and other public international institutions; 21 percent is owed to national governments; and the remaining 62 percent is owed to commercial banks and other private lenders. (See Figure 8–2.) In some heavily indebted countries, such as Zambia, debt-service payments now consume as much as 40 percent of total government expenditures. These excessive interest payments are siphoning away resources that could otherwise be spent on much-needed social and environmental programs, from HIV prevention and treatment to access to clean water and sanitation.[37]

One development of the 1990s that was unanticipated at Rio was the rapid growth of private capital flows to some parts of the developing world. In 1992, private capital flows to developing and former Eastern bloc nations added up to $115.7 billion (in 2000 dollars), more than 60 percent of the funds flowing into the developing world. These funds climbed rapidly over the next five years, peaking at $315 billion in 1997, 88 percent of the total. (See Figure 8–3.) They then declined precipitously in the wake of the Asian financial crisis, rebounded somewhat, and declined again in 2001 in the face of an uncertain world economic and political outlook. The general category of "private flows" covers several different kinds of finance. In 2000, private investment by multinational corporations (foreign direct investment, or FDI) accounted for nearly 70 percent of the total, while stock and bond transactions made up most of the remaining 30 percent.[38]

The impact of private capital on sustainable development is a debated topic. On the positive side, foreign direct investment and stock market investments do not need to be repaid, unlike World Bank loans or commercial bank lending and bond offerings. FDI infusions, in particular, can provide needed investment capital and also facilitate technology transfer. For example, joint ventures with western companies have helped China to become the world's largest producer of efficient compact fluorescent light bulbs and India to become a major wind power producer. Joint ventures

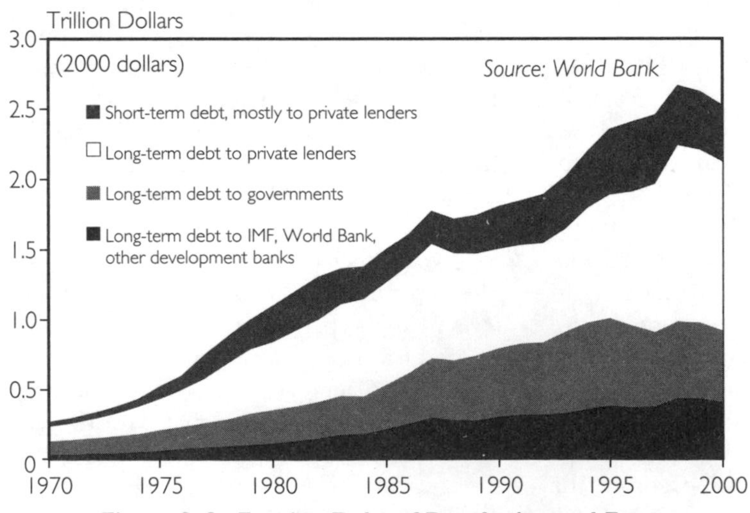

Figure 8–2. Foreign Debt of Developing and Former Eastern Bloc Nations, 1970–2000

can also be a significant source of local employment. On the other hand, critics argue that international corporations can threaten the viability of locally owned businesses, and drain capital out of the country as profits are repatriated. (See Chapter 5 for a discussion of this problem in the tourism sector.)[39]

Even more controversial than FDI are stock and bond investments and commercial bank loans that can move in and out of countries with destructive rapidity. In the aftermath of the Asian financial crisis in 1997, bond financing in developing countries fell from $41 billion in 1998 to $25 billion a year later, while commercial bank lending flows (disbursements of new loans less repayments of old ones) plunged from $50 billion to minus $25 billion over this same period. In addition to these longer-term flows, some $2 trillion worth of foreign exchange occurs every day, up from $400 billion in 1985. Although some of this facilitates legitimate international commerce and investment, a sizable share of it is merely money changing hands to take advantage of short-term changes in currency prices and interest rates. Foreign-exchange speculation on a massive scale is one of the factors that is thought to have helped spark the Asian financial crisis of 1997–98, plunging tens of millions of people into abject poverty within the space of a few months.[40]

In addition to aid, debt reduction, and private capital, another potentially lucrative source of income for developing countries is the removal of trade barriers to their exports. *Agenda 21* called for industrial countries to grant greater market access for developing-country products in the context of the negotiations then under way on the Uruguay Round of trade talks. But many of the WTO's rules have had the effect of prying open the emerging markets of the developing world to exports from industrial countries, while leaving intact large barriers to the entry of developing-country products into northern markets. Frustration over this fundamental imbalance led developing-country negotiators to take a hard line in Seattle in late 1999, contributing even more than the demonstrators to the breakdown of plans to launch a "millennium round" of world trade negotiations.[41]

In the wake of Seattle, many people now argue that any new round of trade talks should be a "development round" that would address the fundamental imbalances that continue to tip the interna-

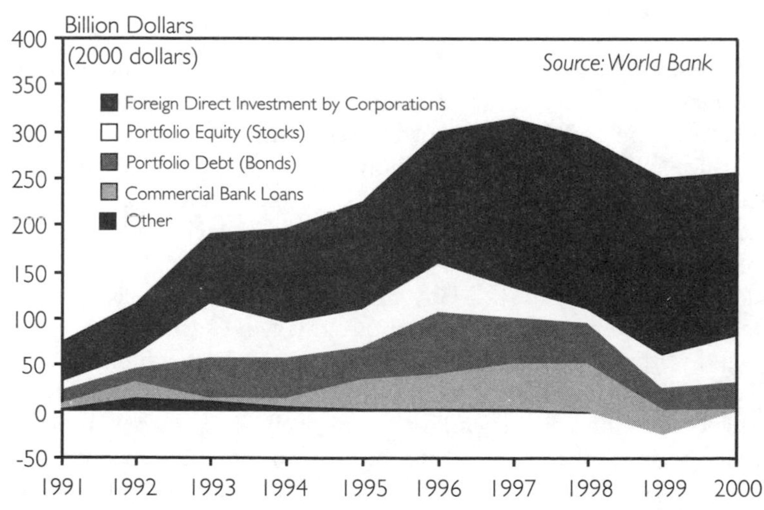

Figure 8–3. Private Capital Flows to Developing Countries, 1991–2000

tional economic scales against the interests of the developing world. Economists estimate that reducing remaining trade barriers could lead to income gains for developing countries in the range of $130 billion a year, although there is no guarantee that this money would be channeled to social or environmental programs. But it is unclear if such a round will have been launched at the WTO's ministerial meeting in November 2001. Many developing-country representatives are leery of jumping into any new negotiations when so many of the commitments made in the Uruguay Round have yet to be implemented. And many of the NGOs that were active in Seattle continue to strongly oppose the launch of a new round, as they believe that the rhetoric about a "development round" is largely a smoke screen for business as usual at the WTO.[42]

The Earth Summit ushered in a new era of global transnational citizen activism that is radically transforming international diplomacy.

One venue where many important financing issues will be raised in the run-up to Johannesburg is the U.N. Financing for Development Summit in Monterrey, Mexico, in March 2002. This meeting will address a daunting array of issues, including the role of domestic financial resources, private capital flows, international trade, development assistance, and debt relief. The summit also plans to tackle broad systemic issues related to governance of the international monetary, financial, and trading systems. Preparations for it brought the World Bank, the IMF, the WTO, and the United Nations to the same negotiating table, which is a notable achievement. In the past, the United Nations has been marginalized from the corridors of international economic power. One of the underlying goals is to infuse some of the environmental and social values that the United Nations embodies into the narrower economic worldview that tends to prevail in the other institutions.[43]

Given the important links between the issues to be addressed in Monterrey and Johannesburg, it is disappointing that the Financing for Development Summit preparations have so far shied away from using the word "sustainable" to modify "development." Nonetheless, some useful initiatives may emerge from Monterrey that could lay the groundwork for success in Johannesburg. One proposal being widely discussed is the Tobin tax, first proposed in 1978 by Yale economist and Nobel laureate James Tobin. Under this proposal, a small tax would be levied on foreign-exchange transactions that would be sizable enough to put a damper on frequent short-term speculative trading but not large enough to discourage longer-term investment and commerce. Although not intended primarily for raising revenue, this type of tax has the potential to generate substantial resources for environmental and social programs. With the enormous volume of daily currency transactions, a tax of only 0.1 percent of the total could generate as much as $400 billion per year. In comparison, in 1999 the U.N. budget, including the activities of its specialized agencies, added up to $10.6 billion.[44]

Many practical problems would need to be worked out in levying such a tax, including both how the revenues would be collected and who would be responsible for spending them well. But these obstacles need not be insurmountable, and political support for a Tobin tax seems to be building. Several national parliaments have held debates or hearings on the idea within the last few years, and the Canadian parliament passed a motion by a two to one margin in March 1999 favoring such a tax. In June 2000, the more than 160 governments present at the Copenhagen+5 meeting in Geneva (on the fifth anniversary of the U.N. Summit on Social Development) agreed to study the feasibility of implementing such a tax. And a number of NGO coalitions are actively advocating a Tobin tax, including the Canadian-based Halifax Initiative and the French-based Association for the Taxation of Transactions to Aid Citizens.[45]

Other ways to generate financial resources for sustainable development and to pay for "global public goods" such as peace and environmental stability will also be on the table in Monterrey and Johannesburg. One idea being discussed is small taxes on the use of the global commons. Levies on international air travel or on emissions of carbon dioxide, for instance, would help countries meet the goals and targets of the Kyoto Protocol while also raising substantial sums that could be used to finance investments in meeting broader sustainable and human development goals.[46]

New Global Actors

As the Earth Summit began in June 1992, some 15,000 representatives of nongovernmental organizations converged in Rio from around the world. Nearly 1,500 of them were accredited to participate in the official conference, where they observed and reported on the negotiations, lobbied delegates, and interpreted developments for the thousands of journalists present. Many more attended the Global Forum, a parallel event that was the scene of numerous exhibits and panel discussions as well as the venue where NGOs negotiated a set of alternative treaties that were intended to hold governments' feet to the fire by enunciating the bold steps needed in order to reverse ecological decline.[47]

In retrospect, it is clear that the Earth Summit ushered in a new era of global transnational citizen activism that is radically transforming the landscape of international diplomacy. Once the staid province of diplomats, U.N. negotiating sessions now attract a diverse and colorful crowd of participants—from NGOs and business representatives to farmers and local officials. Innovative new forms of global governance have emerged since Rio that tap into the dynamism of these different groups.[48]

The number of NGOs operating across international borders grew rapidly over the last century, climbing from 176 in 1909 to more than 24,000 in 2000, according to estimates by the Brussels-based Union of International Organizations. Prominent among them are groups devoted to human rights, peace, women's rights, environmental issues, and Third World development.[49]

Although many NGOs have become vocal critics of the current globalization path in the years since Rio, they have also become adept at using the new tools of the information age to organize themselves into effective cross-border alliances. The Climate Action Network, for instance, has been a powerful and tenacious player in the international climate negotiations for more than a decade. And the Third World Network has helped developing-country NGOs have a voice in international deliberations in diverse arenas, from the annual World Economic Forum held at Davos, Switzerland, to the United Nations and the WTO.[50]

But NGOs are not the only sector to have become increasingly effective at working together across international borders: the international business community is also active at the global level. The number and scale of transnational corporations has climbed rapidly in recent decades, rising from 7,000 in 1970 to an estimated 60,000 today, with some 800,000 foreign affiliates and over $15 trillion in annual sales. The global reach of transnational corporations poses a challenge to regulators, who, unlike corporations, continue to operate mainly at the national and local levels. Corporations who find local or national regulations too onerous can pressure regulators to relax them by threatening to move their operations to other parts of the world.[51]

This regulatory gap has led to calls for a binding code of conduct for transnational corporations to be negotiated under the auspices of the United Nations. Efforts to negotiate such a code in the 1970s and 1980s ran aground because of opposition from corporations themselves as well as from free-market governments in the United States, the United Kingdom, and elsewhere. Although international corporations generally oppose binding codes of conduct, they have taken part in a wide range of voluntary codes and other initiatives aimed at improving corporate environmental and social performance—and, skeptics would say, burnishing public images. (See Table 8–3.)[52]

Most prominently, U.N. Secretary-General Kofi Annan unveiled the Global Compact in 2000, which calls on participating companies to "embrace, support, and enact" nine core values within their operations related to human rights, labor standards, and environmental protection. So far, more than 300 companies have signed on, and several NGOs have also participated in its meetings, including the World Wide Fund for Nature and Amnesty International. But the Compact has also run into a firestorm of criticism from other NGOs that charge it is merely giving cover to "bad actors" while requiring little in the way of specific actions and providing no effective monitoring of implementation or compliance.[53]

If the proliferating number of industry codes of conduct are to make a meaningful difference in shifting the world onto a more environmentally and socially responsible course, they will need to become both more specific and more verifiable. National governments could play an important role in creating the incentive structures that would help to make these instruments more effective. For example, Susan Aaronson of the Washington-based National Policy Association suggests requiring firms that wish to bid on government contracts to demonstrate that they are abiding by terms of designated codes and guidelines, such as OECD's Guidelines for Multinational Enterprises.[54]

One of the most commonly expressed fears about globalization is that it will cause a race to the bottom in environmental and labor standards. But recent history with the Kyoto Protocol offers some hope that the pressures of doing business in a global marketplace can instead spur a race to the top. After the July 2001 agreement on the protocol, hopes have risen that the rest of the world will move ahead with greenhouse gas emissions trading and other provisions of Kyoto despite the current U.S. position. (See Chapter 2.)

Some U.S. companies are worried about their government's lack of involvement, as this will make it difficult for them to participate in the emissions trading provisions. These companies are also concerned about not being included when the rules that govern these instruments are refined. And in the age of globalization, U.S.-based companies operate in many countries, so they will be held to the terms of the Kyoto Protocol in some of their operations, whether or not

Table 8–3. Selected Environmentally Focused Business Codes of Conduct

OECD Guidelines for Multinational Enterprises, 1976
Originally formulated in 1976 and updated in 2000, the Guidelines cover a wide range of topics, with chapters on employment and industrial relations, transparency and anti-corruption, consumer protection, human rights, and the environment. The Guidelines are nonbinding, but the 30 OECD members and 3 nonmember states (Argentina, Brazil, and Chile) have pledged to adhere to them. This is a unique approach, signing on governments rather than individual businesses. Many groups were involved in the process of revising the Guidelines, including NGOs, nonmember governments, and representatives from both labor and business groups.
CERES Principles, 1989
The Coalition for Environmentally Responsible Economies (CERES), formed by an alliance of socially responsible business leaders and environmental activists, released the Principles shortly after the Exxon Valdez oil spill. There are currently 57 companies signed on, including 13 Fortune 500 companies. Some of the largest and best-known endorsers are American Airlines, Ford, Bethlehem Steel, Coca-Cola, and Bank of America. While other codes use general language advising companies to respect the environment, these 10 principles detail specific aspects of the environment that companies must address (such as the sustainable use of natural resources; the reduction, recycling, and safe disposal of wastes; energy conservation and efficiency measures). One notable feature is a commitment to compensate for injuries and damages to the environment and to take steps to restore the environment in case of harm.
The Natural Step, 1989
The Natural Step takes a holistic approach to guiding both society and the environment toward sustainability, setting out four "system conditions" for a sustainable society. It states that nature should be protected from physical degradation, from increased levels of human-made substances, and from rising levels of resources extracted from Earth's crust. The final condition is that "all human needs should be met." The guidance offered on fulfillment of these goals is broad, instructing groups to engage in "systems thinking" to create a vision statement and proceed from there. Some organizations that have adopted this framework are IKEA, Nike, Starbucks, Home Depot, the U.S. Marine Corps, and the municipality of Whistler, British Columbia.
Business Charter for Sustainable Development, 1991
This initiative of the International Chamber of Commerce specifically targets the environment and has 16 principles designed to promote environmental stewardship and high standards for human health and safety. The Charter emphasizes the need for companies to improve their environmental performance continually, incorporating new technologies and information and the precautionary principle into their business strategy. The principles are considered a list of "best practices" and companies are not required to demonstrate adherence.
Eco-Management and Audit Scheme (EMAS), 1993
A voluntary program of the European Community, the project recognizes companies that surpass the minimum environmental standards set by the Community. Businesses must develop an Environmental Management System, submit independently verified environmental reports, and demonstrate continuous environmental improvement. Companies meeting these requirements may display the EMAS logo, a signal to consumers and other businesses that a company is environmentally friendly.

(continued)

Table 8–3. Selected Environmentally Focused Business Codes of Conduct

(continued)

ISO 14000, 1996

A series of standards developed and administered by the International Organization for Standardization, ISO 14001 and ISO 14004 outline principles of environmental management systems and are followed up with detailed directions on implementation. The body of rules is geared toward helping companies develop internal environmental policies and goals, but does not specify performance standards. The series has issued nearly 23,000 certifications in 98 countries, and continues to grow quickly. Japan is the leading ISO 14000 country, with over 5,000 certifications, while Sweden, the United States, and Germany follow with over 1,000 each. The certification process is costly and is most often pursued by large companies, but implementation has proved to be cost-effective in many cases due to waste reduction, energy efficiency, and other measures encouraged by the standards. Unlike other codes, under ISO 14001 independent verification of compliance is required.

Sustainability Reporting Guidelines, 1999

The Global Reporting Initiative (GRI), a collaboration between UNEP and CERES, was founded in 1997. Following the sustainable development framework, its Guidelines are grouped according to economic, social, and environmental indicators. These Guidelines are unique among the codes in laying out detailed rules for reporting information. GRI aims to make sustainability reporting routine, reliable, and credible. More than 60 companies currently participate, including British Airways, Bristol-Myers Squibb, Chiquita, Ford, General Motors, Nokia, and Sunoco.

Standards of Corporate Social Responsibility, 1999

The Social Venture Network is a coalition of entrepreneurs dedicated to the idea that the business community can and should effect positive social change. The Standards cover three topics—ethics, accountability, and governance—and address six sets of stakeholders, ranging from employees and communities to the environment. Meticulously organized, each of the nine chapters is broken down into principle, practices, measures, and resources. The emphasis of the regime is concrete steps that affect the daily operations and management of businesses. The Standards have not yet been widely adopted outside the network's membership.

U.N. Global Compact, 2000

Introduced by U.N. Secretary-General Kofi Annan, the Compact was constructed through collaboration between the United Nations, business, labor, and civil society groups. It has nine principles organized into three categories: human rights, labor, and environment. The Compact also foresees the creation of a network of linked businesses, NGOs, labor groups, and intergovernmental organizations that will help facilitate implementation of shared goals. So far, over 300 companies have pledged their support, from major multinationals such as Royal Dutch/Shell, Nike, and DuPont to small and medium-sized companies in the developing world. Some civil society groups have voiced support, while others are more skeptical, noting that some of the companies that have signed on have been implicated in polluting and harsh labor practices in the past.

SOURCE: See endnote 52.

the U.S. government is formally part of the accord. Over time, the pressures of operating in the global marketplace are thus likely to make it impossible for the United States to remain outside of the Kyoto Protocol forever.[55]

In addition to the rise of both transnational NGO and industry initiatives, recent years have seen growing interest in the concept of "global public policy networks"—joint initiatives involving NGOs, businesses, national governments, and international institutions in which some or all of the parties come together to forge international guidelines or standards for specific activities in which they have relevant knowledge and a large stake in the outcome. These transnational networks are by definition flexible and loose gatherings of experts that are roughly modeled on the decentralized nature of the new information economy, making them fundamentally different in character from traditional top-down intergovernmental treaties and institutions.[56]

A prominent example of such a network is the World Commission on Dams, which was established in 1998 as an outgrowth of a meeting on large dams that was jointly convened by the World Bank and the World Conservation Union–IUCN. The group had 12 commissioners representing all sides in what had become an increasingly polarized debate about the environmental and social impact of large dams. For example, both Medha Patkar, a leader in the fight against the Narmada Dam in India, and Goran Lindahl of ABB Ltd., then one of the world's largest construction companies, were members of the Commission. The group was charged with reviewing the effectiveness of large dams at spurring economic development; assessing alternatives for water resources and development; and developing criteria, guidelines, and standards for the construction, operation, and decommissioning of large dams.[57]

The Commission conducted an extensive program of research on the experience to date with large dams around the world, convened four regional consultations, and considered over 900 submissions from interested individuals, groups, and institutions. The process itself was widely heralded as a new model for international decisionmaking, raising expectations for the final report when it was released in November 2000.[58]

Reactions to the report's recommendations varied. The response from the NGO community was largely positive. The International Rivers Network, for one, welcomed it "as a major contribution to the debate on dams and to the management of water and energy resources in general," and called on its recommendations to be implemented by all financiers and dam builders. Industry representatives, on the other hand, took a dimmer view, noting that "the overall tone of the report is negative in regards to the role of dams" and that "if all of the guidelines and recommendations on the report were implemented they would essentially take decisions away from local governments." Many national governments with large dam-building programs also expressed concern about the final report.[59]

Perhaps the most disappointing reaction came from the World Bank, which had helped initiate the process in the first place and often cited it as a model of its new, more participatory way of doing business. After remaining involved with the extended research and report development process, the Bank decided it had reservations about the results and announced that it would not adopt the Commission's guidelines as Bank policy but would instead use them only as a reference point.[60]

Despite this lukewarm reception, the World Bank has nonetheless thrown its weight behind a somewhat similar multi-stakeholder initiative aimed at reviewing the Bank's role in extractive industries such as oil, gas, and mining. NGO representatives have welcomed this process in principle, but warn that it will only have credibility if it is conducted with complete independence and if the World Bank agrees to abide by the report's recommendations.[61]

Democratizing Global Governance

Although globalization has made a range of nongovernmental transnational actors more powerful, including environmental groups, labor unions, and private corporations, it has by no means made governments obsolete. Forging a sustainable development path—the goal governments committed themselves to at Rio—will require a range of reforms in the way governments go about their business at the global, national, and local levels.

One argument of the anti-globalization protesters that has resonated strongly with the public at large is the notion that today's increasingly powerful institutions of international governance suffer from a profound "democratic deficit." Protesters have focused worldwide attention on the fact that decisions that affect peoples' daily lives—from the safety of the food they eat to the amount their government spends on environmental protection or social welfare—are often being made by remote international institutions such as the WTO and the IMF that are not subject to elections, freedom of information, or public review and comment. Critics charge that these procedural flaws leave the institutions susceptible to capture by narrow commercial interests. Restoring legitimacy to global governance will require addressing these due process concerns head-on.[62]

The procedures of the WTO have come in for particularly sharp criticism, especially the closed-door nature of dispute resolution panels that can issue binding rulings on the consistency of national laws with WTO rules. Several national environmental and consumer laws have been successfully challenged as unfair trade barriers at the WTO since it was established in 1995, including a European Union law that bans the sale of beef produced with growth hormones and a U.S. one that aims to protected endangered sea turtles by restricting imports of shrimp caught in nets without a turtle excluder device.[63]

When a national law is found to violate WTO rules, governments are required to either amend or rescind it—or be subjected to retaliatory trade sanctions. In the shrimp-turtle case, the United States changed the way it was administering the law to bring it into conformity with the WTO ruling. In the beef hormone case, on the other hand, the European Union has so far held its ground even though the United States has imposed over $100 million worth of trade sanctions against EU goods in retaliation. The judges who rule on these national laws are appointed by the WTO, usually based on their background in the field of international trade rather than any environmental or social expertise. And unlike in most national court systems and many international tribunals, WTO panels meet in secret, and government submissions and other key documents are typically confidential.[64]

Recent years have seen the WTO take some limited steps to open itself to greater public participation, including convening symposiums for NGOs and allowing outside groups to submit "friend of the Court" *Amicus* briefs in certain disputes. But more far-reaching steps are needed, such as allowing NGOs to observe and contribute to the deliberations of dispute resolution panels and the meetings of the numerous committees that administer the WTO's extensive and complicated rules.[65]

The World Bank and the International Monetary Fund face similar challenges regarding transparency and accountability. Despite a new public disclosure policy at the World Bank that makes more information available to the public, important documents such as project proposals and country assistance strategies are still confidential until after key decisions have been made. Board meetings at both the World Bank and the IMF are closed to the public, and no minutes are made available. Former World Bank Chief Economist and Nobel laureate Joseph Stiglitz explained the dangers of this lack of transparency in an April 2000 article in the *New Republic*: "Bad economics was only a symptom of the real problem: secrecy. Smart people are more likely to do stupid things when they close themselves off from outside criticism and advice."[66]

The United Nations has generally been more open to the active participation of NGOs in its deliberations and its activities. Their influential role was particularly evident during the global conferences of the 1990s—from the 1992 Rio Earth Summit to the 1993 Vienna human rights conference, the 1994 Cairo population conference, the 1995 Copenhagen social development summit, the 1995 Beijing conference on women, and the 1996 Istanbul conference on cities. At all of these, NGOs joined together in powerful transnational networks to influence both the formal agreements and the informal cooperation resulting from the conference. In more recent years, NGOs have even joined forces with sympathetic governments to promote new international agreements, such as the 1997 Ottawa treaty that banned antipersonnel landmines and the 1998 agreement in Rome to create a U.N. International Criminal Court.[67]

Today's increasingly powerful institutions of international governance suffer from a profound "democratic deficit."

One U.N. institution that has been a pioneer in engaging nongovernmental actors in its work is the U.N. Commission on Sustainable Development, created at Rio as a forum for overseeing the implementation of the agreements reached there. Following the model forged at Rio itself, the CSD has encouraged NGO participation through straightforward accreditation procedures and other steps. The number of NGO representatives from around the world at its annual forums has risen steadily, from 200–300 in 1993 to 700–800 in 2000. High-level government ministers, local officials, business organizations, farmers, and indigenous peoples, among others, have all taken part in the CSD meetings.[68]

The annual CSD meetings in recent years have included "multistakeholder dialogues" dedicated to specific issues, such as tourism, agriculture, and energy. At these sessions, representatives from diverse sectors convene at the United Nations to share their experiences and forge common ground. The multistakeholder model has been incorporated into the preparatory process for the Johannesburg Summit, and is likely to figure prominently in activities there and in any structures or processes that stem from it.[69]

All of these steps at the international level to democratize decisionmaking processes are to the good. But if global governance for sustainable development is to be effective, it must be built on the foundations of good national and local governance. International institutions, after all, are composed of representatives of nation-states, and their policies largely reflect the collective will of these governments, which should themselves represent the collective will of their people (at least in the case of democracies). And when governments make commitments in international forums, whether at the WTO or through U.N. processes, these international commitments must be translated into domestic laws and actions that are implemented and enforced by national and local governmental agencies and judicial systems.

The documents that emerged from the Earth Summit underscored the need to translate abstract global commitments into action at the national and local levels. *Agenda 21* called on all nations to devise national sustainable development strategies, and since the Earth Summit some 70 countries have created National Councils on Sustainable Development or similar organizations charged, among other things, with encouraging the implementation of the Rio agreements at the national level. There is also a growing movement worldwide to create sustainable cities and communities, with many cities and towns adopting local versions of *Agenda 21.* A 2001 survey by the Toronto-based International Council for Local Environmental Initiatives found that more than 4,000 local governments in 63 countries have initiated local *Agenda 21* processes—double the number identified in a survey done in 1997. Successful programs are under way in cities as diverse as Jinja in Uganda, Leicester in the United Kingdom, Porto Alegre in Brazil, and Korolev in Russia.[70]

The Earth Summit also broke new ground by officially recognizing the importance of public participation in environmental decisionmaking at the national level. Principle 10 of the Rio Declaration on Environment and Development stipulates that individuals are entitled to access to information and judicial proceedings, as well as the chance to be involved in decisionmaking. Six years later, this concept was enshrined in a legally binding form in the June 1998 Aarhus Convention on Access to Information, Public Participation in Decision-making and Access to Justice, negotiated under the auspices of the U.N. Economic Commission for Europe. Other regional initiatives on public participation are under way in Latin America and in East Africa, and it is likely that issues of public participation and democratic governance more generally will figure prominently on the agenda for Johannesburg. Along these lines, the U.N. Centre for Human Settlements (Habitat) has already launched a campaign on "good urban governance" that is beginning to have some success. (See Box 8–1.)[71]

Paralleling the need for more democratic governance structures within countries is the importance of more equitable relations among nations. During the first half of 2001, the Bush administration not only pulled out of the Kyoto Protocol but attempted to derail a range of U.N. deliberations on everything from the control of small arms and biological weapons to the well-being of the world's children. But if any good can rise from the rubble and loss in the terrible events of September 11th, it is that they have begun to awaken the American public and the current U.S. administration to the need to cooperate with other countries.[72]

In a stinging rebuke to the earlier policies of his son's administration, former U.S. President George Bush told an audience in Boston in the days immediately following the terrorist attacks that "just as Pearl Harbor awakened this country from the notion that we could somehow avoid the call to duty...in World War II, so, too, should this most recent surprise attack erase the concept in some quarters that America can somehow go it alone in the fight against terrorism or in anything else for that matter." Governments from all corners of the globe have stepped up to the plate to cooperate in the anti-terrorism coalition, and they will now look to the United States to do likewise on issues such as climate change and poverty alleviation.[73]

As the world struggles to respond to the catastrophic events of September 11th and their aftermath, we are understandably dis-

BOX 8–1. GOOD URBAN GOVERNANCE

As the world's population becomes more urban and national governments shift some responsibilities to towns and cities, local authorities are becoming increasingly important. In 1999, Habitat—the U.N. agency responsible for human settlements—launched a global campaign that aims to help people have a voice in local government. It is developing consensus on seven tenets of good urban governance:

- sustainability: balancing the social, economic, and environmental needs of present and future generations—for example, by drafting a Local Agenda 21 action plan for environment and development;
- subsidiarity: decentralizing responsibility and resources to the lowest appropriate level;
- equality: ensuring that all citizens have equal access to decisionmaking;
- efficiency: managing local revenue in a cost-effective way;
- accountability: making local authorities accountable to their citizens, such as by improving public access to government information;
- participation: promoting civic engagement and citizenship—for instance, by making use of public hearings and surveys; and
- security: striving to maintain safe public spaces—for example, by involving citizens in crime and conflict prevention and disaster preparedness, or developing a public awareness campaign to encourage tolerance of diversity.

A better environment is one benefit of translating these sorts of principles into concrete action. In Bangalore, India, an NGO called the Public Affairs Centre surveyed citizens in 1993 to prepare a "report card," and found widespread dissatisfaction and rampant corruption in municipal offices. The corruption was bad for the environment: some people wasted money on bribes that they could have spent on food and education, while those who could not afford the bribes were denied access to needed water, sanitation, and shelter. The survey gave people ammunition to press for improvements in municipal services, and prompted some offices to reform. Inspired by that success, citizens and citizens' groups around the world have devised report cards for other cities. "Good governance isn't so much about technical capability," says Habitat's Paul Taylor, "as it is about the hearts and minds" of the public.

— *Molly O'Meara Sheehan*

SOURCE: See endnote 71.

tracted from preparations for Johannesburg. But recent events only strengthen the need for the World Summit on Sustainable Development. It is now more evident than ever that the persistence of extreme poverty in the face of unprecedented plenty poses grave ethical and moral challenges for international society and calls into question the durability of our current globalization path. Johannesburg offers us an opportunity to shift the course of the global economy and the institutions that underpin it away from destruction and toward ecological and social integrity. We must seize the moment.

WORLD SUMMIT PRIORITIES ON GOVERNANCE

For International Institutions

- Strengthen and streamline the U.N. system's diverse environmentally related agencies and programs.
- Promote more cooperation and coherence between the United Nations, the World Bank, the International Monetary Fund, and the World Trade Organization.
- Promote transparency by making information available and opening negotiations to NGO observers and participants.

For Governments

- Prepare and adopt national and local *Agenda 21s*.
- Implement Rio commitments on freedom of environmental information, public participation, and access to justice.
- Ratify and implement environmental treaties.
- Honor funding pledges from Rio.

For NGOs

- Monitor government and corporate compliance with international norms and standards.
- Strengthen transnational NGO networking and collaboration.
- Forge partnerships with businesses, governments, and international institutions.
- Advocate for strong environmental policies and transparent governmental processes at the global, national, and local levels.

For Business

- Participate in the U.N.'s Global Compact and other corporate codes of conduct, and accept independent monitoring and verification of compliance with them.
- Respect the goals and provisions of international environmental, human rights, and labor treaties and standards.
- Forge partnerships with NGOs, governments, and international institutions.

Notes

Preface

1. Janet Larsen, "Wetlands Decline," in Worldwatch Institute, *Vital Signs 2001* (New York: W.W. Norton & Company, 2001), pp. 96–97; Ashley Mattoon, "Bird Species Threatened," in ibid., pp. 98–99; access to clean water from Robert Engelman et al., *People in the Balance: Population and Natural Resources in the New Millennium* (Washington, DC: Population Action International, 2000), pp. 8–9.

2. J. R. McNeill, *Something New Under the Sun* (New York: W.W. Norton & Company, 2000), p. xxii.

3. Ann Hwang, "AIDS Erodes Decades of Progress," in Worldwatch Institute, op. cit. note 1, pp. 78–79; Janet Larsen, "Hydrological Poverty Worsening," in ibid., pp. 94–95.

4. J. T. Houghton et al., eds., *Climate Change 2001: The Scientific Basis*, Contribution of Working Group I to the Third Assessment Report of the Intergovernmental Panel on Climate Change (Cambridge, U.K.: Cambridge University Press, 2001); Lisa Mastny, "World's Coral Reefs Dying Off," in Worldwatch Institute, op. cit. note 1, pp. 92–93; Lester R. Brown, "World Economy Expands," in ibid., pp. 56–57; World Bank, *World Development Report 2000* (New York: Oxford University Press, 2000).

5. Gary Gardner, "Population Increases Steadily," in Worldwatch Institute, op. cit. note 1, pp. 74–75; World Bank, op. cit. note 4.

Chapter 1. The Challenge for Johannesburg: Creating a More Secure World

1. "A Nation Challenged: President Bush's Address on Terrorism Before a Joint Meeting of Congress," *New York Times*, 21 September 2001.

2. Ibid.

3. Botswana from UNAIDS, "Botswana: Epidemiological Fact Sheets on HIV/AIDS and Sexually Transmitted Infections," 2000 Update (revised), at <www.unaids.org/hivaidsinfo/statistics/fact_sheets/pdfs/Botswana_en.pdf>, viewed 31 October 2001; extinction from G. Tyler Miller, *Living in the Environment* (Pacific Grove, CA: Brooks/Cole Publishing Company, 2000), p. 150; inequality from U.N. Development Programme (UNDP), *Human Development Report 1998* (New York: Oxford University Press, 1998), p. 30.

4. BP Amoco, *Statistical Review of World Energy June 2000* (London: Group Media & Publications, June 2000), pp. 9, 25, 33.

5. Dollar a day from World Bank, *World Development Report 2000/2001* (New York: Oxford University Press, 2000), p. 3.

6. UNDP, *Human Development Report 2001* (New York: Oxford University Press, 2001), p. 9.

7. "A Nation Challenged," op. cit. note 1.

8. Robert Costanza et al., "The Value of the World's Ecosystem Services and Natural Capital," *Nature*, May 1997, p. 253.

9. Warning in 1996 from J. T. Houghton et al., eds., *Climate Change 1995: The Science of Climate Change*, Contribution of Working Group 1 to the Second Assessment Report of the Intergovernmental Panel on Climate Change (IPCC) (Cambridge, U.K.: Cambridge University Press, 1996); idem, *Climate Change 2001: The Scientific Basis*, Contribution of Working Group I to the Third Assessment Report of the IPCC, (Cambridge, U.K.: Cambridge University Press, 2001), p. 10.

10. G. Marland, T. A. Boden, and R. J. Andres, Carbon Dioxide Information Analysis Center, Oak Ridge National Laboratory, "Global, Regional, and National Annual CO_2 Emissions from Fossil-Fuel Burning, Cement Production, and Gas Flaring: 1751–1998 (revised July 2001)," at <cdiac.esd.ornl.gov/ndps/ndp030.html>, viewed 6 November 2001; BP, *BP Statistical Review of World Energy 2001* (London: Group Media & Publications, June 2001).

11. Annette des Iles, "The Alliance of Small Island States Looks Towards Kyoto," *The Courier ACP-EU*, May–June 1997, p. 56.

12. Commission on Sustainable Development (CSD), *Comprehensive Assessment of the Freshwater Resources of the World* (New York: United Nations, 1997).

13. Sandra Postel, *Pillar of Sand* (New York: W.W. Norton & Company, 1999), p. 80.

14. Ibid.

15. Low income living in water-stressed countries from CSD, op. cit. note 12; consequences from Peter Gleick, "The Human Right to Water," *Water Policy*, vol. 1, no. 5 (1999), p. 2.

16. Peter H. Gleick, "The Changing Water Paradigm: A Look at Twenty-first Century Water Resources Development," *Water International*, vol. 25, no. 1 (2000), p. 129.

17. Ibid.

18. World Commission on Dams, *Dams and Development: A New Framework for Decision-Making* (London: Earthscan, November 2000), pp. xxxiii, 15, 16–17.

19. The year of removal is available for 377 of the nearly 500 small dams removed. This smaller dataset is graphed. Data and Figure 1–1 from Peter H. Gleick, *The World's Water 2000–2001* (Washington, DC: Island Press, 2000), pp. 275–86; Idaho from American Rivers, "Congressman McDermott Introduces Bill to Save Endangered Salmon in Columbia and Snake Rivers," press release, at <www.amrivers.org/snakeriver/snakepress7.19.01.htm>, viewed 2 November 2001; Adam Werbach, President, Sierra Club, testimony before the Subcommittees on National Parks and Public Lands and on Water and Power Resources, U.S. House of Representatives, Washington, DC, 23 September 1997; 80,000 dams and reservoirs from Peter H. Gleick, *The World's Water 1998–1999* (Washington, DC: Island Press, 1998), p. 6.

20. Sandra L. Postel and Aaron T. Wolf, "Dehydrating Conflict," *Foreign Policy*, September–October 2001.

21. Threatened plants from Kerry S. Walter and Harriet J. Gillett, *1997 IUCN Red List of Threatened Plants* (Gland, Switzerland: World Conservation Union–IUCN, 1998), p. xvii; other threatened groups from Jonathan Baillie and Brian Groombridge, *1996 IUCN Red List of Threatened Animals* (Gland, Switzerland: IUCN, 1996), p. 26; C. Hilton-Taylor, *2000 IUCN Red List of Threatened Species* (Gland, Switzerland: IUCN, 2000).

22. Assessment from U.N. Food and Agriculture Organization, at <www.fao.org/forestry/fo/fra/index_tables.jsp>; Emily Matthews, *Understanding the FRA 2000* (Washington, DC: World Resources Institute, March 2001), pp. 2, 3.

23. Matthews, op. cit. note 22, p. 5; 1.7 billion people from Tom Gardner-Outlaw and Robert Engelman, *Forest Futures* (Washington, DC: Population Action International, 1999), p. 9.

24. Gardner-Outlaw and Engelman, op. cit. note 23, p. 19.

25. Ibid., p. 21.

26. Share without access to paper from ibid., p. 36; consumption and recycling from Janet N. Abramovitz and Ashley T. Mattoon, *Paper Cuts: Recovering the Paper Landscape*, Worldwatch Paper 149 (Washington, DC: Worldwatch Institute, December 1999), p. 11.

27. Clive Wilkinson, *Status of the Coral Reefs of the World: 2000* (Townsville, Australia: Global Coral Reef Monitoring Network, 2000), p. 1.

28. Ibid.

29. Dirk Bryant et al., *Reefs at Risk: A Map-Based Indicator of Threats to the World's Coral Reefs* (Washington, DC: World Resources Institute, 1998), pp. 8, 9–10.

30. Clive Wilkinson, "Status of Coral Reefs of the World: 2000," Australian Institute of Marine Science, at <www.aims.gov.au/pages/research/coral-bleaching/scr2000/scr-00gcrmn-report.html>, viewed 16 August 2001.

31. Reduction of output from Hilary French, *Vanishing Borders* (New York: W.W. Norton & Company, 2000), p. 90; 180 nations from U.N. Environment Programme, "Status of Ratification/Accession/Acceptance/Approval of the Agreements on the Protection of the Stratospheric Ozone Layer," at <www.unep.org/ozone/ratif.shtml>, viewed 3 November 2001.

32. Principle 1, "Rio Declaration of Environment and Development," U.N. Environment Programme, at <www.unep.org/unep/rio.htm>, viewed 21 September 2001.

33. Don Noah and George Fidas, "The Global Infectious Disease Threat and Its Implication for the United States," National Intelligence Estimate 99-17D, January 2000, at <www.cia.gov/cia/publications/nie/report/nie99-17d.html>, viewed 8 August 2001.

34. Share of deaths from Cesar G. Victora et al., "Reducing Deaths from Diarrhoea Through Oral Rehydration Therapy," *WHO Bulletin*, vol. 78, no. 10 (2000), p. 1252. Table 1–1 from the following: Christopher J. L. Murray and Alan D. Lopez, eds., *The Global Burden of Disease* (Cambridge, MA: Harvard University Press, 1996); World Health Organization (WHO), *World Health Report 2001* (Geneva: 2001); resistance from David L. Heymann, Executive Director for Communicable Diseases, WHO, Testimony before the Committee on International Relations, U.S. House of Representatives, Washington, DC, 29 June 2000, pp. 29–32.

35. Victora et al., op. cit. note 34, pp. 1246, 1252.

36. WHO, "Tuberculosis," Fact Sheet No. 104, revised April 2000, at <www.who.int/inf-fs/en/fact104.html>, viewed 7 August 2001; India from WHO, "Health a Key to Prosperity," at <www.who.int/inf-new/tuber3.htm>, viewed 29 September 2001.

37. Figure 1–2 based on data from Neff Walker, UNAIDS, e-mail to Liza Rosen, Worldwatch Institute, 7 August 2001; infection rates from UNAIDS, "Table of Country-Specific HIV/AIDS Estimates and Data," at <www.unaids.org/epidemic_update/report/Final_Table_Eng_Xcel.xls>, viewed 31 October 2001. Throughout *State of the World*, sub-Saharan Africa is used to describe all African countries except those bordering on the Mediterranean.

38. UNDP, *Human Development Report 2000* (New York: Oxford University Press, 2000), p. 148; UNAIDS, "HIV/AIDS and Development," fact sheet produced for the U.N. Special Session on HIV/AIDS, New York, 25–27 June 2001, pp. 3-4.

39. John Eyles and Ranu Sharma, "Infectious Diseases and Global Change," *Aviso* (monograph), (Victoria, BC, Canada: Global Environmental Change and Human Security Project, June 2001).

40. Noah and Fidas, op. cit. note 33.

41. Patrice Trouiller and Piero L. Olliaro, "Drug Development Output from 1975 to 1996: What Proportion for Tropical Diseases?" *International Journal of Infectious Diseases*, winter 1998–99, p. 61.

42. Heymann, op. cit. note 34, pp. 25, 26.

43. WHO, *World Health Report 1998* (Geneva: 1998), pp. 56–57. Table 1–2 complied from the following: 1990 statistics from Christopher Murray and Alan Lopez, *Global Burden of Disease and Injury Series*, Vols. I and II (Geneva: WHO, 1996); 2000 statistics from WHO, op. cit. note 34, Annex Table 2; major cancers include cancers of the mouth and oropharynx, esophagus, stomach, colon/rectum, liver, pancreas, trachea/ bronchus/lung, melanoma and other skin cancers, breast, cervix, corpus uteri, ovary, prostate, bladder, lymphoma, and leukemia.

44. Overweight in Europe from International Obesity Task Force (IOTF), at <www.iotf.org>, viewed 1 November 2001; U.S. overweight from Centers for Disease Control and Prevention, "A Public Health Epidemic: Overweight and Obesity Among Adults," at <www.cdc.gov/nccdphp/dnpa/obesity/epidemic.htm>, viewed 1 November 2001; obesity in Europe from IOTF, op. cit. this note; obesity in United States (for 1991–98) from American Society of Beriatric Physicians, "Medical Journal Reports Describe Obesity Facts," at <www.asbp.org/overweight.htm>, viewed 30 July 2001.

45. Drugs from IMS Health, at <www.imshealth.com/public/structure/navcontent>, viewed 30 October 2001.

46. WHO, "HIV, TB, and Malaria—Three Major Infectious Diseases Threats," Backgrounder No. 1 (Geneva: July 2000); Elizabeth Olson, "Red Cross Says Three Diseases Kill Many More than Disasters," *New York Times*, 29 June 2000; cost of overweight from IOTF, op. cit. note 44; obesity from Graham Colditz, Harvard School of Public Health, unpublished manuscript.

47. Enrolment from UNESCO, *Education for All—Year 2000 Statistical Assessment*, Statistical Document, at <unescostat.unesco.org/en/pub/pub_p/stat.html>, viewed 26 February 2001, p. 10; share of budget from ibid., pp. 10–11; students per teacher from idem, *World Education Report 2000* (Paris: 2000), pp. 152–55; "Literacy Up by 10 pc Over Last 6 Years: President," *The Times of India*, 9 September 1999.

48. Benefits of education, especially for women, from Kevin Watkins, *Education Now: Break the Cycle of Poverty*, Executive Summary (London: Oxfam International, 1999); illiteracy in developing countries based on data in UNESCO, *Statistical Yearbook 1999 Database*, at <unescostat.unesco.org/en/stats/stats0.htm>, viewed 26 February 2001.

49. UNICEF, *The State of the World's Children 1999* (New York: 1999), pp. 15, 25.

50. Ibid., pp. 80–81.

51. Box 1–1 based on the following: UNDP, op. cit. note 6, p. 14; happiness data from David G. Meyers, "Does Economic Growth Improve Human Morale?" Center for a New American Dream, at <www.newdream.org/newsletter/myers.html>, viewed 27 September 2001.

52. European experience from David Roodman, "Environmental Tax Shifts Multiplying," in Lester R. Brown, Michael Renner, and Brian Halweil, *Vital Signs 2000* (New York: W.W. Norton & Company, 2000), p. 139; computer modeling from Organisation for Economic Co-operation and Development, *OECD Environmental Outlook* (Paris: 2001), p. 263.

53. *European Organic Farming Statistics*, University of Wales, Aberystwyth, at <www.organic.aber.ac.uk/stats.shtml>, viewed 18 September 2001; Christopher Flavin, "Wind Energy Growth Continues," in Worldwatch Institute, *Vital Signs 2001* (New York: W.W. Norton & Company, 2001), p. 44; Seth Dunn, "King Coal's Weakening Grip on Power," *World Watch*, September/October 1999.

54. Social Investment Forum, "Socially Responsible Investing in U.S. Tops Two Trillion Dollar Mark; One Out of Every $8 Under Management Now Invested Responsibly," press release (Washington, DC: 4 November 1999).

55. Grameen clientele from Grameen Bank, *Annual Report 1999* (Dacca, Bangladesh: 1999), p. 39; 58 countries and 94 percent of borrowers from Grameen Bank, at <www.grameen-info.org/bank/index.html>, viewed 4 March 2001; global growth and share of borrowers who are poorest from Microcredit Summit, "Empowering Women with Microcredit: 2000 Microcredit Summit Campaign Report," at <www.microcreditsummit.org/campaigns/report00.html>, viewed 26 February 2001; Kate Druschel, Jennifer Quigley, and Cristina Sanchez, *State of the Microcredit Summit Campaign Report 2001* (draft) (Washington, DC: Microcredit Summit, October 2001).

56. Loans and savings from Consultative Group to Assist the Poorest, "About & History," at <www.cgap.org/html/mi_about_history.html>, viewed 5 March 2001; 70 percent from UNDP, *Human Development Report 1995* (New York: Oxford University Press, 1995), p. 4; 2005 goal from David S. Gibbons and Jennifer W. Meehan, CASHPOR Financial and Technical Services, "The Microcredit Summit's Challenge: Working Towards Institutional Financial Self-Sufficiency while Maintaining a Commitment to Serving the Poorest Families," unpublished draft, June 2000, pp. 3–4; 1.2 billion from World Bank, *World Development Report 2000/2001* (New York: Oxford University Press, 2001), p. 3.

57. International Factor 10 Club, *Statement to Governments and Business Leaders* (Carnoules, France: 1997). The 90-percent goal was also advocated by several government agencies and intergovernmental organizations during the 1990s.

58. Government of Australia, "No Waste by 2010: A Waste Management Strategy for Canberra," at <www.act.gov.au/nowaste/wastestrategy/index.htm>, viewed 15 October 2001; City of Toronto, "Task Force 2010 Seeks Made-in-Toronto Solutions for Waste," press release, at <www.grrn.org/zerowaste/articles/toronto_zerowaste.html>, viewed 15 October 2001; Zero Waste New Zealand, "National Campaign: Zero Waste Councils," at <www.zerowaste.co.nz/integration/basepage.cfm?thepageid=59#Map>, viewed 16 October 2001; take-back policies from Raymond Communications, "Recycling Policy News," at <www.raymond.com/subscribe/rli.html>, viewed 18 October 2001. Table 1–4 from the following: German packaging from U.S. Environmental Protection Agency, Office of Solid Waste, "Product Stewardship: International Initiatives," at <www.epa.gov/epr/products/pintern.html>, viewed 12 October 2001; European Parliament, "European Parliament and Council Directive 94/62/EC of 20 December 1994 on Packaging and Packaging Waste," *Official Journal of the European Communities*, 31 December 1994, Article 6; Sarah D. Hanson and Tetsuo Hamamoto, "U.S. Exports Affected by Japan's New Recycling Law," *FAS Online*, at <www.fas.usda.gov/info/agexporter/2000/Apr/us.htm>, viewed 14 October 2001; European Parliament, "Council Directive 1999/31/EC of 26 April 1999 on the Landfill of

Waste," *Official Journal of the European Communities*, 16 July 1999, Article 5; European Parliament, "Directive 2000/53/EC of the European Parliament and of the Council of 18 September 2000 on End of Life Vehicles," *Official Journal of the European Communities*, 21 October 2000, Articles 4 and 7; Embassy of Japan, "Household Electric Appliance Recycling Law Takes Effect from April—Major Step Toward Age of Full-scale Recycling," Japan Brief 0106, Foreign Press Center, 21 February 2001; Raymond Communications, "European Electronics Directives Will Create New Patchwork," *Recycling Policy News*, press release, 23 May 2001.

59. Edward Cohen-Rosenthal and Thomas N. McGalliard, "Eco-Industrial Development: The Case of the United States," paper issued by the Work and Environment Initiative, Cornell University, at <www.jrc.es/iptsreport/vol27/english/COH1E276.htm>.

60. Consumers International, *Green Testing: Recyclability, Repairability, and Upgradability—A Practical Handbook for Consumer Organizations* (London: 1999), pp. 8–10; Xerox Corporation, *2001 Environment, Health & Safety Progress Report* (Stamford, CT: 2001), p. 19.

61. "Manufacturing: Once Is Not Enough," *Business Week Online*, 16 April 2001.

62. European Car Sharing, "How Does Car Sharing Work?" at <www.carsharing.org>, viewed 25 October 2001.

63. European studies from Bealtaine, Ltd. "Pay As You Drive Carsharing: Final Report," (Cranfield, U.K.: International Ecotechnology Research Centre at Cranfield University, circa 1997), p. 10; idle cars from Rens Meijkamp and Roger Theunissen, "Car Sharing: Consumer Acceptance and Changes on Mobility Behavior," in *Transport Policy and Its Implementation*, Proceedings of Seminar B of the PTRC European Transport Forum, Brunel University, U.K., 2–6 September 1996, p. 1.

64. North America from Car-sharing.net, at <www.carsharing.net>, viewed 30 October 2001; global total from EcoPlan and the Commons, *CarSharing 2001* (Paris: 2001), p. 6; Ford quoted in Terry Slavin, "The Motown Missionary," *The Observer* (London), 12 November 2000.

65. Richard W. Stevenson, "Bush Advisers Say Congress Must Act Now on Economy," *New York Times*, 31 October 2001.

66. Paul H. Ray and Sherry Ruth Anderson, *Cultural Creatives: How 50 Million People Are Changing the World* (New York: Harmony Books, 2000).

67. Geoffrey Ryan, Bureau of Public Affairs, Department of Environmental Protection, City of New York, discussion with author, 19 October 2001.

68. Receptivity and society-wide value change from Ronald Inglehart, *Modernization and Postmodernization: Cultural, Economic, and Political Change in 43 Societies* (Princeton, NJ: Princeton University Press, 1997), p. 33; communication strategies from Gerald T. Gardner and Paul C. Stern, *Environmental Problems and Human Behavior* (Boston: Allyn and Bacon, 1996), pp. 205–52, and from Doug McKenzie-Mohr and William Smith, *Fostering Sustainable Behavior* (Gabriola Island, BC, Canada: New Society Publishers, 1999), pp. 95–99.

Chapter 2. Moving the Climate Change Agenda Forward

1. Eric Pianin, "U.S. Aims to Pull Out of Warming Treaty," *Washington Post*, 28 March 2001.

2. Andrew C. Revkin, "178 Nations Reach A Climate Accord; U.S. Only Looks On," *New York Times*, 24 July 2001.

3. J. T. Houghton et al., eds., *Climate Change 2001: The Scientific Basis*, Contribution of Working Group I to the Third Assessment Report of the Intergovernmental Panel on Climate Change (IPCC) (Cambridge, U.K.: Cambridge University Press, 2001). Box 2–1 based on the following: Michael Grubb, Christian Vrolijk, and Duncan Brack, *The Kyoto Protocol: A Guide and Assessment* (London: Royal Institute of International Affairs/Earthscan, 1999); U.N. Framework Convention on Climate Change (UN FCCC), "Implementation of the Buenos Aires Plan of Action," Conference of the Parties, Bonn, Germany, 24 July 2001; "Summary of the Resumed Sixth Session of the Conference of the Parties to the UN Framework Convention to Climate Change: 16–27 July 2001," *Earth Negotiations Bulletin*, 30 July 2001; UN FCCC, "Bonn Decisions Promise to Speed Action on Climate Change," press release (Geneva: 27 July 2001); idem, "Governments Adopt Bonn Agreement on Kyoto Protocol Rules," press release (Geneva: 23 July 2001).

4. Houghton et al., op. cit. note 3, p. 12.

5. Second IPCC report cited in ibid., p. 10.

6. Ibid., pp. 2–4; Figure 2–1 based on James Hansen, Goddard Institute for Space Studies, "Global Temperature Anomalies in .01 C, 1867–2000," at <www.giss.nasa.gov/data/update/gistemp>, viewed 13 August 2001.

7. Houghton et al., op. cit. note 3, pp. 5–10; C. D. Keeling and T. P. Whorf, "Atmospheric CO_2 Concentrations—Mauna Loa Observatory, Hawaii, 1958–2000 (revised August 2001)," Scripps Institution of Oceanography, La Jolla, CA, 13 August 2001; Figure 2–2 based on G. Marland, T. A. Boden, and R. J. Andres, Carbon Dioxide Information Analysis Center, Oak Ridge National Laboratory (ORNL), "Global, Regional, and National Annual CO_2 Emissions from Fossil-Fuel Burning, Cement Production, and Gas Flaring: 1751–1998 (revised July 2001)," at <cdiac.esd.ornl.gov/ndps/ndp030.html>, viewed 13 August 2001, and on BP, *BP Statistical Review of World Energy* (London: Group Media & Publications, June 2001).

8. Houghton et al., op. cit. note 3, pp. 12–16.

9. Ibid., p. 15.

10. James J. McCarthy et al., eds., *Climate Change 2001: Impacts, Adaptation, and Vulnerability*, Contribution of Working Group II to the IPCC (Cambridge, U.K.: Cambridge University Press, 2001), pp. 3–4.

11. Ibid., pp. 5–6.

12. Ibid., pp. 5–7.

13. Ibid., p. 6.

14. Ibid., p. 6–8; Robert T. Watson et al., eds., *The Regional Impacts of Climate Change: An Assessment of Vulnerability*, A Special Report of IPCC Working Group II (Cambridge, U.K.: Cambridge University Press, 1999); National Assessment Synthesis Team, *Climate Change Impacts on the United States; The Potential Consequences of Climate Variability and Change*, Report for the U.S. Global Change Research Program (Cambridge, U.K.: Cambridge University Press, 2001); Committee on the Science of Climate Change, *Climate Change Science: An Analysis of Some Key Questions* (Washington, DC: National Research Council, 2001).

15. Houghton et al., op. cit. note 3, pp. 75–76.

16. Bert Metz et al., eds., *Climate Change 2001: Mitigation*, Contribution of Working Group III to the Third Assessment Report of the IPCC (Cambridge, U.K.: Cambridge University Press, 2001), pp. 3–5.

17. Ibid., pp. 5–8.

18. Ibid.

19. Ibid., p. 8; Figure 2–3 based on Marland, Boden, and Andres, op. cit. note 7, on BP, op. cit. note 7, and on International Monetary Fund, *World Economic Outlook* (Washington, DC: October 2000), p. 197.

20. Metz et al., op. cit. note 16, p. 9.

21. Ibid.; Luis Cifuentes et al., "Hidden Health

Benefits of Greenhouse Gas Mitigation," *Science*, 17 August 2001, pp. 1257–59; Giulio A. DeLeo et al., "The Economic Benefits of the Kyoto Protocol," *Nature*, 4 October 2001, pp. 478–79.

22. Metz et al., op. cit. note 16, p. 10; Interlaboratory Working Group, *Scenarios for a Clean Energy Future* (Oak Ridge, TN, and Berkeley, CA: Oak Ridge National Laboratory and Lawrence Berkeley National Laboratory, November 2000); European Climate Change Programme, *European Climate Change Programme Report—June 2001* (Brussels: June 2001); "European Union Says Enhanced Efficiency, Conservation Could Make Reductions Easier," *International Environment Reporter*, 20 June 2001, p. 513.

23. Metz et al., op. cit. note 16, p. 10; Peter N. Spotts, "Less Costly Views of Climate Pact," *Christian Science Monitor*, 19 July 2001.

24. Metz et al., op. cit. note 16.

25. Ibid.; Jae Edmonds, Joseph M. Roop, and Michael J. Scott, *Technology and the Economics of Climate Change Policy* (Arlington, VA: Pew Center on Global Climate Change, September 2000).

26. Stephen J. DeCanio et al., *New Directions in the Economics and Integrated Assessment of Global Climate Change* (Arlington, VA: Pew Center on Global Climate Change, October 2000).

27. Metz et al., op. cit. note 16, p. 11.

28. Ibid.

29. Ibid.

30. Ibid.

31. Ibid., pp. 11–12.

32. Ibid., p. 12.

33. Ibid.

34. Ibid., pp. 12–13.

35. Worldwatch estimates based on Marland, Boden, and Andres, op. cit. note 7, and on BP, op. cit. note 7.

36. Table 2–1 based on Marland, Boden, and Andres, op. cit. note 7, and on BP, op. cit. note 7.

37. Figure 2–4 based on Marland, Boden, and Andres, op. cit. note 7, and on BP, op. cit. note 7.

38. Marland, Boden, and Andres, op. cit. note 7; BP, op. cit. note 7; Jonathan Sinton and David Fridley, "Hot Air and Cold Water: The Unexpected Fall in China's Energy Use," *China Environment Series,* Issue 4 (Washington, DC: Woodrow Wilson Center, 2001), pp. 3–20; John Pomfret, "Research Casts Doubt on China's Pollution Claims," *Washington Post*, 15 August 2001.

39. Worldwatch estimates based on Marland, Boden, and Andres, op. cit. note 7, and on BP, op. cit. note 7.

40. Worldwatch estimates based on Marland, Boden, and Andres, op. cit. note 7; BP, op. cit. note 7.

41. Table 2–2 based on International Energy Agency (IEA), *Dealing With Climate Change: Policies and Measures in IEA Countries* (Paris: Organisation for Economic Co-operation and Development (OECD)/IEA, 2000), pp. 24–25; Worldwatch assessment of good practices is based on ibid.

42. IEA, op. cit. note 41; "Economic Man, Cleaner Planet," *Economist*, 29 September 2001, pp. 73–75; OECD, *Environmentally Related Taxes in OECD Countries: Issues and Strategies* (Paris: October 2001).

43. IEA, op. cit. note 41, pp. 26–33; OECD, op. cit. note 42.

44. IEA, op. cit. note 41, pp. 33–37.

45. U.K. Department for Environment, Food & Rural Affairs, "£215m Scheme Offers UK Firms Chance to Be World Leaders," press release (London: 15 August 2001); idem, *Framework for the UK Emissions Trading Scheme* (London: August 2001).

46. IEA, op. cit. note 41, pp. 38–40.

47. Ibid.; U.N. Environment Programme (UNEP) and World Energy Council, "Up to Two Billion Tons of Carbon Dioxide Saved by Cleaner Energy Schemes by 2005; Industry Acting to Fight Global Warming Despite Political Disagreements Over Kyoto," press release (Nairobi and London: 29 June 2001).

48. IEA, op. cit. note 41, pp. 40–41; cuts from local governments from International Council for Local Environmental Initiatives (ICLEI), *Local Government Implementation of Climate Protection: Case Studies* (Toronto: 1997); John J. Fialka, "As the Federal Government Shies Away, States Step Up Efforts to Curb Pollution," *Wall Street Journal*, 11 September 2001; Gary Polakovic, "States Taking the Initiative to Fight Global Warming," *Los Angeles Times*, 7 October 2001; New England Governors/Eastern Canadian Premiers, *Climate Change Action Plan 2001*, prepared by the Committee on the Environment and the Northeast International Committee on Energy of the Conference of New England Governors and Eastern Canadian Premiers, 29–30 March 2001 (Fredericton, N.B., Canada: 28 August 2001). Box 2–2 based on the following: ICLEI, op. cit. this note; current number of cities in campaign from ICLEI, at <www.iclei.org/co2>, viewed 12 October 2001; U.S. and Philippines from Nancy Skinner, "Energy Management in Practice: Communities Acting to Protect the Climate," *UNEP Industry and Environment*, January–June 2000, pp. 43–48, and from Nancy Skinner, International Director, ICLEI Cities for Climate Protection Campaign, Berkeley, CA, e-mail to Molly Sheehan, Worldwatch Institute, 28 August 2001.

49. Kevin A. Baumert and Nancy Kete, *The U.S., Developing Countries, and Climate Protection: Leadership or Stalemate?* Issue Brief (Washington, DC: World Resources Institute, June 2001).

50. Kimberly O'Neill Packard and Forest Reinhardt, "What Every Executive Needs to Know About Global Warming," *Harvard Business Review*, July–August 2000, pp. 128–36.

51. Forest Reinhardt and Kimberly O'Neill Packard, "A Business Manager's Approach to Climate Change," in Eileen Claussen, Vicky Arroyo Cochran, and Debra P. Davis, eds., *Climate Change: Science, Strategies, and Solutions* (Boston: Brill, 2001), pp. 269–79; Packard and Reinhardt, op. cit. note 50, p. 135.

52. Reinhardt and Packard, op. cit. note 51; John Palmisano, "Hedging Corporate Risks Associated With Greenhouse Gas Controls," *Evolution Markets Executive Brief*, 27 August 2001.

53. Davignon quoted in Michael Mann, "Corporate Chiefs Give Backing to Kyoto Pact," *Financial Times*, 11 July 2001.

54. Fred L. Smith and Robert Crandall, Cato Institute, "CO_2 Controls are a Bad Idea, Voluntary or Not" (op-ed), *Wall Street Journal*, 31 July 2001; Ross Gelbspan, *The Heat Is On: The Climate Crisis, the Coverup, the Prescription* (Cambridge, MA: Perseus Publishing, September 1998); Brad Knickerbocker, "Businesses Take Greener Stand on Global Warming," *Christian Science Monitor*, 24 January 2000; Lester R. Brown, "The Rise and Fall of the Global Climate Coalition," Issue Alert (Washington, DC: Worldwatch Institute, 25 July 2000); Laurent Belsie, "Firms Climb Toward Climate-Neutral," *Christian Science Monitor*, 20 August 2001.

55. Thomas L. Friedman, "A Tiger by the Tail," *New York Times*, 1 June 2001; Dick Cheney et al., *National Energy Policy: Reliable, Affordable, and Environmentally Sound Energy for America's Future*, Report of the National Energy Policy Development Group (Washington, DC: U.S. Government Printing Office, 16 May 2001).

56. Royal Dutch/Shell Group of Companies, "Shell Looks to Energy Futures," press release (New York: 3 October 2001); Philip Watts, Chairman of the Committee of Managing Directors, Royal Dutch/Shell Group, "Remarks at the Launch of *Energy Needs, Choices, and Possibilities—Scenarios to 2050*" (New York: 3 October 2001); Rick Popely, "H2 Oh: Carmakers Move Toward Hydrogen Fuel Cells," *Chicago Tribune*, 12 August 2001.

57. John Browne, Chief Executive, BP, "Addressing Global Climate Change," speech delivered at Stanford University, Palo Alto, CA, 19 May 1997; Sir John Browne, Group Chief Executive, BP, "Leading a Global Company: The Case of BP," speech delivered at Yale School of Management, New Haven, CT, 18 September 1998; Charles C. Nicholson, Group Senior Advisor, BP, "The Corporate Commitment to Kyoto," remarks at Chatham House Conference, Delivering Kyoto: Could Europe Do It? Royal Institute of International Affairs, London, 1 October 2001; Neela Banerjee, "Can Black Gold Ever Flow Green?" *New York Times*, 12 November 2000; BP, "BP Launches New Ad Campaign to Engage Public In Energy Issues," press release (London: 5 September 2001).

58. Innovest Strategic Value Advisors, *Climate Change and Shareholder Value: Case Study of BP* (New York: 2001), pp. 3, 6.

59. UNEP, "Impact of Climate Change to Cost the World $US 300 Billion a Year," press release (Nairobi: 3 February 2001); idem, "Financial Sector Responding to Climate Change," press release (Nairobi: 18 July 2001); Vanessa Houlder, "Raising the Temperature," *Financial Times*, 18 April 2001.

60. Natsource from Rana Foorohar, "The New Green Game," *Newsweek*, 27 August 2001; World Bank from "Global Trade in CO_2 Emissions Reaches $100 Million—Study," *Reuters*, 3 August 2001; co2e.com in Carl Frankel, "The Sky's the Limit," *Tomorrow*, May–June 2000, pp. 24–28.

61. Sulfur figure from Ricardo Bayon, "Trading Futures in Dirty Air," *Washington Post*, 5 August 2001; Sandor quoted in Foorohar, op. cit. note 60; Joyce Foundation, "U.S. Voluntary Carbon Trading Emerging," press release (Chicago: 30 May 2001); Nikki Tait, "Greenhouse Gas Trade Scheme Gets Go Ahead," *Financial Times*, 30 May 2001.

62. Christopher Loreti, William Wescott, and Michael Isenberg, "Taking Inventory of Greenhouse Gas Emissions," in Claussen, Cochran, and Davis, op. cit. note 51, pp. 331–56; Christopher P. Loreti, Scot A. Foster, and Jane E. Obbagy, *An Overview of Greenhouse Gas Verification Issues* (Arlington, VA: Pew Center on Global Climate Change, 2001); Shell from Royal Dutch/Shell, "The Shell Tradeable Emission Permit System: An Overview," and "Recent Progress on Climate Change," at <www.shell.com>, viewed 5 October 2001; DuPont from Keith Bradsher and Andrew C. Revkin, "A Pre-Emptive Strike on Global Warming; Many Companies Cut Gas Emissions to Head Off Tougher Regulations," *New York Times*, 15 May 2001. Table 2–3 based on the following: Michael Margolick and Doug Russell, *Corporate Greenhouse Gas Reduction Targets* (Arlington, VA: Pew Center on Global Climate Change, November 2001); Environmental Defense, "Global Corporations and Environmental Defense Partner to Reduce Greenhouse Gas Emissions," press release (Washington, DC: 17 October 2000); Eric Pianin, "Mexican Company Agrees to Reduce Emissions," *Washington Post*, 5 June 2001; Sarah Wade, Environmental Defense, Washington, DC, discussion with Seth Dunn, 7 September 2001; World Wildlife Fund (WWF), "Nike Partners With World Wildlife Fund and the Center for Energy & Climate Solutions to Reduce Greenhouse Gas Emissions," press release (Washington, DC: 2 October 2001); idem, "What is the Climate Savers Program?" at <www.wwfus.org/climate>, viewed 5 October 2001; Alcan, "Alcan Inc. to Reduce Greenhouse Gas Emissions by 500,000 Tonnes over Four Years," press release (Montreal, PQ, Canada: 9 October 2001); "Kyoto Spurned Is Mixed Bag for U.S. Firms" (news roundup), *Wall Street Journal*, 25 July 2001. Note that some corporate targets address greenhouse gas emissions indirectly through improvements in energy efficiency or the use of renewable energy. Some targets may also imply the use of tree planting and other forms of carbon sequestration.

63. Environmental Defense, op. cit. note 62; WWF, "What is the Climate Savers Program?" op. cit. note 62; idem, "Nike Partners," op. cit. note 62; World Resources Institute and World Business Council for Sustainable Development, *The Greenhouse Gas Protocol: A Corporate Accounting and Reporting Standard* (Washington, DC: 2001).

64. Eileen Claussen, "Industry Leaders Accept the Climate Challenge," *Climate Policy*, January 2001, pp. 135–36; "Business Environment Leadership Council," at <www.pewclimate.org/belc/index.cfm>, viewed 5 October 2001; Taryn Fransen, Pew Center on Global Climate Change, e-mail to Seth Dunn, 5 October 2001.

65. "Business Council for Sustainable Energy," at <www.bcse.org>, viewed 5 October 2001; "European Business Council for a Sustainable Energy Future," at <www.e5.org>, viewed 5 October 2001; Social Venture Network, "Business Leaders Call for U.S. Leadership on Global Warming," *New York Times* (advertisement), 21 June 2001; Mann, op. cit. note 53.

66. Lisa Hymas, "An Ocean of Difference," *Tomorrow*, August 2001, pp. 20–23; Geoff Winestock, "The Very Flexible Business of Cutting Emissions," *Wall Street Journal*, 27 July 2001; Jeffrey Ball, "Environmentalists Highlight Firms' Rifts on International Global-Warming Treaty," *Wall Street Journal*, 27 August 2001.

67. U.S. Council in Bradsher and Revkin, op. cit. note 62; Jacob quoted in Vanessa Houlder, "Winners and Losers Wait for Fog to Clear Around Kyoto," *Financial Times*, 25 July 2001; "Kyoto Spurned," op. cit note 62; energy executive quoted in Andrew Revkin, "Some Energy Executives Urge U.S. Shift on Global Warming," *New York Times*, 1 August 2001.

68. "Analysis—CO_2 Emissions Trading Long Way Off Despite Kyoto Deal," *Reuters*, 27 July 2001; Vanessa Houlder, "Carbon Trading Plans May Be Hampered by the Politics of Pollution," *Financial Times*, 14 August 2001.

69. Box 2–3 based on UN FCCC, "Kyoto Protocol: Status of Ratification (As of 28 September 2001)," at <www.unfccc.de>, viewed 1 October 2001, on UN FCCC, "Marrakesh Climate Talks to Finalize Kyoto Rulebook," press release (Marrakesh: 9 October 2001), and on Michael Grubb, Jean-Charles Hourcade, and Sebastian Oberthur, *Keeping Kyoto: A Study of Approaches to Maintaining the Kyoto Protocol on Climate Change* (London: Climate Strategies, July 2001).

70. UN FCCC, "Marrakesh Climate Talks," op. cit. note 69; UN FCC, "Bonn Decisions," op. cit. note 3; "Summary of the Resumed Sixth Session," op. cit. note 3; "Kyoto Protocol May Be Alive, But Battle Is Far From Over," *International Environment Reporter*, 1 August 2001, pp. 635–36.

71. "In Morocco, Experts Cross the T's of Kyoto Treaty" (news roundup), *Wall Street Journal*, 31 October 2001; Hisane Masaki, "Japan Ready to Seek Ratification of Kyoto Protocol," *Japan Times*, 12 October 2001; Winestock, op. cit. note 66; "Kyoto Spurned," op. cit. note 62.

72. "Japanese Study Predicts Limited Impact on Economy if Kyoto Accord Implemented," *International Environment Reporter*, 20 June 2001, pp. 512–13.

73. Text and Japanese carmaker market share in Yasuhiro Murota, Shonan Environmental Research Force, "Will Ratification of the Kyoto Protocol Result in Economic Loss?" commissioned by World Wide Fund for Nature, 28 June 2001; Toyota, "Prius Vision," at <prius.toyota.com>, viewed 5 October 2001; Honda, "American Honda Vehicle Sales for September, 2001," at <www.honda2001.com/news>, viewed 5 October 2001.

74. Murota, op. cit. note 73.

75. Text and Blok quote from Mirjam Harmelink et al., ECOFYS Energy and Environment, "Kyoto Without the U.S.: Costs and Benefits of EU Ratification of the Kyoto Protocol," commissioned by World Wide Fund for Nature, July 2001.

76. Jon E. Hilsenrath, "Eco-Economists Back Bush on Kyoto Pact," *Wall Street Journal*, 7 August 2001; David G. Victor, *The Collapse of the Kyoto Protocol and the Struggle to Slow Global Warming* (Princeton, NJ: Princeton University Press, 2001).

77. Michael Grubb, "A Constructive Scheme Unraveled?" *Nature*, 12 April 2001, pp. 750–01.

78. Pronk quote in "Summary of the Resumed Sixth Session," op. cit. note 3, p. 13; Benito Muller, Axel Michaelova, and Christian Vrolijk, *Rejecting Kyoto: A Study of Proposed Alternatives to the Kyoto Protocol*, pre-publication copy for general distribution (London: Climate Strategies, July 2001).

79. Grubb, Hourcade, and Oberthur, op. cit. note 69; Climate Strategies, "New International Research Organisation Argues that the Kyoto Protocol Can and Must be Rescued by the EU, Japan, and Russia—and That the US Will Then Come Back In," press release (London: 10 July 2001).

80. Grubb, Hourcade, and Oberthur, op. cit. note 69.

81. Edward A. Parson, "Moving Beyond the Kyoto Impasse" (op ed), *New York Times*, 31 July 2001.

82. Ibid.; Multilateral Fund from Richard Eliot Benedick, *Ozone Diplomacy: New Directions in Safeguarding the Planet* (Cambridge, MA: Harvard University Press, 1991).

83. David E. Rosenbaum, "Senate Committee United in Seeking Bush Action on Emissions," *New York Times*, 2 August 2001; Henry Lee, Vicky Arroyo Cochran, and Manik Roy, "US Domestic Climate Change Policy," *Climate Policy*, August–September 2001, pp. 381–95; Vicki Arroyo Cochran and Manik Roy, "Climate Change Legislation in the United States," in Claussen, Cochran, and Davis, op. cit. note 51, pp. 361–71; Congressional Budget Office, *An Evaluation of Cap-and-Trade Programs for Reducing U.S. Carbon Emissions* (Washington, DC: June 2001); Eileen Claussen and Eliot Diringer, "The Climate Challenge Begins At Home" (op ed), *Washington Post*, 19 August 2001.

84. "Summary of the Resumed Sixth Session," op. cit. note 3; "Koizumi Tells Cabinet to Draft GHG Strategy; Japanese Firm Inks Emissions Trade Accord," *International Environment Reporter*, 15 August 2001; Masaki, op. cit. note 71.

85. Joseph Nye, "The Limits of Change," *Financial Times*, 14 September 2001; "Summary of the Resumed Sixth Session," op. cit. note 3; Vanessa Houlder, "Warming to the Kyoto Protocol," *Financial Times*, 24 July 2001; Joseph E. Aldy, Peter R. Orszag, and Joseph E. Stiglitz, "Climate Change: An Agenda for Global Collective Action," prepared for the conference on The Timing of Climate Change Policies, Pew Center on Global Climate Change, Arlington, VA, October 2001; Pew Center on Global Climate Change, "Nobel Prize Recipient Dr. Joseph E. Stiglitz Calls for Immediate Action Against Climate Change," press release (Arlington, VA: 11 October 2001).

Chapter 3. Farming in the Public Interest

1. Roland Bunch, COSECHA, Tegucigalpa, Honduras, discussion with author, 5 May 2001.

2. Roland Bunch, "Increasing Productivity Through Agroecological Approaches: Experiences from Hillside Agriculture in Central America," in Norman Uphoff, ed., *Agroecological Innovations: Increasing Food Production with Participatory Development* (London: Earthscan, forthcoming); Roland Bunch and Gabinó Lòpez, *Soil Recuperation in Central America:*

Sustaining Innovation After Intervention, Gatekeeper Series No. 55 (London: International Institute for Environment and Development (IIED), 1995), p. 12.

3. Bunch, op. cit. note 2.

4. Ibid.

5. Ibid.; Zelaya quote from Jules Pretty, University of Essex, Colchester, U.K., discussion with author, 24 September 2001.

6. World Neighbors, *Lessons from the Field, Reasons for Resiliency: Toward a Sustainable Recovery after Hurricane Mitch* (Oklahoma City, OK: 2000).

7. Agriculture is the focus of chapters 14 and 32 of *Agenda 21*, the blueprint for change adopted at the U.N. Conference on Environment and Development, though issues related to agriculture are mentioned throughout the document; synergies between agricultural policies in Jules Pretty and Rachel Hine, *Reducing Food Poverty with Sustainable Agriculture: A Summary of New Evidence*, Executive Summary (Colchester, U.K.: SAFE-World Research Project, University of Essex, February 2001), p. 8.

8. Figure 3–1 from U.N. Food and Agriculture Organization (FAO), *FAOSTAT Statistics Database*, at <apps.fao.org>, updated 2 May 2001; food prices from International Monetary Fund (IMF), *International Financial Statistics Yearbooks* (Washington, DC: various years).

9. J. N. Pretty et al., "An Assessment of the Total External Costs of UK Agriculture," *Agricultural Systems*, August 2000, pp. 113–36; health care costs associated with poor diets in the United Kingdom are conservatively estimated at over $3 billion, according to Mike Rayner, Department of Public Health, Oxford University, e-mail to author, 5 October 2001.

10. Cheap food policies from Tim Lang, "The Complexities of Globalization: The UK as a Case Study of Tensions Within the Food System and the Challenge to Food Policy," *Agriculture and Human Values*, June 1999, pp. 169–85; Green Revolution from Lori Ann Thrupp, *Cultivating Diversity: Agrobiodiversity and Food Security* (Washington, DC: World Resources Institute, 1998), p. 21.

11. Jeffrey A. McNeely and Sara J. Scherr, *Common Ground, Common Future: How Ecoagriculture Can Help Feed the World and Save Wild Biodiversity* (Gland, Switzerland: World Conservation Union–IUCN, May 2001); David Tilmen et al., "Forecasting Agriculturally Driven Global Environmental Change," *Science*, 13 April 2001, p. 281.

12. Calorie intake provided by 30 plants from FAO, *The State of the World's Plant Genetic Resources for Food and Agriculture* (Rome: 1997), p. 14; China from ibid., p. 34, with present-day update from Zhonghu He, International Maize and Wheat Improvement Center, e-mail to author, 10 October 2001.

13. Landscape structures from Organisation for Economic Co-operation and Development (OECD), *Environmental Indicators for Agriculture: Methods and Results, Executive Summary* (Paris: 2000), p. 42; farmland birds from Pretty et al., op. cit. note 9, p. 125.

14. Cropland in just two species from U.S. Department of Agriculture (USDA), National Agricultural Statistics Service, *1997 Census of Agriculture* (Washington, DC: February 1999).

15. Nitrate and pesticide levels in groundwater from U.S. Geological Survey (USGS), *The Quality of Our Nation's Waters—Nutrients and Pesticides* (Reston, VA: 1999); soil quality from Phil Barak, University of Wisconsin-Madison, "Acidification from Fertilizer Use Linked to Soil Aging," press release, 3 March 1999; share of our crop lost to pests from Montague Yudelman et al., *Pest Management and Food Production*, Food, Agriculture, and the Environment Discussion Paper 25 (Washington, DC: International Food Policy Research Institute (IFPRI), September 1998), pp. 7, 13; Figure 3–2 from FAO, *Fertilizer Yearbook* (Rome: various years) and from K.G. Soh and M. Prud'homme, *Fertilizer*

Consumption Report: World and Regional Overview and Country Reports (Paris: International Fertilizer Industry Association (IFA), December 2000); Figure 3–3 from Rob Bryant, Agranova, letter to author, 17 July 2001, adjusted for inflation using U.S. Department of Commerce, Bureau of Economic Analysis, *U.S. Implicit GNP Price Deflator*, <www.bea.doc.gov>.

16. R. Alexander et al., "Effect of Stream Channel Size on the Delivery of Nitrogen to the Gulf of Mexico," *Nature*, 24 February 2000, pp. 758–60.

17. McNeely and Scherr, op. cit. note 11, p. 7; Robert J. Diaz, "Hypoxia: A Global Perspective" (manuscript), Virginia Institute of Marine Science, College of William and Mary, Williamsburg, VA, sent to author, 22 August 2000.

18. Irrigated land for 1950 and share of world's grain provided by irrigated land from Sandra Postel, *Pillar of Sand* (New York: W.W. Norton & Company, 1999), pp. 41–42; irrigated land for 1999 from FAO, op. cit. note 8, updated 10 July 2001; continued expansion of irrigated area from Postel, op. cit. this note.

19. Water table drop from Liu Yonggong and John B. Penson, Jr., "China's Sustainable Agriculture and Regional Implications," paper presented to the symposium on Agriculture, Trade and Sustainable Development in Pacific Asia: China and Its Trading Partners, Texas A&M University, College Station, TX, 12–14 February 1998, and from Michael Ma, "Northern Cities Sinking as Water Table Falls," *South China Morning Post*, 11 August 2001; 10 percent of the world's grain harvest from Postel, op. cit. note 18, p. 80; wetlands from McNeely and Scherr, op. cit. note 11, p. 7.

20. Share of poor in rural areas and higher rural rates of poverty from International Fund for Agricultural Development (IFAD), *Rural Poverty Report 2001* (New York: Oxford University Press, 2001), pp. 15, 21; U.S. rural poverty from "Farm Families Flock to Food Banks," *AgJournal*, 29 July 2001; rural indicators and hunger concentrated in the countryside from IFAD, op. cit. this note, pp. 72, 91, 106, 107; 800 million estimate based on calculating available food supplies from FAO, *The State of Food Insecurity in the World* (Rome: 2001); Gary Gardner and Brian Halweil, in *Underfed and Overfed: The Global Epidemic of Malnutrition*, Worldwatch Paper 150 (Washington, DC: Worldwatch Institute, March 2000), provide an estimate of 1.1 billion based on surveys of body weight around the world; number and levels of hunger in sub-Saharan Africa from United Nations, Administrative Committee on Coordination, Sub-Committee on Nutrition, *Fourth Report on the World Nutrition Situation* (Geneva: January 2000).

21. Share of food dollar going to farmer from Stewart Smith, University of Maine, letter to author, 20 August 2000.

22. Goal for 1996 and 2001 declaration from FAO, op. cit. note 20; 1974 declaration from FAO, at <www.fao.org/wfs/index_en.htm>.

23. Norman Uphoff, "Challenges Facing World Agriculture in Our New Century," in Uphoff, op. cit. note 2; 1.8 billion from John Pender and Peter Hazell, "Promoting Sustainable Development in Less-Favored Areas, Overview," in John Pender and Peter Hazell, eds., *Promoting Sustainable Development in Less-Favored Areas*, 2020 Vision, Focus 4, Brief 1 (Washington, DC: IFPRI, November 2000), and from Peter Hazell, IFPRI, discussion with author, 19 June 2001.

24. Bunch, op. cit. note 1; Uphoff, op. cit. note 23. Box 3–1 based on the following: 100–140 million children from World Health Organization (WHO), "Micronutrient Deficiencies: Combating Vitamin A Deficiency," <www.who.int/nut/vad.htm>, viewed 10 October 2001; amount of Golden Rice needed for necessary vitamins from Greenpeace, "Genetically Engineered 'Golden Rice' is Fool's Gold," press release (Manila/Amsterdam, 9 February 2001); Richard Lewontin, "Genes in the Food!" *The New York Review*, 21 June 2001; concentration of transgenic crops from Clive James, *Global*

Review of Commercialized Transgenic Crops: 2000, ISAAA Brief No. 21 (Ithaca, NY: International Service for the Acquisition of Agri-Biotech Applications, 2000); U.N. Development Programme, "Although Controversial, GMOs Could Be Breakthrough Technology for Developing Countries," press release for *Human Development Report 2001* (New York: 10 July 2001).

25. Uphoff, op. cit. note 23.

26. Pretty and Hine, op. cit. note 7, pp. 10, 48; Jules Pretty, University of Essex, Colchester, U.K., e-mail to author, 26 September 2001.

27. Nutrient depletion rates from Julio Henao and Carlos Baanante, *Nutrient Depletion in the Agricultural Soils of Africa* (Washington, DC: IFPRI, October 1999), and from Stanley Wood et al., *Agroecosystems: Pilot Analysis of Global Ecosystems* (Washington, DC: IFPRI and World Resources Institute, 2000), p. 52; improved fallow from Pedro Sanchez, "Benefits from Agroforestry in Africa, with Examples from Kenya and Zambia," in Uphoff, op. cit. note 2, and from Pedro Sanchez, International Centre for Research in Agroforestry (ICRAF), discussion with author, 16 February 2001.

28. Norman Uphoff, "Opportunities for Raising Yields by Changing Management Practices: The System of Rice Intensification in Madagascar," in Uphoff, op. cit. note 2.

29. Arie Kuyvenhoven and Ruerd Ruben, "Economic Considerations for Sustainable Agricultural Intensification," in Uphoff, op. cit. note 2; SRI from Norman Uphoff, Cornell University, discussion with author, 23 March 2001.

30. Sahelian project described in Pretty and Hine, op. cit. note 7, pp. 124–25, and in Amadou Makhtar Diop, "Organic Input Management to Increase Food Production in Senegal," in Uphoff, op. cit. note 2; more than half the children malnourished from FAO, op. cit. note 20; more stable levels of productivity from Arie Kuyvenhoven and Ruerd Ruben, "Economic Considerations for Sustainable Agricultural Intensification," in Uphoff, op. cit. note 2, and from Rick Welsh, *The Economics of Organic Grain and Soybean Production in the Midwestern United States*, Henry A. Wallace Institute for Alternative Agriculture, Policy Studies Report No. 13 (Greenbelt, MD: May 1999).

31. Neglect of arid areas and advantages of smaller schemes from IFAD, op. cit. note 20, pp. 92, 94, and from Sandra Postel, Global Water Policy Project, discussion with author, 23 August 2001; focus on irrigated areas from Peter Hazell, IFPRI, discussion with author, 19 June 2001; *tassas* example from Pretty and Hine, op. cit. note 7, pp. 123–24.

32. Gal Oya from C. M. Wijayaratna and Norman Uphoff, "Farmer Organization in Gala Oya: Improving Irrigation Management in Sri Lanka," in Anirudh Krishna et al., eds., *Reasons for Hope: Instructive Experiences in Rural Development* (Bloomfield, CT: Kumarian Press, 1997), pp. 166–83; Norman Uphoff and C. M. Wijayaratna, "Demonstrated Benefits from Social Capital: The Productivity of Farmer Organizations in Gal Oya, Sri Lanka," *World Development*, November 2000.

33. Uphoff and Wijayaratna, op. cit. note 32; Postel, op. cit. note 18, p. 11.

34. Underappreciation of women in agriculture from Jules Pretty, Ruerd Ruben, and Lori Ann Thrupp, "Policies and Institutional Changes," in Uphoff, op. cit. note 2; female-headed households in India and migration from Bina Agarwal, "Disinherited Peasants, Disadvantaged Workers: A Gender Perspective on Land and Livelihood," *Economic and Political Weekly*, March 1998, pp. 2–14, and from Ruth Meinzen-Dick et al., *Gender, Property Rights, and Natural Resources*, FCND Discussion Paper No. 29 (Washington, DC: IFPRI, May 1997), p. 27; 2 percent of land worldwide from Katherine Spengler, "Expansion of Third World Women's Empowerment: The Emergence of Sustainable Development and the Evolution of International Economic Strategy," *Colorado Journal of International Environmental Law and Policy*, summer 2001, p. 320; biases against women from ibid., and from IFAD, op. cit. note 21, p. 87; five African nations from FAO, "Women Feed the World," prepared for World Food Day, 16 October 1998 (Rome: 1998).

35. Liz Alden Wily, *Land Tenure Reform and the Balance of Power in Eastern and Southern Africa*, ODI Natural Resource Perspectives 58 (London: Overseas Development Institute, June 2000); Liz Alden Wily, Overseas Development Institute, e-mail to author, 7 June 2001; yield increases of 20 percent from A. Quisumbing, "Male-Female Differences in Agricultural Productivity," *World Development*, October 1996, pp. 1579–95; weeding technique from D. Elson, "Gender Awareness in Modeling Structural Adjustment," *World Development*, November 1995, pp. 1851–68.

36. Families without land ownership rights from Roy L. Prosterman and Tim Hanstad, *Land Reform: A Revised Agenda for the 21st Century, Rural Development Institute*, RDI Reports on Foreign Aid and Development No. 108 (Seattle, WA: Rural Development Institute, July 2000), p. 1; effects of lack of ownership from ibid., from IFAD, op. cit. note 20, p. 71, and from Arie Kuyvenhoven and Ruerd Ruben, "Economic Considerations for Sustainable Agricultural Intensification," in Uphoff, op. cit. note 2; improved fallow from F. Ksweige et al., "The Effect of Short Rotation *Sesbania sesban* Planted Fallows on Maize Yield," *Forest Ecology and Management*, April 1994, pp. 199–208. Table 3–1 from the following: Zimbabwe from Jon Jeter, "Africa's Racial Land Divide," *Washington Post*, 21 February 2001; South Africa and Namibia from Rachel L. Swarns, "The West Sees One Mugabe, but Africa Sees Another," *New York Times*, 6 August 2000; Stephen Buckley, "Brazil Unveils Agrarian Reform Program," *Washington Post*, 5 July 2000; Department of Agriculture and Co-operation, Ministry of Agriculture, Government of India, "Agricultural Statistics at a Glance," <agricoop.nic.in/statistics/hold1.htm>, viewed 4 September 2001; U.S. data are for 1998 from USDA, National Agricultural Statistics Service, U.S. and State Data, "Farm Numbers and Land in Farms," at <www.nass.usda.gov:81/ipedb>, viewed 4 September 2001; worldwide from International Labour Organisation, *Sustainable Agriculture in a Globalized Economy*, Report for discussion at the Tripartite Meeting on Moving to Sustainable Agricultural Development through the Modernization of Agriculture and Employment in a Globalized Economy (Geneva: 2000), p. 22.

37. Uphoff, op. cit. note 29; Chris Reij and Ann Waters-Bayer, eds., *Farmer Innovation in Africa: A Source of Inspiration for Agricultural Development* (London: Earthscan, October 2001); Ann Waters Bayer, ETC Ecoculture Netherlands, discussion with author, 5 April 2001; Jules Pretty, "Social and Human Capital for Sustainable Agriculture," in Uphoff, op. cit. note 2.

38. San Martin from Bunch, op. cit. note 2; farmer field schools in Latin America and general importance of this approach to farmer learning from Pretty and Hine, op. cit. note 7, pp. 15, 65.

39. FAO, op. cit. note 8; nearly 40 percent includes cropland and rangeland.

40. Youyung Zhu et al., "Genetic Diversity and Disease Control in Rice," *Nature*, 17 August 2000, pp. 718–21; rye and nitrogen from Paul Porter, Department of Agronomy and Plant Genetics, University of Minnesota, "Precipitation and Runoff in the Cottonwood River Watershed: an Historical Perspective with a Future Vision of a Cropping System Involving a Cereal Rye Cover Crop," at <www.rrcnet.org/~porterp/COTTON.htm>.

41. "Algeria to Convert Large Cereal Land to Tree-Planting," *Reuters*, 8 December 2000.

42. Jeffrey A. McNeely, IUCN, Gland, Switzerland, discussion with author, 4 June 2001; McNeely and Scherr, op. cit. note 11, pp. 10, 13.

43. Ademir Calegari, "The Spread and Benefits of No-Till Agriculture in Paraná State, Brazil," in Uphoff, op. cit. note 2; half of state in no-till from Pretty and Hine, op. cit. note 7, p. 67. Box

3–2 from the following: Michael Janofsky, "Plague of Crickets Does $25 Million Damage to Crops in Utah," *New York Times*, June 18, 2001; "China Suffers Worst Locust Attack in Years," *Agence France Presse*, 18 June 2001; FAO, Committee on Agriculture, 16th Session, "Climate Variability and Change: A Challenge for Sustainable Agricultural Production," Rome, 26–30 March 2001, p. 3; amounts of carbon stored in temperate and tropical soils and trees from Pretty and Hine, op. cit. note 7, p. 16, and from John O. Niles et al., "Potential Carbon Mitigation and Income in Developing Countries from Changes in Use and Management of Agricultural and Forest Lands," Centre for Environment and Society Occasional Paper 2001–04, University of Essex, U.K., July 2001; three times from Pedro A. Sanchez, "Linking Climate Change Research with Food Security and Poverty Reduction in the Tropics," *Agriculture, Ecosystems and Environment*, December 2000, p. 378; Chiapas example from McNelly and Scherr, op. cit. note 11, p. 20.

44. Soil Association, *The Biodiversity Benefits of Organic Farming* (Bristol, U.K.: May 2000).

45. Ibid.

46. Pretty and Hine, op. cit. note 7, p. 75; Dave Brubaker from "Global Resource Center for the Environment (GRACE) Calls for a Tax on Factory Farm Meat," press release, at <www.factoryfarm.org>, 14 June 2001; fastest growing form of meat production from Cees de Haan et al., "Livestock & the Environment: Finding a Balance," report of a study coordinated by FAO, U.S. Agency for International Development, and World Bank (Brussels: 1997), p. 53.

47. Pretty and Hine, op. cit. note 7, p. 131.

48. Ibid., p. 74; OECD, *Agricultural Policies in OECD Countries: Monitoring and Evaluation 2001* (Paris: 2001), pp. 25, 178, 183–84; barriers to diversifying from Thomas L. Dobbs and Jules N. Pretty, *The United Kingdom's Experience with Agri-environmental Stewardship Schemes: Lessons and Issues for the United States and Europe* (Colchester, U.K.: South Dakota State University Economics Staff Paper 2001-1 and University of Essex Centre for Environment and Society Occasional Paper 2001-1, March 2001), p. 7.

49. Pretty et al., op. cit. note 9, p. 131; food safety connection from Dan Bilefsky, "EU's Women Farm Ministers Espouse 'Green' Agriculture," *Financial Times*, 7 March 2001, and from Renate Künast, "The Magic Hexagon," *The Ecologist*, April 2001; distribution of farm payments from OECD, *Agricultural Policies in OECD Countries: Monitoring and Evaluation 2000* (Paris: 2000), p. 47.

50. OECD, op. cit. note 13, p. 15; 85 percent from Dave Serfling, Land Stewardship Project, testimony to Hearing on Conservation on Working Lands, Agriculture Committee, U.S. Senate, Washington, DC, 2 August 2001.

51. "Greening the edges" is an expression used by Dobbs and Pretty, op. cit. note 48, p. 6; room for change under existing laws from Franz Fischler, European Commissioner for Agriculture, "A Three-Pronged Reform: The Common Agricultural Policy Should be Adapted to Changing Consumer Needs—But Not Scraped," *Financial Times*, 8 May 2001; funding for U.S. conservation from Jules Pretty et al., "Policy Challenges and Priorities for Internalising the Externalities of Modern Agriculture," *Journal of Environmental Planning and Management*, forthcoming; details on France from Dobbs and Pretty, op. cit. note 48, pp. 1, 17.

52. Nicolas Lampkin, University of Wales, Aberystwyth, U.K., "Organic Farming in the European Union—Overview, Policies and Perspectives," paper presented at the EU conference Organic Farming in the European Union—Perspectives for the 21st Century, Vienna, 27–28 May 1999; Figure 3–4 from University of Wales, Aberystwyth, European Organic Farming Statistics, at <www.organic.aber.ac.uk/stats.shtml>, viewed 18 September 2001.

53. Pretty and Hine, op. cit. note 7, p. 74; Kristen Corselius et al., *Sustainable Agriculture:*

Making Money, Making Sense (Minneapolis, MN: Institute for Agriculture and Trade Policy, March 2001), pp. 9, 32; Welsh, op. cit. note 30, p. 41.

54. Michael Lipton, University of Sussex, "Rural Poverty Reductions: Are We Winning the War?" presentation at the World Bank, Washington, DC, 31 October 2000; aid to agriculture from IFAD, op. cit. note 20, pp. 1–2, 229; World Bank commitments from IFAD, *Annual Report* (Rome: 2001), p. 9.

55. Reduction of services from Deborah Bryceson, ed., *Disappearing Peasantries? Rural Labour in Africa, Asia, and Latin America* (London: Intermediate Technology Publications, 2000), pp. 54, 304–05; Rafael Mariano, Peasant Movement of the Philippines, e-mail to author, 5 August 1999.

56. National Farmers Union (Canada), "The Farm Crisis, EU Subsidies, and Agribusiness Market Power," presentation to the Senate Standing Committee on Agriculture and Forestry, Ottawa, ON, Canada, 17 February 2000. Table 3–2 from the following: pesticide and seed market from "Globalization, Inc., Concentration in Corporate Power: The Unmentioned Agenda," *Communique* (Winnipeg, MN, Canada: ETC Group (formerly RAFI), 5 September 2001); vegetable seeds from "The Gene Giants: Update on Consolidation in the Life Industry," *Communique* (Winnipeg, MN, Canada: Rural Advancement Foundation International, 30 March 1999); trade statistics and retailers in Europe from Fileman Torres et al., "Agriculture in the Early XXI Century: Agrodiversity and Pluralism as a Contribution to Address Issues on Food Security, Poverty, and Natural Resource Conservation" (draft) (Rome: Global Forum on Agricultural Research, April 2000), p. 14; chicken purchases in Central America and retail sector in Brazil from William Vorley and Julio Berdegué, "The Chains of Agriculture," *World Summit on Sustainable Development Opinion* (London: IIED, May 2001); beef and pork packing from William Heffernan, University of Missouri, Columbia, "Consolidation in the Food and Agriculture System," Report to the National Farmers Union, 5 February 1999; Hong Kong retail from Tim Lang, Thames Valley University, London, discussion with author, 14 June 2001.

57. Lipton, op. cit. note 54; West Africa from Christopher Delgado et al., "Agricultural Growth Linkages in Sub-Saharan Africa," IFPRI Research Report 107 (Washington, DC: IFPRI, December 1998), p. xii; Japan, South Korea, and Taiwan from Peter Rosset, "The Multiple Functions and Benefits of Small Farm Agriculture," Policy Brief No. 4 (Oakland, CA: Foodfirst/Institute for Food and Development Policy, September 1999), pp. 12–13.

58. IFAD, op. cit. note 20, p. 2; Brazil from Rosset, op. cit. note 57, pp. 11–12.

59. Higher labor requirements of ecological agriculture from Corselius et al., op. cit. note 53, p. 33, and from Arie Kuyvenhoven and Ruerd Ruben, "Economic Considerations for Sustainable Agricultural Intensification," in Uphoff, op. cit. note 2; India examples from Pretty and Hine, op. cit. note 7, p. 61.

60. Prosterman and Hanstad, op. cit. note 36, pp. 8, 11; general discussion of land rights included in Klaus Deininger, "Land Tenure, Investment, and Land Values: Evidence from Uganda," World Bank, 5 March 2001, unpublished paper, pp. 2–3; India and China from IFAD, op. cit. note 20, pp. 74–76.

61. Bina Agarwal, *A Field of One's Own: Gender and Land Rights in South Asia* (Cambridge: Cambridge University Press, 1994), see especially p. 487; IFAD, op. cit. note 20, p. 171.

62. Pretty and Hine, op. cit. note 7, pp. 10, 17; Jules Pretty et al., "Policies and Institutional Changes," in Uphoff, op. cit. note 2; Jim Cheatle, Director, Association for Better Land Husbandry, Nairobi, Kenya, discussion with author, 12 February 2001; IFAD, op. cit. note 20, pp. vi, 161–73.

63. Diane Carney, *Approaches to Sustainable Livelihoods for the Rural Poor*, ODI Poverty

Brief (London: Overseas Development Institute, 2 January 1999); gap between rural and urban areas and greater impact of extra investment in rural areas from IFAD, op. cit. note 20, pp. 6, 101–02, 105–06; Xiabo Zhang and Shenggen Fan, *Public Investment and Regional Inequality in Rural China*, Environment and Production Technology Division (Washington, DC: IFPRI, December 2000), p. 23.

64. Peter Uvin, *The International Organization of Hunger* (London: Kegan Paul International, 1994); FAO, "Issues and Options in the Forthcoming WTO Negotiations from the Perspective of Developing Countries, Paper No 3: Synthesis of Country Case Studies," FAO Symposium on Agriculture, Trade and Food Security, Geneva, 23–24 September 1999, p. 4; Global Trade Watch, *Down on the Farm: NAFTA's Seven-Years War on Farmers and Ranchers in the U.S., Canada, and Mexico* (Washington, DC: Public Citizen, June 2001).

65. Sophia Murphy, *Trade and Food Security: An Assessment of the Uruguay Round Agreement on Agriculture* (London: Catholic Institute for International Relations, 1999); increased barriers in industrial nations from OECD, op. cit. note 48, p. 11; Michael Windfuhr, FIAN, discussion with author, 11 February 2001.

66. "The Pleasures of Eating," in Wendell Berry, *What Are People For?* (New York: North Point Press, 1990); Scott Kilman, "Monsanto Co. Shelves Seed That Turned Out To Be A Dud Of A Spud," *Wall Street Journal*, 21 March 2001.

67. Renate Künast, "The Magic Hexagon," *The Ecologist*, April 2001, p. 48; Dan Bilefsky, "EU's Women Farm Ministers Espouse 'Green' Agriculture," *Financial Times*, 7 March 2001.

68. OECD, op. cit. note 49, pp. 11, 165–66.

69. Don Wyse, Department of Agronomy and Plant Genetics, University of Minnesota, discussion with author, 23 May 2001.

70. German water companies from Jules Pretty, *The Living Land* (London: Earthscan, 1998), p. 283; "PCC Farmland Fund Rescues Its First Farm: Fund Created by PCC Natural Markets Saves Strategic Growing Area for Organics," press release (Seattle, WA: PCC Farmland Fund, 11 April 2000); Scott Hayes, Landcare Policy and Program Section, Agriculture, Fisheries and Forestry Australia, e-mail to author, 17 July 2001.

71. Ernesto Mendez, Agroecology Program, University of California, Santa Cruz, discussion with author, 20 June 2001.

72. William Vorley, *Agribusiness and Power Relations in the Agri-Food Chain* (draft), background paper (London: IIED, June 2000), p. 21.

73. Based on a statement developed at the International Federation for Alternative Trade Annual General Meeting in May 1999, at <www.ifat.org/fairtrade-defin.html>; $400 million from Fair Trade Federation, "Fair Trade Facts," at <www.fairtradefederation.com/ab_facts.html>, viewed 1 October 2001.

74. WHO in collaboration with U.N. Environment Programme, *Public Health Impact of Pesticides Used in Agriculture* (Geneva: 1990); pesticide-intensive nature of export crops from Lori Ann Thrupp, *Bittersweet Harvest for Global Supermarkets: Challenges in Latin America's Agricultural Export Boom* (Washington, DC: World Resources Institute, 1995).

75. Lang, op. cit. note 10; hormone dispute from "EU and US Claim Beef 'Victory,'" *Financial Times*, 16 January 1998, and from "Beef Hormones: EU Pushes for Compensation Deal with US," *Wall Street Journal*, 14 September 2000.

76. Lang, op. cit. note 10; Slow Food statistics from <www.slowfood.com>, viewed 21 August 2001.

77. Transportation and storage as big energy users and sources of greenhouse gas emissions from Andrew Jones, Sustain, London, e-mail to author, 23 June 2001; United Kingdom from *Food Miles—Still on the Road to Ruin?* (London: Sustain–The Alliance for Better Food and Farming, October 1999), p. 6; distance to Chicago market from Rich Pirog et al., *Food, Fuel, and Freeways: An Iowa Perspective on How Far Food Travels, Fuel Usage, and Greenhouse Gas Emissions* (Ames, IA: Leopold Center for Sustainable Agriculture, Iowa State University, 2001), p. 1; eight times as much energy from Daniel B. Wood, "Coming Soon to City Near You: A Farm," *Christian Science Monitor*, 3 January 2001.

78. Iowa Project from Pirog et al., op. cit. note 77, pp. 1–2 (the ranges depend on the system and truck type used to move the food).

79. USDA, Agricultural Marketing Service, *Innovative Marketing Opportunities for Small Farmers: Local Schools as Customers* (Washington, DC: February 2000); Rich Pirog, Leopold Center for Sustainable Agriculture, Iowa State University, Ames, IA, discussion with author, 20 June 2001.

Chapter 4. Reducing Our Toxic Burden

1. U.N. Environment Programme (UNEP), *Report of the Intergovernmental Negotiating Committee for an International Legally Binding Instrument for Implementing International Action on Certain Persistent Organic Pollutants on the Work of Its Fifth Session,* fifth session, held in Johannesburg, 4–9 December 2000 (Geneva: UNEP Chemicals, 26 December 2000).

2. John Buccini, Chair of UNEP Negotiations on Persistent Organic Pollutants (POPs) and Interim Chair of the Stockholm Convention on POPs, presentation at the World Bank, Washington, DC, 21 June 2001; Stockholm Convention on POPs, at <www.chem.unep.ch/pops/POPs_Inc/dipcon/meetingdocs/25june2001/conf4_finalact/en/FINALACT-English.doc>.

3. Mary O'Brien, *Making Better Environmental Decisions: An Alternative to Risk Assessment* (Cambridge, MA: The MIT Press, 2000); Carolyn Raffensperger and Joel Tickner, eds., *Protecting Public Health & the Environment: Implementing the Precautionary Principle* (Washington, DC: Island Press, 1999).

4. Chapter 19, "Environmentally Sound Management of Toxic Chemicals, Including Prevention of Illegal International Traffic in Toxic and Dangerous Products," in *Agenda 21*, at <www.un.org/esa/sustdev/agenda21chapter19.htm>.

5. David Kriebel et al., "The Precautionary Principle in Environmental Science," *Environmental Health Perspectives*, September 2001, pp. 871–76.

6. Terry Collins, "Toward Sustainable Chemistry," *Science*, 5 January 2001, pp. 48–49; Kenneth Geiser, *Materials Matter: Toward a Sustainable Materials Policy* (Cambridge, MA: The MIT Press, 2001), pp. 1–3, 55–81.

7. Sales in 1998 from Organisation for Economic Co-operation and Development (OECD), *OECD Environmental Outlook for the Chemicals Industry* (Paris: 2001), p. 10.

8. Energy from OECD, op. cit. note 7, p. 42; water use from ibid., pp. 45–46.

9. U.S. Environmental Protection Agency (EPA), Office of Environmental Information, "TRI On-site and Off-site Reported Releases (in pounds), All Chemicals, By Industry, U.S., 1999," *TRI Explorer*, 12 October 2001; Figure 4–1 based on Hemamala Hettige et al., *The Industrial Pollution Projection System*, Policy Research Working Paper No. 1431 (Part 2) (Washington, DC: World Bank, December 1994).

10. Pesticide usage of 2.5 million tons from David Pimentel, "Protecting Crops," in Wallace C. Olsen, ed., *The Literature of Crop Science* (Ithaca, NY: Cornell University Press, 1995), p. 50; magnitude from Polly Short and Theo Colborn, "Pesticide Use in the U.S. and Policy Implications: A Focus on Herbicides," *Toxicology and Industrial Health*, vol. 15, nos. 1–2 (1999), p. 240; 71 percent from Environmental

Defense Fund (EDF), *Toxic Ignorance: The Continuing Absence of Basic Health Testing for Top-Selling Chemicals in the United States* (New York: 1997); less than 10 percent from Geiser, op. cit. note 6, p. 179; Figure 4–2 from OECD, op. cit. note 7, pp. 115–16.

11. Factors in developing countries from OECD, op. cit. note 7, pp. 26–27; specialty chemicals from ibid., pp. 36–37.

12. Production of 25 million tons from Chemical Market Associates Inc., "Polyvinyl Chloride," *PVC Insight*, 21 January 2000, p. 1, from "Chemical Market Associates Inc.'s Industry Report," *PVC Insight*, 10 April–1 May 2000, p. 1, and for Asia from Aida M. Jebens, "Polyvinyl Chloride (PVC) Resins," *CEH* [Chemical Economics Handbook] *Marketing Research Report* (Zurich: SRI International, 1997), p. 580.1880G.

13. Figure of 40 percent from U.N. Food and Agriculture Organization (FAO), *FAOSTAT Statistics Database*, at <apps.fao.org>, viewed 5 October 2000; 35 tons from Janet N. Abramovitz and Ashley T. Mattoon, *Paper Cuts Recovering the Paper Landscape*, Worldwatch Paper 149 (Washington, DC: Worldwatch Institute, December 1999), p. 51; global production from Ashley T. Mattoon, "Paper Piles Up," in Lester R. Brown, Michael Renner and Brian Halweil, *Vital Signs 2000* (New York: W.W. Norton & Company, 2000), p. 78; Latin America from FAO, op. cit. this note, and from Mark Payne, "Latin America Aims High for Next Century," *Pulp and Paper International*, August 1999; Asia from Hou-Main Chang, "Economic Outlook for Asia's Pulp and Paper Industry," *TAPPI Journal*, January 1999; current Asia production from FAO, op. cit. this note.

14. Hilary French, *Vanishing Borders* (New York: W.W. Norton & Company, 2000), pp. 83–86.

15. Oil byproducts from OECD, op. cit. note 7, pp. 22–23, and from Geiser, op. cit. note 6, pp. 67–70, 74–76.

16. Industrial ecology from Tim Jackson, *Material Concerns: Pollution, Profit, and Quality of Life* (New York: Routledge, 1996); Joe Thornton, *Pandora's Poison: Chlorine, Health, and a New Environmental Strategy* (Cambridge, MA: The MIT Press, 2000), pp. 229–58.

17. Thornton, op. cit. note 16; 60 percent from Robert Ayres, "The Life-Cycle of Chlorine, Part I: Chlorine Production and the Chlorine-Mercury Connection," *Journal of Industrial Ecology*, vol. 1, no. 1 (1997), p. 83; Sterns quote from Ivan Amato, "The Crusade to Ban Chlorine," *Garbage*, summer 1994.

18. EDF, op. cit. note 10; European Environment Agency (EEA) and UNEP Regional Office for Europe, *Chemicals in the European Environment: Low Doses, High Stakes?* (Copenhagen: EEA, September 1998), pp. 8–10.

19. Figure 4–3 from Geiser, op. cit. note 6, p. 339; general discussion from ibid., pp. 336–40; P. Barry Ryan, Kelly A. Scanlon, and David L. MacIntosh, "Analysis of Dietary Intake of Selected Metals in the NHEXAS-Maryland Investigation," *Environmental Health Perspectives*, February 2001, p. 121; Daniel Pruzin, "UN/ECE Draft Protocol on Heavy Metals, Persistent Organic Pollutants Concluded," *International Environment Reporter*, 18 February 1998; EEA, *European Environment Assessment 1998* (Copenhagen: 1998), pp. 109–29; Michiel H. H. Hötte, Jaap van der Vlies, and Wim A. Hafkamp, "Levy on Surface Water in the Netherlands," in Robert Gale and Stephan Barg, eds., *Green Budget Reform: An International Casebook of Leading Practices* (Winnipeg, MN, Canada: International Institute for Sustainable Development, 1995), pp. 220–30.

20. Mary O. Amdur, John Doull, and Curtis D. Klaassen, eds., *Cassarett and Doull's Toxicology: The Basic Science of Poisons*, fourth ed. (New York: Pergamon Press, 1991), p. 623.

21. History from Jerome O. Nriagu, "Tales Told in Lead," *Science*, 11 September 1998, pp. 1621–22, from Josef Eisinger, "Sweet Poison," *Natural History*, July 1996, pp. 50–51, from

Seth Dunn, "King Coal's Weakening Grip on Power," *World Watch*, September/October 1999, pp. 10–19, and from Nicola Pirrone et al., "Historical Atmospheric Mercury Emissions and Depositions in North America Compared to Mercury Accumulations in Sedimentary Records," *Atmospheric Environment*, vol. 32, no. 5 (1998), pp. 929–40; U.S. metals and wood products from Grecia Matos, U.S. Geological Survey, Reston, VA, e-mail to Payal Sampat, Worldwatch Institute, 17 October 2001; ratios from J. O. Nriagu, "A Global Assessment of Natural Sources of Atmospheric Trace Metals," *Nature*, 2 March 1989, pp. 47–49.

22. Geiser, op. cit. note 6, pp. 7–8.

23. Jamie Lincoln Kitman, "The Secret History of Lead," *The Nation*, 20 March 2000, pp. 11–44; U.S. production from Geiser, op. cit. note 6, p. 8; Japan from John R. McNeil, *Something New Under the Sun: An Environmental History of the Twentieth-Century World* (New York: W.W. Norton & Company, 2001), p. 62; 90 percent from "Bromide Baron," Rap Sheet No. 2, at <www.corpwatch.org/trac/feature/bromide/greatlakes.html>, viewed 14 October 2001.

24. Current blood lead levels from Shilu Tong, Yasmin E. von Schirnding, and Tippawan Prapamontol, "Environmental Lead Exposure: A Public Health Problem of Global Dimensions," *Bulletin of the World Health Organization*, vol. 78, no. 9 (2000), p. 1069; all living things from Ted Schettler et al., *Generations at Risk: Reproductive Health and the Environment* (Cambridge, MA: The MIT Press, 1999), p. 52; free lead from Eisinger, op. cit. note 21, p. 52.

25. Ethylene dibromide from Kitman, op. cit. note 23; development toxic from Environmental Defense, "Methyl Bromide," at <www.scorecard.org/chemical-profiles/summary.tcl?edf_substance_id=74%2d83%2d9#hazards>, viewed 12 October 2001; World Meterological Organization, "Executive Summary," *Scientific Assessment of Ozone Depletion: 1994* (Geneva: 1994), p. 9.

26. F. G. Hank Hilton, "Income, Liberties, Idiosyncracies, and the Decline of Leaded Gasoline, 1972 to 1992," *Journal of Environment & Development*, March 1999, pp. 49–69.

27. OECD and UNEP, *Phasing Lead Out of Gasoline: An Examination of Policy Approaches in Different Countries* (Paris: 1999), p. 3.

28. Carolyn Stephens et al., "Box 1. 'L'Ethyl' Freedoms?—Canada Fails in Opposition to U.S. Corporate Power," in "Health, Sustainability and Equity: Global Trade in the Brave New World," *Global Change & Human Health*, vol. 1, no. 1 (2000), p. 48; "Ethyl Corp. v. EPA," 51 F.3d 1053, 1054, 1055 (citing Administrator Browner's statements) (D.C. Cir. 1995); EDF, "Producers of Over 70% of US Gasoline Not Using MMT," press release (Washington, DC: 18 March 1996).

29. Lead sources from Tong, von Schirnding, and Prapamontol, op. cit. note 24, p. 1070; Jackson quoted in "Tightrope Walker," *Environmental Forum*, May/June 2001, p. 58.

30. Lead added to PVC from Thornton, op. cit. note 16, p. 313; North America from Alexander H. Tullo, "Plastics Additives' Steady Evolution," *Chemical and Engineering News*, 4 December 2000, p. 25.

31. Jozef M. Pacyna and Elisabeth G. Pacyna, "An Assessment of Global and Regional Emissions of Trace Metals to the Atmosphere from Anthropogenic Sources Worldwide," *Environmental Reviews* (in press), p. 64 in draft.

32. N. Pirrone, G. J. Keeler, and J. O. Nriagu, "Regional Differences in Worldwide Emissions of Mercury to the Atmosphere," *Atmospheric Environment*, vol. 30 (1996), pp. 2981–87; Asia from E. G. Pacyna and J. M. Pacyna, "Global Emission of Mercury from Anthropogenic Sources in 1995," *Water, Air, and Soil Pollution* (in press); Robert B. Finkelman, Harvey E. Belkin, and Baoshan Zheng, "Health Impacts of Domestic Coal Use in China," *Proceedings of the National Academy of Sciences*, March 1999, pp. 3427–31; source of exposure from Arnold Schecter, "Exposure Assessment: Measurement of Dioxins and Related Chemicals in Human Tissues," in Arnold Schecter, ed., *Dioxins and Health* (New York: Plenum Press, 1994), p. 449.

33. Amdur, Doull, and Klaassen, op. cit. note 20, p. 647; Schettler et al., op. cit. note 24, pp. 57–63; global estimate from Pacyna and Pacyna, op. cit. note 31; lake analogy from "Dangerous Levels of Mercury Found in New England Rain," *National Wildlife*, December/January 2001, p. 62.

34. EPA, "Update: National Listing of Fish and Wildlife Advisories," fact sheet (Washington, DC: April 2001); EPA, "Mercury Update: Impact on Fish Advisories," fact sheet (Washington, DC: June 2001), p. 4; U.S. Food and Drug Administration, Center for Food Safety and Applied Nutrition, "An Important Message for Pregnant Women and Women of Childbearing Age Who May Become Pregnant About the Risks of Mercury in Fish," *Consumer Advisory*, March 2001; "Pregnant Women Warned Not to Eat Shark, Swordfish, Mackerel," *Washington Post*, 13 January 2001; Faroe Islands and New Zealand from National Academy of Sciences, *Toxicological Effects of Methylmercury* (Washington, DC: National Research Council, July 2000). Box 4–1 based on the following: gold price from Kitco Precious Metals Inc., online database, at <www.kitco.com/charts/historical gold.html>; mercury use from Roger Moody, "The Lure of Gold—How Golden is the Future?" Panos Media Briefing No. 19 (London: Panos Institute, May 1996), and from Eric Taylor, *Mercury, Mining and Mayhem: Slow Death in the Amazon,* Report for the People's Gold Summit (Berkeley, CA: Project Underground, May 1999); Ed Susman, "The Price of Gold: Indians at Risk in French Guiana," *Environmental Health Perspectives*, May 2001, p. A225; D. Cleary and I. Thornton, "The Environmental Impact of Gold Mining in the Brazilian Amazon," in R. E. Hester and R. M. Harrison, eds., *Mining and its Environmental Impact* (Cambridge, U.K.: Royal Society of Chemistry, 1994), p. 19; EPA, "Toxic Releases Inventory 1999: Executive Summary," April 2001, at <www.epa.gov/tri/tri99/pdr/index.htm>.

35. Hatmakers from Geiser, op. cit. note 6, p. 112; Akio Mishima, *Bitter Sea: The Human Cost of Minamata Disease* (Tokyo: Kosei Publishing Company, 1992); Schettler et al., op. cit. note 24, p. 61; Bernard Weiss, "Vulnerability of Children and the Developing Brain to Neurotoxic Hazards," *Environmental Health Perspectives*, June 2000, pp. 375–81.

36. Tong, von Schirnding, and Prapamontol, op. cit. note 24, p.1071; "Blood and Hair Mercury Levels in Young Children and Women of Childbearing Age—United States, 1999," *Morbidity and Mortality Weekly Report*, 2 March 2001, pp. 140–43; David S. Salkever, "Updated Estimates of Earnings Benefits from Reduced Exposure of Children to Environmental Lead," *Environmental Research*, vol. 70 (1995), pp. 1–6; five IQ points from "Tightrope Walker," op. cit. note 29, p. 59.

37. Tong, von Schirnding, and Prapamontol, op. cit. note 24, pp. 1072–73; EPA, *America's Children and the Environment: A View of Available Measures* (Washington, DC: 2000); Clean Water Action, *Time To Act: Preventing Harm to Our Children* (Boston, MA: 12 May 2001), p. 3.

38. Other factors and leaded gasoline in Africa from Tong, von Schirnding, and Prapamontol, op. cit. note 24, pp. 1068–77; "One Child in Five in Beijing Has Excessive Lead in Blood," *Times of India*, 16 October 2001; Reinhard Kaiser et al., "Blood Lead Levels of Primary School Children in Dhaka, Bangladesh," *Environmental Health Perspectives*, June 2001, pp. 563–66.

39. World Health Organization (WHO) quote from Hilton, op. cit. note 26, p. 49; Tong, von Schirnding, and Prapamontol, op. cit. note 24, p. 1073.

40. UNEP Governing Council, *Mercury Assessment* (Nairobi: 9 February 2001); H. K. T. Wong, A. Gauthier, and J. O. Nriagu, "Dispersion and Toxicity of Metals from Abandoned Gold Mine Tailings at Goldenville, Nova Scotia, Canada," *The Science of the Total Environment*, vol. 228 (1999), p. 44.

41. Krimsky quoted in Geiser, op. cit. note 6, p. 187.

42. Not chemically bonded from Thornton, op. cit. note 16, p. 313; particular conditions from Amdur, Doull, and Klaassen, op. cit. note 20, p. 889; health problems in animal studies from Schettler et al., op. cit. note 24, pp. 181–82, 335.

43. Stephen D. Pearson and Lawrence A. Trissel, "Leaching of Diethylhexyl Phthalate from Polyvinyl Chloride Containers by Selected Drugs and Formulation Components," *American Journal of Hospital Pharmacology*, July 1993, pp. 1405–09; Children's Hospital quote from Bette Hileman, "Alert on Phthalates," *C&EN*, 7 August 2000, p. 54; Toxicology Panel quote cited in Janet Raloff, "New Concerns About Phthalates," *Science News*, 2 September 2000, p 154.

44. National Center for Environmental Health, *National Report on Human Exposure to Environmental Chemicals* (Atlanta, GA: Centers for Disease Control and Prevention, March 2001); Benjamin C. Blount et al., "Levels of Seven Urinary Phthalates Metabolites in a Human Reference Population," *Environmental Health Perspectives*, October 2000, pp. 979–82; J. Raloff, "Girls May Face Risks from Phthalates," *Science News*, 9 September 2000, p. 165; Environmental Working Group, *Beauty Secrets: Does a Common Chemical in Nail Polish Pose Risks to Human Health?* (Washington, DC: November 2000).

45. Peter Matthiessen and Peter E. Gibbs, "Critical Appraisal of the Evidence for Tributyltin-Mediated Endocrine Disruption in Mollusks," *Environmental Toxicology and Chemistry*, vol. 17, no. 1 (1998), pp. 37–43; Terry L. Wade, Bernardo Garcia-Romero, and James M. Brooks, "Tributyltin Contamination in Bivalves from United States Coastal Estuaries," *Environmental Science & Technology*, vol. 22, no. 12 (1988), pp. 1488–93.

46. Wade, Garcia-Romero, and Brooks, op. cit. note 45; Joint Group of Experts on the Scientific Aspects of Marine Environmental Protection, *Protecting the Oceans from Land-based Activities*, GESAMP Report No. 71 (Nairobi: 15 January 2001), p. 23.

47. International Conferences on the Protection of the North Sea, *The North Sea: An Integrated Ecosystem Approach for Sustainable Development* (Oslo: Ministry of the Environment, April 1999), p. 11; World Wide Fund for Nature (WWF), "WWF Welcomes the Adoption of a Convention to Eliminate Dangerous Chemicals from the Seas," news release (Gland, Switzerland: 5 October 2001).

48. Terry Collins, Department of Chemistry, Carnegie Mellon University, Pittsburgh, PA, discussion with author, 20 June 2001; Collins, op. cit. note 6; The Royal Society, *Endocrine Disrupting Chemicals* (London: June 2000), p. 5.

49. Quote from Jocelyn Kaiser, "Panel Cautiously Confirms Low-Dose Effects," *Science*, 27 October 2000, p. 695; Royal Society, op. cit. note 48, pp. 1–4.

50. Sohail Khattak et al., "Pregnancy Outcome Following Gestational Exposure to Organic Solvents: A Prospective Controlled Study," *Journal of the American Medical Association*, 24–31 March 1999, pp. 1106–09; Cheryl Siegel Scott and V. James Cogliano, "Trichloroethylene Health Risks: State of the Science," *Environmental Health Perspectives*, Supplement 2, May 2000, pp. 159–60; OSPAR Convention for the Protection of the Marine Environment of the North-East Atlantic, *OSPAR Strategy with Regard to Hazardous Substances* (Sintra, Portugal: 22–23 July 1998).

51. Warren P. Porter, James W. Jaeger, and Ian H. Carlson, "Endocrine, Immune, and Behavioral Effects of Aldicarb (Carbamate), Atrazine (Triazine) and Nitrate (Fertilizer) Mixtures at Groundwater Concentrations," *Toxiciology and Industrial Health*, vol. 15, nos. 1–2 (1999), pp. 133–50.

52. Thornton, op. cit. note 16, pp. 7–19; Geiser, op. cit. note 6, pp. 195–212.

53. "World Bank and UNEP Sign Agreement on Persistent Organic Pollutants," news release (Washington, DC: UNEP, 18 May 2001).

54. UNEP, op. cit. note 1; Buccini, op. cit. note 2.

55. UNEP, op. cit. note 1; Buccini, op. cit. note 2.

56. Barbara Dinham, *PIC—A Tool for Change and a Partner for POPs? NGO Briefing on the Rotterdam Convention on Prior Informed Consent for Participants at the Signing of the POPs Treaty* (London: Pesticide Action Network–UK, 14 May 2001); Clifton Curtis and Cynthia Palmer Olsen, *Targeting Toxic Chemicals on the Way to the Johannesburg Summit* (Washington, DC: WWF, September 2001).

57. Dinham, op. cit. note 56; Curtis and Olsen, op. cit. note 56.

58. Article 14 of the Rotterdam Convention on the Prior Informed Consent Procedure for Certain Hazardous Chemicals and Pesticides in International Trade (September 1998); Dinham, op. cit. note 56; EEA and UNEP, op. cit. note 18, p. 18.

59. Jonathan Krueger, "What's to Become of Trade in Hazardous Wastes? The Basel Convention One Decade Later," *Environment*, November 1999, p. 13; special responsibility from Jim Puckett, Basel Action Network, Seattle, WA, e-mail to author, 16 October 2001.

60. Basel Action Network, "China Ratifies Global Accord to End Waste Dumping," news release (New York: 15 May 2001); Pat Phibbs, "Ratification of Basel Accord Without Ban On Exports Opposed by Environment Coalition," *International Environment Reporter*, 15 August 2001, p. 687.

61. Krueger, op. cit. note 59, pp. 15, 18, 19; Puckett, op. cit. note 59.

62. Texas Center for Policy Studies, *The Generation and Management of Hazardous Wastes and Transboundary Hazardous Waste Shipments between Mexico, Canada and the United States, 1990–2000* (Austin, TX: May 2001).

63. National Resources Council of Maine and Mercury Policy Project (MPP), "India Rejects U.S. Waste Mercury Shipment; Mercury To Be Turned Back to U.S. Port," news release (Burlington, VT: MPP, 25 January 2001); Port Said from Neil Tangri, Essential Action, Washington, DC, discussion with author, 20 June 2001.

64. Steve Gorman and Clif Curtis, "African Stockpile Program (Obsolete Pesticides)," presentation at POPs Day meeting, World Bank, Washington, DC, 21 June 2001; "Chemical Time Bomb in the CIS," *Pesticides News*, September 1999, p. 8; Alemayehu Wodageneh, "Trouble in Store," *Our Planet*, vol. 8, no. 6 (1998), pp. 12–14; Mark Davis, "Picking Up the Poison Bill," *Pesticides News*, June 2001, pp. 3–5.

65. Darryl Luscombe, Toxics Campaign, Greenpeace International, discussion with author, 28 June 2001.

66. Michael Gregory, Mariann Lloyd-Smith, and Carl Smith, "UNEP Global POPs Treaty (INC5/Johannesburg, December 2000): POPs and Right to Know," IPEN Briefing Paper (Washington, DC: International POPs Elimination Network, 2000).

67. U.N. Economic Commission for Europe (UNECE), "Aarhus Convention Starts Countdown to Entry into Force," press release (Geneva: 9 August 2001); Kofi Annan, "Foreword," in UNECE, *The Aarhus Convention: An Implementation Guide* (Geneva: 2000), p. v.

68. EPA, "Chapter 1: Toxics Release Inventory Reporting and the 1997 Public Data Release," *Toxics Release Inventory* (Washington, DC: 1997); EPA, "Toxics Release Inventory," op. cit. note 34; U.S. General Accounting Office, *Toxic Chemicals: EPA's Toxic Release Inventory Is Useful but Can Be Improved* (Washington, DC: June 1991); EPA, "Administration Promotes Rule on Lead Emissions Information," press release (Washington, DC: 17 April 2001).

69. Countries cited in EPA, "Chapter 1," op. cit. note 68, in UNEP Chemicals, *National Inventories of Persistent Organic Pollutants: Selected Examples as Possible Models,* Preliminary Report (Geneva: July 1999), and in OECD, op. cit. note 7, pp. 64–69; UNEP, *National PRTR Activities: Challenges and Experiences* (Nairobi: 1 March 1996); "The Citizens Right to Know" (editorial), *New York Times*, 1 June 1999.

70. Pat Costner, scientist, Greenpeace International, discussion with author, 13 April 2000. For list of participating organizations, see <www.ipen.org/ipen_endorse99.html>.

71. "Denmark to Stop Funding Toxic Waste Project," *Xinhua News Service*, 22 August 2001; U.N. Foundation, "GEF Approves Project to Safely Dispose of PCBs," news release (Washington, DC: 11 May 2001).

72. Lotta Fredholm, "Sweden To Get Tough on Lingering Compounds," *Science*, 1 December 2000, p. 1666.

73. "WWF Calls on the IMO Members to Adhere to Ban on a Highly Toxic Chemical," press release (Gland, Switzerland: WWF, 1 October 2001); "Toxic Toys," *Environmental Health Perspectives*, June 1999, p. A295; Joe Kirwin, "European Commission Seeks Ban on PVC Toys Containing Phthalates," *International Environment Reporter*, 24 November 1999, pp. 937–38.

74. Figure 4–4 from Kees Baas, Central Bureau of Statistics, The Hague, e-mail to David Roodman, Worldwatch Institute, 24 September 1997; Hötte, van der Vlies, and Hafkamp, op. cit. note 19.

75. Olle Pettersson, "Pesticide Use in Swedish Agriculture: The Case of a 75% Reduction," in David Pimentel, ed., *Techniques for Reducing Pesticide Use: Economic and Environmental Benefits* (New York: John Wiley & Sons, 1997), pp. 79–102; OECD and UNEP, op. cit. note 27, pp. 7–10; Odil Tunali, "Lead in Gasoline Slowly Phased Out," in Lester R. Brown et al., *Vital Signs 1995* (New York: W.W. Norton & Company, 1995), pp. 126–27.

76. World Bank, "Greening Industry: Communities, Stock Markets, and Governments Join Hands to Cut Industrial Pollution," press release (Washington, DC: 22 November 1999).

77. Geiser, op. cit. note 6, pp. 193–94; New Jersey official from U.S. General Accounting Office, *Environmental Protection: EPA Should Strengthen Its Efforts to Measure and Encourage Pollution Prevention* (Washington, DC: February 2001), p. 21.

78. Miguel Altieri, "Escaping the Treadmill," *Ceres*, July/August 1995; definitions from FAO, "Organic Agriculture: Item 8 of the Provisional Agenda," Committee on Agriculture, Fifteenth Session, Rome, 25–29 January 1999, 3 February 2000; Carol Kaesuk Yoon, "Simple Method Found to Increase Crop Yields Vastly," *New York Times*, 22 August 2000; Rick Welsh, *The Economics of Organic Grain and Soybean Production in the Midwestern United States*, Policy Studies Report No. 13 (Greenbelt, MD: Henry A. Wallace Institute of Alternative Agriculture, May 1999).

79. Brian Halweil, "Organic Gold Rush," *World Watch*, May/June 2001, p. 22; 25 percent from Rob Bryant, Agranova, e-mail to Liza Rosen, Worldwatch Institute, 18 July 2001.

80. Stockholm Convention, op. cit. note 2; World Wildlife Fund, *Resolving the DDT Dilemma: Protecting Human Health and Biodiversity* (Washington, DC: June 1998); WHO, *WHO Expert Committee on Malaria, Twentieth Report*, Technical Report Series No. 892 (Geneva: 2000), p. 3.

81. Geoffrey A. T. Target and Brian M. Greenwood, "Impregnated Bednets," *World Health*, May–June 1998, pp. 10–11; "Malaria: Waiver by Uganda of Taxes and Tariffs on Insecticide-treated Nets," *Weekly Epidemiological Record,* 21 July 2000, pp. 233–34.

82. Henk Bouwman, "Malaria Control and the Paradox of DDT," *Africa: Environment and*

Wildlife, May 2000, p. 56; Roger Thurow, "In Malaria War, South Africa Turns to Pesticide Long Banned in the West," *Wall Street Journal*, 26 July 2001.

83. Thornton, op. cit. note 16, pp. 383–86; 87 percent from French, op. cit. note 14, p. 90.

84. Massachusetts Toxics Use Reduction Institute, *"The Cost of Changing": Total Cost Assessment of Solvent Alternatives*, Methods/Policy Report No. 6 (Lowell, MA: 1994); A. H. Verschoor and L. Reijnders, "Toxics Reduction in Ten Large Companies, Why and How," *Journal of Cleaner Production*, vol. 8, no. 1 (2000), pp. 69–78; Thornton, op. cit. note 16, pp. 383–86; Geiser, op. cit. note 6, p. 255, 351–54.

85. Geiser, op. cit. note 6, pp. 283–304; National Research Council, *Biobased Industrial Products: Priorities for Research and Commercialization* (Washington, DC: National Academy of Sciences, 2000); "EarthShell: Greening the Drive-Thru," *The Carbohydrate Economy*, spring 2001, p. 11.

86. Gruber cited in Peter Fairley, "Bioprocessing Comes Alive," *Chemical Week*, 14 March 2001, pp. 23–25.

87. Lena Q. Ma et al., "A Fern That Hyperaccumulates Arsenic," *Nature*, 1 February 2001, p. 579; 1 percent from Mike Ferullo, "U.S. Study Finds Fern May Remediate Arsenic Contamination in Soil, Water," *International Environment Reporter*, 23 May 2001, p. 436; detox center parallel from Andrew C. Revkin, "Plants as Pollution Sponges," *New York Times*, 6 March 2001.

88. Revkin, op. cit. note 87.

89. Geiser, op. cit. note 6, p. 237.

90. "Living Free of Pollution Called a Basic Human Right," *Environmental News Service*, 30 April 2001.

Chapter 5. Redirecting International Tourism

1. Jayakumar Chelaton, Thanal Conservation Action and Information Network, Kerala, India, e-mail to author, 9 July 2001.

2. Ibid.

3. Susan C. Stonich, "Political Ecology of Tourism," *Annals of Tourism Research*, vol. 25, no. 1 (1998), p. 25.

4. World Tourism Organization (WTO), "Tourism Knocked Down, But Not Out," press release (Madrid: 17 September 2001).

5. Martha Honey, *Who Owns Paradise? Ecotourism and Sustainable Development* (Washington, DC: Island Press, 1998).

6. WTO, "Tourism is One of the Environment's Best Friends. Secretary-General Addresses Special Session of United Nations," press release (Madrid: 24 June 1997).

7. International tourist arrivals refers to the number of nonresident arrivals who stay at least one night, not to the number of distinct persons. Definitions and 2020 from WTO, *Tourism Market Trends: World Overview and Tourism Topics* (Madrid: May 2001), pp. 14, 47; Figure 5–1 from Rosa Songel, Statistics and Economic Measurement of Tourism, WTO, e-mail to author, 17 April 2001; 4–10 times higher estimate from Geoffrey Wall, Associate Dean, Graduate Studies and Research, Faculty of Environmental Studies, University of Waterloo, ON, Canada, e-mail to author, 27 September 2001.

8. Auliana Poon, "The 'New Tourism' Revolution," *Tourism Management*, vol. 15, no. 2 (1994), pp. 91–92; technologies from World Trade Organization, Council for Trade in Services, "Tourism Services," background note (Geneva: 23 September 1998), pp. 5–6; Travel Industry Association of America, "TIA Releases Two Studies on E-Travel Consumers," press release (Washington, DC: 27 February 2001).

9. Sources of tourists from WTO, op. cit. note 7, p. 33; share of passports derived from U.S. Department of State, discussion with Uta Saoshiro, Worldwatch Institute, 9 October 2001, and from Population Reference Bureau, electronic database, at <www.prb.org/Content/NavigationMenu/Other_reports/2000-2002/sheet1.html>, viewed 9 October 2001; 3.5 and 7 percent from World Trade Organization, op. cit. note 8, p. 5.

10. Motivations from WTO, op. cit. note 7, p. 23; Poon, op. cit. note 8; U.S. study from Honey, op. cit. note 5, p. 9.

11. Regional breakdown and Figure 5–2 from WTO, op. cit. note 7, p. 35, and from Songel, op. cit. note 7; 2020 from WTO, *Tourism 2020 Vision: A New Forecast* (Madrid: 1997); China from WTO, "East Asia/Pacific Arrivals Set to Double in Next Ten Years," press release (Madrid: 9 January 1999).

12. Receipts estimates exclude international transport fares and were deflated using U.S. Implicit GNP Price Deflator, obtained from Virginia Mannering, U.S. Commerce Department, Bureau of Economic Analysis, 27 February 2001; data and Table 5–1 from WTO, op. cit. note 7, pp. 15, 17, 34, 42.

13. Rapid rise from World Wide Fund for Nature (WWF), *Preliminary Assessment of the Environmental & Social Effects of Trade in Tourism* (Gland, Switzerland: May 2001), p. 22; 8 percent and 40 percent from Henryk Handszuh, "Overview of International Trade in Tourism Services, Including Current Statistics and Trends," presentation at WTO Symposium on Tourism Services, Geneva, 22–23 February 2001; top five from WTO, op. cit. note 7, p. 41.

14. Methodology from World Travel & Tourism Council (WTTC), *World Travel & Tourism Council Year 2001 Tourism Satellite Accounting Research: Documentation* (London: June 2001), p. 5; 2000 estimates from idem, "Tourism Satellite Accounting Confirms Travel and Tourism as World's Foremost Economic Activity," press release (London: 11 May 2000).

15. Airlines from Miguel Alejandro Figueras, Advisor to the Minister of Tourism, Cuba, "International Tourism in the Cuban Economy," presentation at WTO Symposium on Tourism Services, Geneva, 22–23 February 2001, p. 7; hotels and tour operators from WTO, op. cit. note 7, p. 40.

16. World Trade Organization, op. cit. note 8, p. 7; WWF, op. cit. note 13.

17. One in five from Honey, op. cit. note 5, p. 8; Asian destinations and growth rates from WTO, op. cit. note 7, pp. 18–22, 46; Cuba from Figueras, op. cit. note 15, p. 2.

18. United Nations Conference on Trade and Development (UNCTAD), "Tourism and Development in the Least Developed Countries," background paper for the Third U.N. Conference on the Least Developed Countries, Las Palmas, Canary Islands, 26–29 March 2001, p. 15.

19. "World Bank Revisits Role of Tourism in Development," *World Bank News*, 18 June 1998; $500 million from Maurice Desthuis-Francis, "Tourists & Cities: Friend or Foe?" *Impact* (International Finance Corporation newsletter), spring-summer 2000, p. 18; International Monetary Fund from Raymond de Chavez, "Globalization and Tourism: Deadly Mix for Indigenous Peoples," *Third World Resurgence*, March 1999.

20. Deloitte & Touche, International Institute for Environment and Development (IIED), and Overseas Development Institute (ODI), *Sustainable Tourism and Poverty Elimination Study*, a report to the U.K. Department for International Development (UKDFID) (London: April 1999), pp. 9–10; UNCTAD, op. cit. note 18, p. 4; national economy estimates derived from World Bank, *World Development Indicators*, electronic database, accessed 6 August 2001, and from WTO, op. cit. note 7, pp. 91–96.

21. Surplus from World Trade Organization, op. cit. note 8, p. 3; 1999 estimates from Handszuh, op. cit. note 13.

22. Obstacles to development from UNCTAD, op. cit. note 18, pp. 9–13; Brazil from "Wish Your Dollars Were Here," *The Economist*, 16 December 2000, p. 66.

23. WTO estimate cited in David Diaz Benavides, UNCTAD, "The Sustainability of International Tourism in Developing Countries," presentation at the Seminar on Tourism Policy and Economic Growth, Berlin, 6–7 March 2001, pp. 8–9.

24. Estimates of 50–70 percent from World Trade Organization, op. cit. note 8, p. 4; linkage problems from Matthew J. Walpole and Harold J. Goodwin, "Local Economic Impacts of Dragon Tourism in Indonesia," *Annals of Tourism Research*, vol. 27, no. 3 (2000), p. 572.

25. Estimate of 90 percent from U.N. Environment and Development–UK (UNED–UK), *Sustainable Tourism and Poverty Elimination*, report on a workshop held 9 February 1999 in preparation for the Seventh Meeting of the U.N. Commission on Sustainable Development (UNCSD), New York, April 1999, p. 8; Thailand from Michael Doyle, "New Law Retains US Advantage for Now," *Bangkok Post*, 8 January 2000, and from Chatrudee Theparat, "Petition Will Demand Amendment of Laws," *Bangkok Post*, 12 November 2000; crowding out from International Labour Organisation (ILO), *Human Resources Development, Employment and Globalization in the Hotel, Catering and Tourism Sector*, report for discussion at the Tripartite Meeting on Human Resources Development, Employment and Globalization in the Hotel, Catering and Tourism Sector, Geneva, 2001, pp. 43–44.

26. Foreign operators from WWF, op. cit. note 13, p. 15; 80 percent from Erlet Cater, "Ecotourism in the Third World: Problems for Sustainable Tourism Development," *Tourism Management*, April 1993, p. 86; packages from Caroline Ashley, Charlotte Boyd, and Harold Goodwin, "Pro-Poor Tourism: Putting Poverty at the Heart of the Tourism Agenda," *Natural Resource Perspectives* (London: ODI, March 2000).

27. Estimate of 90 percent and inflation from Cater, op. cit. note 26; 65 percent from Polly Pattullo, *Last Resorts: The Cost of Tourism in the Caribbean* (London: Cassell, 1996), pp. 121–23.

28. David L Edgell, Sr., "A Barrier-Free Future for Tourism?" *Tourism Management*, vol. 16, no. 2 (1995), pp. 107–10; de Chavez, op. cit. note 19; WWF, op. cit. note 13, pp. 10–11; 112 countries from World Trade Organization, op. cit. note 8, p. 9; TRIMS from Anita Pleumarom, "Tourism, Globalization and Sustainable Development," *Third World Resurgence*, March 1999.

29. WTTC, *Travel & Tourism's Economic Impact* (London: March 1999); women from Carmen Michard et al., "Women's Employment in Tourism World-Wide: Data and Statistics," in UNED–UK, *Gender & Tourism: Women's Employment and Participation in Tourism*, report prepared for the Seventh Session of the UNCSD, April 1999, pp. 17–34; informal work from Ashley, Boyd, and Goodwin, op. cit. note 26.

30. Service jobs from Lee Pera and Deborah McLaren, Rethinking Tourism project, "Globalization, Tourism & Indigenous Peoples: What You Should Know About the World's Largest 'Industry,'" at <www.planeta.com/ecotravel/resources/rtp/globalization.html>, November 1999; ILO, op. cit. note 25, p. 121; Maggie Black, *In the Twilight Zone: Child Workers in the Hotel, Tourism and Catering Industry* (Geneva: ILO, 1995); 2 million from ILO, "Seeking Socially Responsible Tourism," *World of Work*, June 2001.

31. Dependence from Patricia P. A. A. H. Kandelaars, *A Dynamic Simulation Model of Tourism and Environment in the Yucatan Peninsula* (Laxenburg, Austria: International Institute for Applied Systems Analysis, April 1997), p. 2; Douglas W. Payne, "Letter from Grenada," *The Nation*, 22 March 1999, pp. 22–24; slow season from ILO, "The Social Impact on the Hotel and Tourism Sector of Events Subsequent to 11 September 2001," briefing paper for discussion at the Informal Meeting on the Hotel and Tourism Sector: Social Impacts of Events Subsequent to

11 September 2001, Geneva, 25–26 October 2001.

32. Respect for minorities from Ashley, Boyd, and Goodwin, op. cit. note 26; Deborah Ramer McLaren, "The History of Indigenous Peoples and Tourism," *Cultural Survival Quarterly*, summer 1999, p. 27; Mike Robinson, "Is Cultural Tourism on the Right Track?" *UNESCO Courier*, July–August 1999, p. 22–23; Bali from Tony Wheeler, "Philosophy of a Guidebook Guru," *UNESCO Courier*, July–August 1999, p. 55; Rachel Proctor, "Tourism Opens New Doors, Creates New Challenges, for Traditional Healers in Peru," *Cultural Survival Quarterly*, winter 2001, pp. 14–16; Myra Shackley, "The Himalayas: Masked Dances and Mixed Blessings," *UNESCO Courier*, July–August 1999, p. 28.

33. De Chavez, op. cit. note 19; Peru from Kenneth McCormick, "Can Ecotourism Save the Rainforests?" information sheet (San Francisco: Rainforest Action Network, 1994); contact effects from Sven Wunder, *Promoting Forest Conservation Through Ecotourism Income?* Occasional Paper No. 21 (Jakarta: Center for International Forestry Research, March 1999), pp. 13–14; globalization from Lesley France, ed., *The Earthscan Reader in Sustainable Tourism* (London: Earthscan Publications, Ltd., 1997), p. 8.

34. John S. Akama, "Marginalization of the Maasai in Kenya," *Annals of Tourism Research*, vol. 26, no. 3 (1996), pp. 716–18; Isaac Sindiga, *Tourism and African Development: Change and Challenge of Tourism in Kenya* (Aldershot, Hamps, U.K., and Leiden, Netherlands: Ashgate Publishing Led and African Studies Centre, 1999), pp. 126–45; Tourism Concern, "Burma," at <www.tourismconcern.org.uk/campaigns/frame.htm>, viewed 20 September 2001.

35. Megan Epler Wood, International Ecotourism Society, Burlington, VT, e-mail to author, 3 October 2001; Stephen G. Snow, "The Kuna General Congress and the Statute on Tourism," *Cultural Survival Quarterly*, winter 2001, pp. 17–18.

36. United Nations Environment Programme (UNEP), *Tourism and Environmental Protection*, Addendum C, Contribution of UNEP to the Secretary-General's Report on Industry and Sustainable Tourism for the Seventh Session of the UNCSD, 1999; capacity from Pleumarom, op. cit. note 28.

37. Estimate of 90 percent from International Federation Free Trade Union & Trade Union Advisory Committee, Organisation for Economic Co-operation and Development, *Tourism and Sustainable Development: Workers and Trade Unions in the Web of Tourism*, Background Paper No. 2, prepared for the Seventh Session of the UNCSD, New York, 19–30 April 1999, p. 9; shares of transport are 1998 estimates, per WTO, op. cit. note 7, pp. 24–25; 90 percent air travel estimate from World Trade Organization, op. cit. note 8, p. 5.

38. Fastest growing from "Global Warming Could Hurt Tourism," *Associated Press*, 30 August 1999; Intergovernmental Panel on Climate Change (IPCC), *Special Report: Aviation and the Global Atmosphere* (Geneva: 1999); 57 percent and higher share a 1996 estimate cited in Stefan Gössling, "Tourism—Sustainable Development Option?" *Environmental Conservation*, vol. 27, no. 3 (2000), pp. 223–24.

39. Room estimates from WTO, op. cit. note 7, p. 29; larger hotels from idem, *Tourism Highlights 1999* (Madrid: rev. 19 May 1999), p. 10; Singy Hanyona, "Victoria Falls Marred by Pollution," *Environment News Service*, 28 September 2000.

40. Caribbean from Pattullo, op. cit. note 27, pp. 105–09; "Mexico: Resort's Boom Raises Environmental Concerns," *UN Wire*, 20 August 1999; two ecosystems from Carlos Castilho and Roberto Herrscher, *Ecotourism: Paradise Gained, or Paradise Lost?* Panos Media Briefing No. 14 (London: Panos Institute, January 1995).

41. Expensive resource use from Mitzi Perdue, "Hotels Are Going Green," at <www.eggscape.com/hotels.htm>, viewed 9 December 1998;

Elisabeth Eaves, "Feature—Dead Sea Tourism Threatens Fragile Environment," *Reuters*, 2 September 1998; Caroline Hawley, "Dead Sea 'to Disappear by 2050,'" *BBC News Online*, 3 August 2001.

42. Grenada from Payne, op. cit. note 31, pp. 22–24; Malaysia from Annette Groth, "Sustainable Tourism and the Environment," *Connect* (UNESCO International Science, Technology & Environmental Education Newsletter), vol. 25, no. 1 (2000), p. 1; Philippines from Maurice Malanes, "Tourism Killing World's Eighth Wonder," at Third World Network, <www.twnside.org.sg/title/mm-cn.htm>, viewed 18 September 2001.

43. UNEP, op. cit. note 36, p. 2; Frederick Noronha, "Goa's Tourist Boom Backfires With Ugly Smell," *Environment News Service*, 27 December 2000.

44. Obstacles to treatment and Caribbean from Pattullo, op. cit. note 27, pp. 112–13; Thailand from James E. N. Sweeting, Aaron G. Bruner, and Amy B. Rosenfeld, *The Green Host Effect*, CI Policy Papers (Washington, DC: Conservation International, 1999), p. 10.

45. Doubling from WTO, op. cit. note 7, p. 59; Kira Schmidt, *Cruising for Trouble: Stemming the Tide of Cruise Ship Pollution* (San Francisco: Bluewater Network, March 2000); 90,000 tons from Honey, op. cit. note 5, p. 40; "Royal Caribbean Sentenced for Ocean Dumping," *Reuters*, 4 November 1999.

46. New vessels from WTO, op. cit. note 7, p. 59; floating cities from "Cruise Ship Dumping Sparks Interest," *Environmental News Network*, 3 December 1999; passengers and crew from Douglas Frantz, "Gaps in Sea Laws Shield Pollution by Cruise Lines," *New York Times*, 3 January 1999; S. H. Smith, "Cruise Ships: A Serious Threat to Coral Reefs and Associated Organisms," *Ocean and Shoreline Management*, vol. 11 (1998), pp. 231–48; Cayman Islands from Pattullo, op. cit. note 27, p. 110.

47. Ill-equipped sites from Paul F. J. Eagles, *International Trends in Park Tourism and Ecotourism* (Waterloo, ON, Canada: University of Waterloo, Department of Recreation and Leisure Studies, 31 August 1999), pp. 28–32; "Cambodia's Tourist Boom Seen Threatening Angkor," *Reuters*, 21 June 2000; trash from "Tourists Soil Malaysia Rain Forest," *Associated Press*, 23 March 1999.

48. Jean M. Blane and Reiner Jaakson, "The Impact of Ecotourism Boats on the St Lawrence Beluga Whales," *Environmental Conservation*, autumn 1994, pp. 267–69; safari from Dilys Roe, Nigel Leader-Williams, and Barry Dalal-Clayton, *Take Only Photographs, Leave Only Footprints: The Environmental Impacts of Wildlife Tourism* (London: IIED, October 1997), p. 44; ivory from Danna Harman, "Former Stray Dogs Join Fight to Save Africa's Elephants," *Christian Science Monitor*, 2 July 2001.

49. B. R. Tershy et al., "A Survey of Ecotourism on Islands in Northwestern México," *Environmental Conservation*, vol. 26, no. 3 (1999), pp. 212–17; 66 percent from Samantha Newport, "Oil Spill Highlights Hazards of Isles' Growth," *Washington Post*, 27 January 2001; doubling from Larry Rohter, "Isles Rich in Species Are Origin of Much Tension," *New York Times*, 27 January 2001.

50. UNEP, op. cit. note 36, p. 3; Sanjay Nepal, "Tourism in Protected Areas: The Nepalese Himalaya," *Annals of Tourism Research*, vol. 27, no. 3 (2000), pp. 661–81; fuelwood loss from Castilho and Herrscher, op. cit. note 40; Nicodemus Odhiambo, "Millennium Celebrants Swarm to Mt. Kilimanjaro," *Environment News Service*, 20 September 1999.

51. Dive numbers from International Coral Reef Initiative and UNEP, "Tourism and Coral Reefs," information sheet (Nairobi: UNEP, undated); Julie P. Hawkins and Callum M. Roberts, "The Growth of Coastal Tourism in the Red Sea: Present and Future Effects on Coral Reefs," *Ambio*, December 1994, p. 506; Sakanan Plathong, Graeme J. Inglis, and Michael E. Huber, "Effects of Self-Guided Snorkeling Trails on Corals in a Tropical Marine

Park," *Conservation Biology*, December 2000, pp. 1821–30; Bonaire from Julie P. Hawkins et al., "Effects of Recreational Scuba Diving on Caribbean Coral and Fish Communities," *Conservation Biology*, August 1999, pp. 888–97; looting from Pattullo, op. cit. note 27, p. 110.

52. David Viner and Maureen Agnew, "Climate Change and Its Impacts on Tourism," report prepared for WWF–UK (Norwich, U.K.: University of East Anglia's Climatic Research Unit, July 1999); James J. McCarthy et al., *Climate Change 2001: Impacts, Adaptation, and Vulnerability*, Contribution of Working Group II to the Third Assessment Report of the IPCC (Cambridge, U.K.: Cambridge University Press, 2001), pp. 843–75; 85 percent derived from World Bank, op. cit. note 20, and from WTO, op. cit. note 7, p. 90; Black Forest and Adriatic from Kandelaars, op. cit. note 31, p. 2; developing world from Desthuis-Francis, op. cit. note 19, p. 17; Shawn W. Crispin and G. Pierre Goad, "Bangkok Renaissance," *Far Eastern Economic Review*, 30 September 1999, p. 62.

53. International Ecotourism Society, "Ecotourism Statistical Fact Sheet," information sheet (Burlington, VT: 2000); United Nations Economic and Social Council, "Resolution 1998/40–Declaring the Year 2002 as the International Year of Ecotourism," agreed to at the 46th Plenary Meeting, New York, 30 July 1998.

54. Honey, op. cit. note 5, p. 5; Cater, op. cit. note 26, p. 88; Peru from Laurent Belsie, "Treading Lightly," *Christian Science Monitor*, 1 February 2001.

55. International Ecotourism Society cited in Mike Tidwell, "No Glaciers in Glacier National Park?" *Washington Post*, 9 September 2001; 7 percent from WTO, op. cit. note 7; 1992 study from Fern Fillion et al., cited in International Ecotourism Society, op. cit. note 53.

56. Pamela A. White, "North American Ecotourists: Market Profile and Trip Characteristics," *Journal of Travel Research*, spring 1996; WTO, "WTO Picks Hot Tourism Trends for 21st Century," press release (Madrid: 4 June 1998).

57. Increases from Gerard S. Dharmaratne et al., "Tourism Potentials for Financing Protected Areas," *Annals of Tourism Research*, vol. 27, no. 3 (2000), pp. 590–610; South Africa from Eagles, op. cit. note 47, pp. 18–19; Central America from Elizabeth Boo, *Ecotourism: The Potentials and Pitfalls* (Washington DC: World Wildlife Fund, 1990), p. 47.

58. Financing problems from Eagles, op. cit. note 47, p. 10–18, and from Dharmaratne et al., op. cit. note 57, pp. 606–07. Box 5–1 based on the following: Bonaire from John A. Dixon and Tom van't Hof, "Conservation Pays Big Dividends in Caribbean," *Forum for Applied Research and Public Policy*, spring 1997, p. 46; Costa Rica from McCormick, op. cit. note 33; Komodo and Galapagos from Matthew J. Walpole, Harold J. Goodwin, and Kari G. R. Ward, "Pricing Policy for Tourism in Protected Areas: Lessons from Komodo National Park, Indonesia," *Conservation Biology*, February 2001, pp. 223, 219; corruption from Karen Archabald and Lisa Naughton-Treves, "Tourism Revenue-Sharing Around National Parks in Western Uganda: Early Efforts to Identify and Reward Local Communities," *Environmental Conservation*, vol. 28, no. 2 (2001), pp. 135–49; Mexico from Tershy et al., op. cit. note 49, pp. 212–17; U.S. survey from Castilho and Herrscher, op. cit. note 40; gorillas from Ian Fisher, "Victims of War: The Jungle Gorillas and Tourism," *New York Times*, 31 March 1999; Central African Republic from David S. Wilkie and Julia F. Carpenter, "Can Nature Tourism Help Finance Protected Areas in the Congo Basin?" *Oryx*, vol. 33, no. 4 (1999), pp. 332–38.

59. "Freelance Conservationists," *The Economist*, 23 August 2001, p. 62; Costa Rica and Belize from Boo, op. cit. note 57, p. xvii; Jeff Langholz, "Economics, Objectives, and Success of Private Nature Reserves in Sub-Saharan Africa and Latin America," *Conservation Biology*, vol. 10, no. 1 (1996), pp. 271–80.

60. Mark B. Orams, "Towards a More Desirable Form of Ecotourism," *Tourism Management*, vol. 16, no. 1 (1995), p. 3; Marilyn Bauer, "Eco-resort Owner Fights to Save the Ti Ti

Monkey," *Environmental News Network*, 21 July 2000.

61. Pleumarom, op. cit. note 28; Ed Stoddard, "Interview—S. Africa Game Parks to Woo Private Sector," *Reuters*, 25 November 1999; John Pomfret, "Privatizing China's Parks," *Washington Post*, 5 July 2001.

62. Cater, op. cit. note 26, pp. 86–88; Honey, op. cit. note 5, p. 25; Geoffrey Wall, "Is Ecotourism Sustainable?" *Environmental Management*, vol. 2, no. 4 (1997), pp. 9–12.

63. David B. Weaver, "Magnitude of Ecotourism in Costa Rica and Kenya," *Annals of Tourism Research*, vol. 26, no. 4 (1999), pp. 809–11; "Ecotourism: 'Hordes Of Visitors' Put Costa Rica At Risk," *UN Wire*, 10 May 1999; Bas Amelung et al., "Tourism in Motion: Is the Sky the Limit?" in Pim Martens and Jan Rotmans, eds., *Transitions in a Globalising World* (Maastricht, Netherlands: International Centre for Integrative Studies, forthcoming), p. 95; 700,000 from Costa Rica Tourist Board, "Resume 2000," at <www.tourism-costarica.com>, viewed 13 September 2001.

64. Keith W. Sproule, "Community-Based Ecotourism Development: Identifying Partners in the Process," in *The Ecotourism Equation: Measuring the Impacts*, Bulletin Series No. 99 (New Haven, CT: Yale University School of Forestry and Environmental Studies, 1996), pp. 233–50; revenue sharing from Archabald and Naughton-Treves, op. cit. note 58; Huaorani from Sylvie Blangy, "Ecotourism Without Tears," *UNESCO Courier*, July–August 1999, p. 32.

65. Wunder, op. cit. note 33, pp. 17–19; Archabald and Naughton-Treves, op. cit. note 58, pp. 144–45; quote from Mountain Agenda, "Mountains of the World: Tourism and Sustainable Mountain Development, Part 1," prepared for the Seventh Session of the UNCSD, New York, 19–30 April 1999; CAMPFIRE from Castilho and Herrscher, op. cit. note 40.

66. Resentment from Cater, op. cit. note 26, p. 88; conflict from Marie Jose Fortin and Christiane Gagnon, "An Assessment of Social Impacts of National Parks on Communities in Quebec, Canada," *Environmental Conservation*, vol. 26, no. 3 (1999), p. 201; Galapagos from Castilho and Herrscher, op. cit. note 40.

67. Partnerships from Sproule, op. cit. note 64, pp. 233–50; 200 guides and 92 percent from Beth Trafk, RARE Center for Tropical Conservation, Arlington, VA, discussion with Uta Saoshiro, Worldwatch Institute, 21 September 2001; The Conference Board, "Business Enterprises for Sustainable Travel's First BEST Practices Highlights Tour Operator's Model Philanthropic Program: Lindblad Guests Have Contributed More than $500,000 to Galapagos Conservation Fund," press release (New York: 12 June 2000).

68. Castilho and Herrscher, op. cit. note 40; 1980 numbers from Nepal, op. cit. note 50, p. 669; 63,000 from WWF Nepal, "Conservation and Sustainable Development through Tourism in Nepal," *Ecocircular* (WWF Nepal Program newsletter), March–April 2001, p. 8.

69. Honey, op. cit. note 5, p. 17; U.S. Agency for International Development, *Win-Win Approaches to Development and the Environment: Ecotourism and Biodiversity Conservation* (Washington, DC: Center for Development Information and Evaluation, July 1996); Shekhar Singh and Claudio Volonte, *Biodiversity Program Study* (Washington, DC: Global Environment Facility Monitoring and Evaluation Unit, 11 April 2001), pp. 2–3.

70. WTO, "International Year of Ecotourism (IYE) 2002," concept paper (Madrid: undated); Anita Pleumarom, "Do We Need the International Year of Ecotourism?" (Bangkok: Tourism Investigation & Monitoring Team, November 2000).

71. Wall, op. cit. note 62, p. 12; Boo, op. cit. note 57.

72. Definition from WTO, "Concepts & Definitions," at <www.world-tourism.org/frameset/frame_sustainable.html>; "Malé Declaration on Tourism & Sustainable Development (1997)," at <www.eco-tour.org/info/w_10194_en.html>; "Berlin Declaration: Biological Diversity and Sustainable Tourism," at <www.eco-tour.org/info/w_10016_en.html>; "CSD-7 to Focus on Tourism, Oceans, and SIDS," *CSD Update*, August 1998; WTO, "Tourism Sector Takes Steps to Ensure Future Growth: Global Code of Ethics Adopted at WTO Summit," press release (Santiago: 1 October 1999); WTTC, WTO, and Earth Council, *Agenda 21 for the Travel & Tourism Industry* (London: 1995).

73. Deloitte & Touche, IIED, and ODI, op. cit. note 20; balance from France, op. cit. note 33, p. 23; reconciling perspectives from Stewart Moore and Bill Carter, "Ecotourism in the 21st Century," *Tourism Management*, April 1993, p. 127.

74. U.S. survey reported in International Ecotourism Society, "U.S. Ecotourism Fact Sheet," information sheet (Burlington, VT: 1999); U.K. survey from MORI, cited in Green Globe, "Green Globe: Securing the Future for Travel and Tourism," information packet (London: undated).

75. Honey, op. cit. note 5, p. 20; Wolf Michael Iwand, "The Ecological Programme of the TUI," brochure (Hannover, Germany: Touristic Union International, undated).

76. "Lessening the Environmental Impacts of Leisure Facilities," *Green Hotelier*, April 1999, pp. 18–24; Green Hotels Association, "Ways Hotels are Helping to Save Our Planet," <www.greenhotels.com/grnideas.htm>, viewed 9 December 1998; water savings from Jennifer Bogo, "Breathing Easy in America's First Environmentally-Smart Hotel," *E Magazine*, January/February 2000, p. 47; value from Perdue, op. cit. note 41.

77. International Hotels Environment Initiative (IHEI), <www.ihei.org>, viewed 10 October 2001; 111 countries from United Nations, "Governments, Tour Industry Seek Plan at United Nations to Cut Environmental and Social Impacts," press release (New York: February 1999). Table 5–2 based on the following: IHEI, "Proving Green Hotel Business is Good Business," information brochure (London: undated); idem, information sheet (London: undated); Bogo, op. cit. note 76, pp. 46–47; Paula Diperna, "Tourism is Only As Sustainable As the Tourism Industry Allows," *Earth Times*, 5 April 1999; Rebecca Hawkins, "Environmental Reviewing for the Hotel Sector: the Experience of Inter-Continental Hotels and Resorts," Tourism Focus No. 5, *UNEP Industry and Environment*, April–June 1996.

78. Recycling from Mary B. Uebersax, "Indecent Proposal: Cruise Ship Pollution in the Caribbean," August 1996, at <www.planeta.com/planeta/96/0896cruise.htm>, viewed 26 October 2001; newer vessels from "Cruise Line Cleans Up Its Wake," *Environmental News Network*, 6 August 2001, and from "Cruise Companies Aim for Cleaner Waste," *Associated Press*, 31 July 2000; International Council of Cruise Lines, "ICCL Industry Standard E-01-01: Cruise Industry Waste Management Practices and Procedures," adopted 11 June 2001, at <www.iccl.org/policies/environmental standards.pdf>.

79. Guide programs from Thomas B. Lawrence and Deborah Wickins, "Managing Legitimacy in Ecotourism," *Tourism Management*, vol. 18, no. 5 (1997), p. 313; UNEP, "UNEP, WTO and Tour Operators Join Forces to Promote Sustainable Tourism Development," press release (Nairobi: 19 April 1999); members from <www.toinitiative.org/initiative.htm>, viewed 17 September 2001.

80. Australia from R. C. Buckley and G. F. Araujo, "Environmental Management Performance in Tourism Accommodation," *Annals of Tourism Research*, vol. 24, no. 2 (1997), pp. 465–70; UNEP, op. cit. note 36, pp. 5–6, 10; monitoring from Cater, op. cit. note 26, p. 90.

81. UNEP, *Environmental Codes of Conduct for Tourism* (Nairobi: 1995); International Association of Antarctic Tour Operators, at <www.iaato.org>, viewed 10 October 2001; "Antarctic Tourism Tests Fragile Ecosystem,"

Reuters, 23 February 1999.

82. UNEP, *Ecolabels in the Tourism Industry* (Nairobi: October 1998); Tanja Mihali, "Ecological Labelling in Tourism," in Lino Briguglio et al., eds., *Sustainable Tourism in Island & Small States: Issues and Policies* (London: Pinter, 1996), pp. 197–205; Ron Mader, "Eco-yardsticks for Hotels—Whose to Use?" *EcoAmericas*, December 1999, p. 8. Table 5–3 based on the following: Green Globe 21 from United Nations, op. cit. note 77, from Green Globe, op. cit. note 74, and from Synergy, "Tourism Certification: An Analysis of Green Globe 21 and Other Tourism Certification Programs," report prepared for WWF (Gland, Switzerland: August 2000); Ecotel from <www.ecotels.org>, viewed 10 September 2001; Blue Flag from <www.blueflag.org>, and from Graham Ashworth, "The Blue Flag Campaign," *Naturopa*, no. 88, (1998), p. 21; Certification for Sustainable Tourism from <www.turismo-sostenible.co.cr>, viewed 10 October 2001, from Diane Jukofsky, "Costa Rica Rates Hotels for Eco-Friendliness," *Environment News Service*, 14 December 1999, and from Beatrice Blake, "Comparing the ICT's Certification of Sustainable Tourism and The New Key to Costa Rica's Sustainable Tourism Rating," August 2001, at <www.planeta.com/planeta/01/0104costa.html>, viewed 7 September 2001; Rainforest Alliance, "About SmartVoyager," at <www.rainforest-alliance.org/programs/sv/index.html>, viewed 21 August 2001; Jorge Peraza-Breedy, Sustainable Tourism Program, Rainforest Alliance, San José, Costa Rica, e-mail to author, 21 August 2001; Green Leaf Foundation, "Profile: Green Leaf Program," informational brochure (Bangkok: undated); number of hotels with Green Leaf from UNEP, Regional Office for Asia and the Pacific, "Thailand's Green Tourism Initiative Applauded," press release (Bangkok: 2 October 2000).

83. Synergy, op. cit. note 82; Anne Becher and Beatrice Blake, "Reflections on 'Green Ratings,'" *La Planeta Platica*, August 1998.

84. Erlet Cater, "Ecotourism in the Third World—Problems and Prospects for Sustainability," in Erlet Cater and Gwen Lowman, *Ecotourism: A Sustainable Option?* (Chichester, U.K.: J. Wiley & Sons, 1994), pp. 69–86; Honey, op. cit. note 5, p. 87; Alan Flook, "The Changing Structure of International Trade in Tourism Services, The Tour Operators Perspective," presentation at WTO Symposium on Tourism Services, Geneva, 22–23 February 2001, p. 7.

85. Honey, op. cit. note 5, p. 32; Pleumarom, op. cit. note 28; UNED–UK, op. cit. note 25, p. 5.

86. UNCTAD, op. cit. note 18, pp. 16–17; Barbara Jones and Tanya Tear, "Australia's National Ecotourism Strategy," Tourism Focus No. 1, *UNEP Industry and Environment*, January–March 1995.

87. Trish Nicholson, *Culture, Tourism and Local Strategies Towards Development: Case Studies in the Philippines and Vietnam*, ESCOR Research Report #R6578 (London: UKDFID, 1997).

88. Sproule, op. cit. note 64; Brian Wheeller, "Tourism's Troubled Times: Responsible Tourism is Not the Answer," in France, op. cit. note 33, pp. 63–64; Wunder, op. cit. note 33, pp. 11–17; Cater, op. cit. note 84; Deloitte & Touche, IIED, and ODI, op. cit. note 20; WWF Nepal, op. cit. note 68.

89. Licensing and training from Sproule, op. cit. note 64; Walpole and Goodwin, op. cit. note 24, p. 573; UNCTAD, op. cit. note 18; Honey, op. cit. note 5, p. 83; Deloitte & Touche, IIED, and ODI, op. cit. note 20; Sustainable Travel & Tourism, "Nepal Bans Child Labour in Tourism," *News*, June 2001.

90. Diverse needs from E. Cater and B. Goodall, "Must Tourism Destroy Its Resource Base?" in France, op. cit. note 33, p. 86; oversight from Castilho and Herrscher, op. cit. note 40; Julio Batle, "Rethinking Tourism in the Balearic Islands," *Annals of Tourism Research*, vol. 27, no. 2 (2000), pp. 524–26; countries with laws from UNEP, op. cit. note 36, p. 7;

Nicole Winfield, "Cuba Seeks Right Mix of Ecotourism," at <msnbc.com>, 20 November 2000.

91. Saren Starbridge and Peter Bramwell, "Charging for Rhinos: Making Conservation Pay," *Living Planet*, spring 2001, pp. 64–69; Wouter Schalken, "Where are the Wild Ones? The Involvement of Indigenous Communities in Tourism in Namibia," *Cultural Survival Quarterly*, summer 1999, pp. 40–42; Namibia Community Based Tourism Association, at <www.nacobta.com.na>, viewed 19 September 2001.

92. John Roach, "Peru Puts Limits on Inca Trail Foot Traffic," *Environmental News Network*, 18 May 2000; Kingdom of Bhutan, at <www.kingdomofbhutan.com>, viewed 11 July 2001; Mridula Chettri, "High Altitude Dilemma," *Down to Earth*, 30 September 2000, p. 30; Ecuador from Castilho and Herrscher, op. cit. note 40.

93. Sweeting, Bruner, and Rosenfeld, op. cit. note 44, pp. 64–69; France from UNEP, op. cit. note 36, p. 4.

94. "Spain's Balearics Approve Tourist Ecotax," *Reuters*, 12 April 2001; Chris Brown, "Spain Blames Strikes, Eco-Tax for Tourism Slowdown," *Reuters*, 1 August 2001; Seychelles from UNEP, op. cit. note 36, p. 4, and from Seychelles New Adventures, "Passport, Visa, and Customs," at <www.sey.net/trv_pp.htm>, viewed 10 October 2001; cruise tax and tax holidays from Cater, op. cit. note 84.

95. Ashley, Boyd, and Goodwin, op. cit. note 26; Asian Development Bank, "Helping Cook Islands Manage Waste Will Reduce Health Risks, Protect Environment," press release (Manila: 17 July 2001).

96. WTO, "WTO and the Environment," press release (Madrid: 27 April 1995); International Maritime Organization, at <www.imo.org>; concern from Pleumarom, op. cit. note 28.

97. Maria Kousis, "Tourism and the Environment: A Social Movements Perspective," *Annals of Tourism Research*, vol. 27, no. 2 (2000), pp. 468–89; Chelaton, op. cit. note 1.

98. Wendy Patterson, "Mexican Government Temporarily Revokes License Granted to Build Along Coastal Area," *International Environment Reporter*, 25 April 2001, pp. 322–23; 40 species from "Corporate Corner," *EarthNet News*, 1 July 1999; Tourism Concern, op. cit. note 34.

99. United Nations, "UN Talks With Tourism Industry Spur Plans to Cut Negative Impacts," press release (New York: 3 May 1999); Tourism Concern, "Child Sex Tourism," at <www.tourismconcern.org.uk/useful%20stuff/frame.htm>, viewed 20 September 2001; WTTC from Cynthia Guttman, "Towards an Ethics of Tourism," *UNESCO Courier*, July–August 1999, p. 56.

100. International Ecotourism Society, "'Your Travel Choice Makes a Difference' Campaign," press release (Burlington, VT: 29 February 2000); Conservation International's Ecotravel Center, at <www.ecotour.org>.

101. International Ecotourism Society, "Cultural Impacts," at <www.ecotourism.org/travelchoice/cultural.html>, viewed 20 September 2001.

Chapter 6. Rethinking Population, Improving Lives

1. Robert Engelman, *Plan and Conserve: A Source Book on Linking Population and Environmental Services in Communities* (Washington, DC: Population Action International (PAI), 1998), pp. 64–66; nearly two out of every five women from Alan Guttmacher Institute, *Sharing Responsibility: Women, Society, and Abortion Worldwide* (New York: 1999), p. 42.

2. Population would have topped 8 billion from Patrick Heuveline, "The Global Impact of Mortality and Fertility Transitions, 1950–2000," *Population and Development Review*, December 1999, pp. 681–702; data on fertility and population change from United Nations,

World Population Prospects: The 2000 Revision (New York: 2001).

3. United Nations, op. cit. note 2.

4. United Nations, *Report of the International Conference on Population and Development, Cairo, 5–13 September 1994*, Programme of Action of the International Conference on Population and Development, Annex to the Report (New York: 18 October 1994).

5. Perdita Huston, *Families As We Are: Conversations from Around the World* (New York: The Feminist Press at the University of New York, 2001), p. 334.

6. For the conference documents described in Box 6–1, see <www.un.org/esa/sustdev/agenda21text.htm> and <www.iisd.ca/linkages/Cairo/program/p00000.html>.

7. John Cleland, "Equity, Security and Fertility: A Reaction to Thomas," *Population Studies,* July 1993, p. 351.

8. Figure 6–1 and historical estimates from Robert Engelman, PAI, based on various written works by historians and demographers; for historical estimates, see also see U.S. Bureau of Census, "Historical Estimates of World Population," <www.census.gov/ipc/www/worldhis.html>; United Nations, op. cit. note 2; New York City population from U.S. Census Bureau 2000 census, at <www.ci.nyc.ny.us/html/dcp/html/pop2000.html>, viewed 18 October 2001.

9. Peter G. Peterson, "Gray Dawn: The Global Aging Crisis," *Foreign Affairs*, January/February 1999, pp. 42–55; Phillip J. Longman, "The Global Aging Crisis," *U.S. News & World Report*, 1 March 1999, pp. 30–39.

10. United Nations, op. cit. note 2.

11. International Organization for Migration (IOM), *World Migration Report* (New York: 2000), p. 3; global migration data from United Nations, *International Migration and Development: The Concise Report* (New York: 1997); National Intelligence Council, *Growing Global Migration and Its Implications for the United States* (Washington, DC: 2001). Box 6–2 based on the following: Beth Gardiner, "Trucker Found Guilty in the Suffocation Deaths of 58 Immigrants," *Associated Press*, 5 April 2001; "U.S. Promises Action after Mexican Migrants' Deaths," *The Arizona Republic*, 24 May 2001; James Sterngold, "Devastating Picture of Immigrants Dead in the Arizona Desert," *New York Times*, 24 May 2001; Thomas Homer Dixon, *Environment, Scarcity, and Violence* (Princeton, NJ: Princeton University Press, 1999), p. 110; IOM, op. cit. this note; "Thousands of Afghan Refugees Pour into Pakistan," *Environmental News Service*, 29 September 2001; Roger Cohen, "Europe's Love-Hate Affair with Foreigners," *New York Times*, 24 December 2000.

12. United Nations, op. cit. note 2.

13. Ibid.; estimate of developing-country contraceptive prevalence in the late 1960s from J. Khanna, P. F. A. Van Look and P. D. Griffin, *Reproductive Health: Key to a Brighter Future, World Health Organization Biennial Report 1990–1991* (Geneva: World Health Organization (WHO), 1992), pp. 5–6; modern contraceptive prevalence from Population Reference Bureau, *2001 World Population Data Sheet*, wall chart (Washington, DC: June 2001).

14. Estimate for unmet need from John A. Ross and William L. Winfrey, "Unmet Need in the Developing World and the Former USSR: An Updated Estimate," unpublished manuscript, received 1 November 2001; 350 million from U.N. Population Fund (UNFPA), *The State of World Population 1999* (New York: 1999), p. 2.

15. Nancy E. Riley, "Gender, Power, and Population Change," *Population Bulletin*, May 1997; effects of education in high-fertility countries from United Nations, Population Division, *Women's Education and Fertility Behaviour: Recent Evidence from the Demographic and Health Surveys* (New York: 1995), p. 29.

16. Barbara Mensch, Judith Bruce, and Margaret Greene, *The Unchartered Passage: Girls' Adolescence in the Developing World* (New York: Population Council, 1998), p. 29.

17. Figurc of 75 million from PAI, *Educating Girls: Gender Gaps and Gains* (Washington, DC: 1998); information on women's earnings from United Nations, *The World's Women 2000: Trends and Statistics* (New York: 2000), p. 132. Table 6–1 based on the following: education from United Nations, op. cit. this note, pp. 87, 91, and from UNFPA, *The State of World Population 2001* (New York: 2001), p. 41; women-headed households' vulnerability from International Fund for Agricultural Development, *Rural Poverty Report 2001* (New York: Oxford University Press, 2001), pp. 28–29; Allen Dupree and Wendell Primus, *Declining Share of Children Lived with Single Mothers in the Late 1990s: Substantial Differences by Race and Income* (Washington, DC: Center on Budget and Policy Priorities, 2001); women's earnings from United Nations, op. cit. this note; one third from UNFPA, op. cit. this note, p. 38; 500 largest corporations from United Nations, op. cit. this note, p. 130; International Monetary Fund from UNIFEM, *Progress of the World's Women 2000* (New York: United Nations Development Fund for Women, 2000), p. 32; International Women's Democracy Center, "Women's Political Participation," fact sheet, at <www.iwdc.org/factsheets.htm>, viewed 23 July 2001; Women's Environment and Development Organization (WEDO), "Fact Sheet 3: Women in Government, Get the Balance Right," at <www.wedo.org/fact_sheet_3.htm>, viewed 17 October 2001; civic freedom from David Dollar and Roberta Gatti, *Gender Inequality, Income, and Growth: Are Good Times Good for Women?* (Washington, DC: World Bank Development Research Group, 1999), pp. 5–6.

18. Usable water from Sandra Postel, *Last Oasis* (New York: W.W. Norton & Company, 1997), pp. 27–28; half of usable portion from Sandra L. Postel, Gretchen C. Daily, and Paul R. Ehrlich, "Human Appropriation of Renewable Fresh Water," *Science*, 9 February 1996. Table 6–2 based on the following: fresh water from Robert Engelman et al., *People in the Balance: Population and Natural Resources in the New Millennium* (Washington, DC: PAI, 2000); cropland from ibid.; forests from Theodore Panayotou, "The Population, Environment, and Development Nexus," in Robert Cassen et al., *Population and Development: Old Debates, New Conclusions* (New Brunswick, NJ: Transaction Publishers, 1994), pp. 172–73, from Engelman et al., op. cit. this note, pp. 12–13, and from Tom Gardner-Outlaw and Robert Engelman, *Forest Futures: Population, Consumption and Wood Resources* (Washington, DC: PAI, 1999); Richard P. Cincotta et al., "Human Population in the Biodiversity Hotspots," *Nature*, 27 April 2001, pp. 990–92.

19. Malin Falkenmark and Carl Widstrand, "Population and Water Resources: A Delicate Balance," *Population Bulletin* (Population Reference Bureau), November 1992.

20. Postel, op. cit. note 18, pp. 6, 129.

21. Water scarcity and stress projections from Engelman et al., op. cit. note 18; poor pay more for water from Patrick Webb and Maria Iskandarani, *Water Insecurity and the Poor: Issues and Research Needs*, ZEF-Discussion Papers on Development Policy No. 2 (Bonn: Center for Development Research, 1998), pp. 29–31; numbers without safe water and sanitation from WHO and UNICEF, *Global Water Supply and Sanitation Assessment 2000 Report* (New York: 2000), p. 8.

22. Number of deaths annually from Peter H. Gleick, ed., *Water in Crisis: A Guide to the World's Fresh Water Resources* (Oxford: Oxford University Press, 1992); Sandra L. Postel and Aaron T. Wolf, "Dehydrating Conflict," *Foreign Policy*, September/October 2001, p. 62.

23. United Nations, op. cit. note 2; 25 percent of emissions from G. Marland, T. A. Boden, and R. J. Andres, Carbon Dioxide Information Analysis Center, Oak Ridge National Laboratory, "Global, Regional, and National Annual CO_2 Emissions from Fossil-Fuel Burning, Cement Production, and Gas Flaring: 1751–1998 (revised July 2001)," at <cdiac.esd.ornl.gov/

ndps/ndp030.html>, viewed 13 August 2001; U.S. and Africa comparison based on U.S. Department of Energy, Energy Information Administration, *International Energy Outlook 1998, April 1998* (Washington, DC: 1998), with growth rates for regional and global carbon emissions 2020–50 in Robert T. Watson et al., eds., *Climate Change 1995: Impacts, Adaptations and Mitigation of Climate Change: Scientific-Technical Analyses: Contribution of Working Group II to the Second Assessment Report of the Intergovernmental Panel on Climate Change* (New York: Cambridge University Press, 1996).

24. Janet Abramovitz, *Unnatural Disasters*, Worldwatch Paper 158 (Washington, DC: Worldwatch Institute, October 2001); Eric Chivian, "Environment and Health: 7. Species Loss and Ecosystem Disruption—The Implications for Human Health," *Canadian Medical Association Journal*, 9 January 2001, p. 68; Jon Cohen, "The Hunt for the Origin of AIDS," *Atlantic Monthly*, October 2000, pp. 88–104; "AIDS Wars," *The Economist*, 16 September 2000, pp. 87–88.

25. Canlas quoted in Doris C. Dumlao, "Business Gov't to Adopt 'Population Management Plan,' Says NEDA," *Philippine Daily Inquirer*, 27 June 2001.

26. David E. Bloom and Jeffrey G. Williamson, "Demographic Transitions and Economic Miracles in Emerging Asia," *World Bank Economic Review*, September 1998, pp. 419–55.

27. Lester Brown, Gary Gardner, and Brian Halweil, *Beyond Malthus* (New York: W.W. Norton & Company, 1999); sub-Saharan Africa from ibid., pp. 91–92.

28. United Nations, Population Division, Department of Economic and Social Affairs, *World Urbanization Prospects, The 1999 Revision* (New York: 1999); Martin Brockerhoff and Ellen Brennan, *The Poverty of Cities in the Developing World*, Policy Research Division Working Paper No. 96 (New York: Population Council, 1997), p. 5.

29. Christian G. Mesquida and Neil I. Wiener, "Male Age Composition and the Severity of Conflicts," *Politics in the Life Sciences*, September 1999, pp. 181–89; for population and environmental scarcity connections to conflict generally, see Thomas Homer-Dixon and Valerie Percival, *Environmental Scarcity and Violent Conflict: Briefing Book* (Washington, DC: American Association for the Advancement of Science/University of Toronto, 1996).

30. The Programme of Action of the International Conference on Population and Development, <www.undp.org/popin/icpd/conference/offeng/poa.html>, viewed 23 July 2001.

31. Jodi Jacobson, Executive Director, Center for Health and Gender Equity, Takoma Park, MD, discussion with authors, 2 October 2001.

32. Jane Hughes and Anne P. McCauley, "Improving the Fit: Adolescents' Needs and Future Programs for Sexual and Reproductive Health in Developing Countries," *Studies in Family Planning*, June 1998, pp. 233–45; International NGO Youth Consultation on Population and Development, "Cairo Youth Declaration," 1994, <youth.unesco.or.kr/youth/english/resources>, viewed on 18 September 2001.

33. Research on the impacts of information and guidance from Douglas Kirby, *Emerging Answers: Research Findings on Programs to Reduce Teen Pregnancy* (Washington, DC: National Campaign to Prevent Teen Pregnancy, 2001), and from Hughes and McCauley, op. cit. note 32; UNFPA, "Supporting the Next Generation of Parents and Leaders," <www.unfpa.org/adolescents/index.htm>, viewed 12 October 2001.

34. Rodolfo A. Bulatao, *The Value of Family Planning Programs in Developing Countries* (Washington, DC: Rand, 1998), p. 24; Hantamalala Rafalimanana and Charles F. Westoff, "Potential Effects on Fertility and Child Health and Survival of Birth-spacing Preferences in Sub-Saharan Africa," *Studies in Family Planning*, June 2000, p. 99; Figure 6–2 based on

Nada Chaya et al., *A World of Difference: Sexual and Reproductive Health & Risks (PAI Report Card 2001)*, wall chart and report (Washington, DC: PAI, 2001).

35. Engelman, op. cit. note 1, pp. 22, 34–35, 42; impact of the pill on American women's education from Claudia Golden and Lawrence F. Katz, "On the Pill: Changing the Course of Women's Education," *The Milken Institute Review*, second quarter 2001, pp. 12–21.

36. PAI, "How Family Planning Protects the Health of Women and Children," fact sheet no. 2 in second series (Washington, DC: April 2001).

37. Role of family planning movement in fertility decline from John Bongaarts, W. Parker Mauldin, and James F. Phillips, "The Demographic Impact of Family Planning Programs," *Studies in Family Planning*, November/December 1990; Bulatao, op. cit. note 34, pp. 28–30.

38. UNFPA, *The State of World Population 2000* (New York: 2000), p. 23.

39. United Nations, op. cit. note 2.

40. Number of women using contraception from UNFPA, op. cit. 38, p. 11; one quarter and two thirds of men from ibid., p. 4.

41. Indonesian government declaration from "Government Faces Shortage of Contraceptives for the Poor," *Jakarta Post*, 17 July 2001; increased contraceptive spending needs from Thoraya Ahmed Obaid, Executive Director, UNFPA, opening address at Meeting the Reproductive Health Challenge: Securing Contraceptives, and Condoms for HIV/AIDS Prevention, Istanbul, 3 May 2001, p. 4.

42. United Nations, *Report on the Global HIV/AIDS Epidemic* (Geneva: UNAIDS, 2000); Rachel L. Swarns, "Study Says AIDS Is Now Chief Cause of Death in South Africa," *New York Times*, 17 October 2001.

43. Noeleen Heyzer, Executive Director, UNIFEM, "Women at the Epicentre of the HIV/AIDS Epidemics: The Challenges Ahead," presentation at panel during the UN Special Session on HIV/AIDS, New York, 27 June 2001.

44. Fertility decline from United Nations, op. cit. note 2.

45. United Nations, "Background Note on the Resource Requirements for Population Programmes in the Years 2000–2015," unofficial white paper, New York, 13 July 1994.

46. Commitments at Cairo from Programme of Action, op. cit. note 30; developing-country spending from UNFPA, *Financial Resource Flows for Population Activities in 1998* (New York: 1999), p. i; 40 percent from Shanti R. Conly and Shyami de Silva, *Paying Their Fair Share? Donor Countries and International Population Assistance* (Washington, DC: PAI, 1998), p. 4.

47. U.S. spending goal of $1.9 billion from Conly and de Silva, op. cit. note 46, p. P82; current spending from Public Policy and Strategic Initiatives Department, PAI, discussion with Robert Engelman, 15 October 2001; gag rule from Richard P. Cincotta and Barbara B. Crane, "The Mexico City Policy and U.S. Family Planning Assistance," *Science*, 19 October 2001, pp. 525–26; U.S. appropriations for HIV/ AIDS spending from USAID, at <www.usaid.gov/pop_health/aids/Funding/index.html>, viewed 22 September 2001.

48. Religious opposition to contraception from Oscar Harkavy, *Curbing Population Growth: An Insider's Perspective on the Population Movement* (New York: Plenum Press, 1995), pp. 93, 95, 163.

49. Farzaneh Foudi, "Iran's Approach to Family Planning," *Population Today*, July/August 1999, p. 4; United Nations, op. cit. note 2.

50. "Church Active in Care for Those with AIDS," *Catholic News Service*, 9 July 2001.

51. Marta Lamas, "Standing Fast in Mexico:

Protecting Women's Rights in a Hostile Climate," *NACLA Report on the Americas*, March/April 2001, p. 40; David M. Adamson et al., *How Americans View World Population Issues: A Survey of Public Opinion* (Santa Monica, CA: RAND, 2000), pp. 40, 41, 51, 52.

52. Mizanur Rahman, Julie DaVanzo, and Abdur Razzaque, "Do Better Family Planning Services Reduce Abortion in Bangladesh?" *Lancet*, 29 September 2001, pp. 1051–56.

53. Karen Hardee et al., *Post-Cairo Reproductive Health Policies and Programs: A Comparative Study of Eight Countries*, Policy Papers No. 2 (Washington: The Futures Group International, September 1998); Celia W. Dugger, "Relying on Hard and Soft Sells, India Pushes for Sterilization," *New York Times*, 22 June 2001; China from Sophia Woodman, "Draft Law Fails to Address Real Population Issues," *South China Morning Post*, 9 July 2001; India from Rami Chhabra, "Saying Goodbye to Targets," *People & the Planet*, vol. 6, no. 1 (1997), pp. 14–15, from Leela Visaria, Shireen Jejeebhoy, and Tom Merrick, "From Family Planning to Reproductive Health: Challenges Facing India," *International Family Planning Perspectives*, vol. 25 supplement (1999), pp. S44–49, and from Michael A. Koenig, Gillian H. C. Foo, and Ketan Joshi, "Quality of Care Within the Indian National Family Welfare Programme: A Review of Recent Evidence," *Studies in Family Planning*, March 2000, p. 13.

54. WHO, *Violence Against Women* (Geneva: 1996); Lori Heise, Mary Ellsberg, and Megan Gottemoeller, "Ending Violence Against Women," *Population Reports*, December 1999, p. 5. Box 6–3 from the following: UNICEF, "Domestic Violence Against Women and Girls," *Innocenti Digest*, May 2000, p. 6; Celia W. Dugger, "Modern Asia's Anomaly: The Girls Who Don't Get Born," *New York Times*, 6 May 2001; UNICEF, *Innocenti Digest*, May 2000, p. 3; WHO, "Female Genital Mutilation," fact sheet no. 241 (Geneva: June 2000); UNFPA, op. cit. note 38, p. 29; Molly Moore, "In Turkey, 'Honor Killing' Follows Families to Cities," *Washington Post*, 8 August 2001; Suzan Fraser, "Suicides of Women Rising in Traditional Southeast Turkey," *Washington Post*, 9 November 2000; UNFPA, op. cit. note 38, p. 38.

55. UNFPA, op. cit. note 38, p. 38; Barbara Mensch, Judith Bruce, and Margaret Greene, *The Unchartered Passage: Girls' Adolescence in the Developing World* (New York: Population Council, 1998), p. 46.

56. World Bank, *Engendering Development: Through Gender Equality in Rights, Resources, and Voice* (New York: Oxford University Press, 2001), pp. 152–54; one third share from UNFPA, op. cit. note 38, p. 38.

57. World Bank, op. cit. note 56; nations in sub-Saharan Africa from UNFPA, op. cit. note 38, p. 41; Lisa C. Smith and Lawrence Haddad, *Overcoming Child Malnutrition in Developing Countries: Past Achievements and Future Choices* (Washington, DC: International Food Policy Research Institute, February 2000), p. 44.

58. Brazil study from World Bank, op. cit. note 56, p. 148; Dollar and Gatti, op. cit. note 17, p. 21; relationship between female enrollment and income from UNFPA, op. cit. note 38, p. 40.

59. Inter-Parliamentary Union (IPU), "Women in National Parliaments," at <www.ipu.org/wmn-e/world.htm>, updated 12 October 2001; IPU, *Women in Parliaments 1945–1995: A World Statistical Survey* (Geneva: 1995); sectors of government from Socorro Reyes, "Getting the Balance Right: Strategies for Change Introduction," at <www.wedo.org/5050/introduction2.htm>, March 2001; Lamas, op. cit. note 51, p. 40.

60. WEDO, "Fact Sheet 2: Women Making a Difference," at <www.wedo.org/fact_sheet_2.htm>, viewed 18 July 2001; IPU, "Women in National Parliaments," op. cit. note 59.

61. Rachel Kyte, Senior Specialist, International Finance Corporation, Washington, DC, discussion with authors, 26 July 2001.

62. World Resources Institute, *World Resources:*

1994–95 (New York: Oxford University Press, 1994), p. 53; Agnes Quisumbing, Senior Research Fellow, International Food Policy Research Institute, Washington, DC, discussion with authors, 24 July 2001.

63. Bina Agarwal from International Fund for Agricultural Development, *Rural Poverty Report 2001* (New York: 2001), p. 86; Bina Agarwal, New Delhi, India, discussion with authors, 23 August 2001; World Bank, op. cit. note 56, p. 149.

64. Engelman, op. cit. note 1, pp. 19–21, 34.

65. World Wildlife Fund, *Disappearing Landscapes: The Population/Environment Connection* (Washington, DC: 2001).

66. Richard E. Benedick, *Human Population and Environmental Stresses in the Twenty-first Century*, Environmental Change & Security Project Report (Washington, DC: Woodrow Wilson Center, 2000).

67. Hardee et al., op. cit. note 53.

Chapter 7. Breaking the Link Between Resources and Repression

1. UNICEF quoted by Holger Jensen, "Spoils of War," *Nando Times*, 15 March 2000; Global Witness, *A Crude Awakening* (London: 1999), p. 4; United Nations Development Programme (UNDP), *Human Development Report 2001* (New York: Oxford University Press, 2001), Annex Tables 1, 4, 9.

2. Displaced population and food aid dependence from Blaine Harden, "Africa's Gems: Warfare's Best Friend," *New York Times*, 6 April 2000, and from Fatal Transactions Campaign, "Diamond, a Merciless Beauty," <www.niza.nl/uk/campaigns/diamonds>, viewed 5 July 2001.

3. Table 7–1 based on the following: Colombia from Thad Dunning and Leslie Wirpsa, "Oil Rigged," Resource Center of the Americas, February 2001, at Global Policy Forum, <www.globalpolicy.org/security/natres/oil/2001/0201colo.htm>, and from Project Underground, "Colombia: Oxy's Relationship with Military Turns Deadly," 30 June 2001, at CorpWatch, <www.corpwatch.org/news/2001/0148.html>; Sudan from Christian Aid, *Scorched Earth* (London: 2001), from Dan Connell, "Sudan: Recasting U.S. Policy," *Foreign Policy in Focus*, August 2001, from Amnesty International, "Oil in Sudan—Deteriorating Human Rights," 3 May 2000, from Elisabeth Sköns et al., "Military Expenditure and Arms Production," in Stockholm International Peace Research Institute (SIPRI), *SIPRI Yearbook 2001: Armaments, Disarmament and International Security* (New York: Oxford University Press, 2001), p. 278, from Human Rights Watch, "Sudan: Human Rights Developments," *Human Rights Watch World Report 2001* (New York: 2001), p. 5, and from Sudan Update, "Raising the Stakes: Oil and Conflict in Sudan," <www.sudanupdate.org/REPORTS/OIL/21oc.html>; Chad and Cameroon from "Hotspots!" *New Internationalist*, June 2001, pp. 22–23, from Norimitsu Onishi with Neela Banerjee, "Chad's Wait for Its Oil Riches May Be Long," *New York Times*, 16 May 2001, and from Abid Aslam and Jim Lobe, "Bush-Cheney Energy Plan Could Aggravate Ethnic Conflicts," Crisis Watch, *Foreign Policy in Focus*, <www.fpif.org/selfdetermination/crisiswatch/energy_body.html>, viewed 3 August 2001; Afghanistan from Bertil Lintner, "Taliban Turns to Drugs," *Far Eastern Economic Review*, 11 October 2001, pp. 26–27, from Tim Golden, "Afghan Ban on Growing of Opium Is Unraveling," *New York Times*, 22 October 2001, from Ahmed Rashid, *Taliban: Militant Islam, Oil and Fundamentalism in Central Asia* (New Haven, CT: Yale University Press, 2001), pp. 117–24, from Jane Perlez, "Taliban Continue Trade Through Closed but Porous Border," *New York Times*, 30 October 2001, from Lucian Kim, "Afghanistan's Emerald Heights," *Christian Science Monitor*, 25 July 2000, and from Michael Ross, "Natural Resources and Civil Conflict: Evidence from Case Studies," University of Michigan, Department of Political Science, 11 May 2001; Cambodia from ibid., and from Jamie Doward, "Mineral Riches Fuel War, Not the Poor," *The*

Observer, 18 June 2000. Number of conflicts active in 2000 from Arbeitsgemeinschaft Kriegsursachenforschung, "Das Kriegsgeschehen des Jahres 2000," press release (Hamburg, Germany: Institute for Political Science, University of Hamburg), December 2000; the one-quarter share of all conflicts having a resource dimension is the author's assessment based on existing literature.

4. Global intact forest area assessment from U.N. Environment Programme, *An Assessment of the Status of the World's Remaining Closed Forests* (Nairobi: 2001); forest loss in 1990s from "Forests: Deforestation Continuing Worldwide At High Rate, FAO Warns," *UN Wire*, 3 October 2001. The U.N. Food and Agriculture Organization found in its *State of the World's Forests 2001* assessment that the countries with the highest net loss of forest area between 1990 and 2000 were Argentina, Brazil, the Democratic Republic of Congo, Indonesia, Mexico, Myanmar, Nigeria, Sudan, Zambia, and Zimbabwe.

5. Michael T. Klare, *Resource Wars: The New Landscape of Global Conflict* (New York: Metropolitan Books, 2001), pp. 15, 20–21.

6. Switch from superpower patronage to resource exploitation from Mark Duffield, "Globalization, Transborder Trade, and War Economies," in Mats Berdal and David M. Malone, eds., *Greed and Grievance: Economic Agendas in Civil Wars* (Boulder, CO: Lynne Rienner Publishers, 2000), p. 73, and from Richard Dowden, "War, Money and Survival: Rounding Up," <www.onwar.org/warandmoney/index.html>; alternative revenue streams from Mary Kaldor, *New and Old Wars: Organized Violence in a Global Era* (Stanford, CA: Stanford University Press, 1999), pp. 102–03, and from David Keen, "Incentives and Disincentives for Violence," in Berdal and Malone, op. cit. this note, pp. 29–31.

7. Paul Collier, *Economic Causes of Civil Conflict and Their Implications for Policy* (Washington, DC: World Bank, 2000), pp. 3–4; Paul Collier, "Doing Well Out of War: An Economic Perspective," in Berdal and Malone, op. cit. note 6, pp. 93–97.

8. Simulated attacks and similar tactics were commonplace during the Liberian civil war of 1989–97; see Mats Berdal and David M. Malone, "Introduction," in Berdal and Malone, op. cit. note 6, p. 5.

9. Kaldor, op. cit. note 6, pp. 90, 98–100; Dowden, op. cit. note 6.

10. Keen, op. cit. note 6, pp. 22, 24, 27; David Keen, "The Economic Functions of Violence in Civil Wars," *Adelphi Paper 320* (Oxford: Oxford University Press for the International Institute for Strategic Studies, 1998); Kaldor, op. cit. note 6, pp. 110–11.

11. Ross, op. cit. note 3, p. 10; Kaldor, op. cit. note 6, pp. 110–11.

12. Indra de Soysa, "The Resource Curse: Are Civil Wars Driven by Rapacity or Paucity?" in Berdal and Malone, op. cit. note 6, pp. 120–21, 125–26.

13. William Reno, "Shadow States and the Political Economy of Civil Wars," in Berdal and Malone, op. cit. note 6, pp. 45–46, 56–57; de Soysa, op. cit. note 12; higher Mobutu wealth estimate from Reno, op. cit. this note, p. 46; lower estimate from Jimmy Burns, Mark Huband, and Michael Holman, "Mobutu Built a Fortune of $4bn from Looted Aid," *Financial Times*, 12 May 1997.

14. Reno, op. cit. note 13, pp. 47–53; Ian Smillie, Lansana Gberie, and Ralph Hazleton, *The Heart of the Matter: Sierra Leone, Diamonds and Human Security* (Ottawa, ON, Canada: Partnership Africa Canada, January 2000), p. 15.

15. Kaldor, op. cit. note 6, pp. 92–93; Alex de Waal, "Contemporary Warfare in Africa," *IDS Bulletin*, vol. 27, no. 3 (1996).

16. Project Underground, "Militarization & Minerals Tour," <www.moles.org/ProjectUnderground/mil/intro.html>, viewed 6 July

2001; Kim Richard Nossal, "Bulls to Bears: The Privatization of War in the 1990s," <www.onwar.org/warandmoney/index.html>; Small Arms Survey, *Small Arms Survey 2001* (New York: Oxford University Press, 2001), p. 109; Smillie, Gberie, and Hazleton, op. cit. note 14, p. 12.

17. Occidental from Project Underground, op. cit. note 3; Shell from Reno, op. cit. note 13, p. 52; Talisman from Christian Aid, op. cit. note 3; ExxonMobil from "Exxon 'Helped Torture in Indonesia,'" *BBC News Online*, 22 June 2001; Freeport-McMoRan from Abigail Abrash, "The Amungme, Kamoro & Freeport," *Cultural Survival Quarterly*, spring 2001, p. 40.

18. Ease of use of small arms and other attributes from Michael Renner, *Small Arms, Big Impact: The Next Challenge of Disarmament*, Worldwatch Paper 137 (Washington, DC: Worldwatch Institute, October 1997), pp. 10–12; statistics and estimates from Small Arms Survey, op. cit. note 16, pp. 7–8, 13–14, 59.

19. Renner, op. cit. note 18, pp. 33–34; Small Arms Survey, op. cit. note 16, pp. 107–08.

20. Box 7–1 based on the following: Karl Vick, "Vital Ore Funds Congo's War," *Washington Post*, 19 March 2001; Kristi Essick, "Guns, Money and Cell Phones," *The Standard: Intelligence for the Internet Economy*, 11 June 2001, <www.thestandard.com/article/0,1902,26784,pp.html>; Edward Marek, "Tantalum and War in the Congo," <www.yourdotcomforafrica.com/USPolicy040801.html>, 8 April 2001; Blaine Harden, "The Dirt in the New Machine," *New York Times Magazine*, 12 August 2001, pp. 35–39; United Nations, *Report of the Panel of Experts on the Illegal Exploitation of Natural Resources and Other Forms of Wealth of the Democratic Republic of the Congo* (New York: 12 April 2001), pp. 8, 11.

21. Duffield, op. cit. note 6, p. 84.

22. Christian Aid, op. cit. note 3; Connell, op. cit. note 3; Amnesty International, op. cit. note 3; Human Rights Watch, op. cit. note 3; "Hotspots!" op. cit. note 3; Leon P. Spencer, *Key Points Related to Sudan and Oil* (Washington, DC: The Washington Office on Africa, 18 May 2001).

23. De Beers and U.N. group estimates from United Nations, *Report of the Panel of Experts Appointed Pursuant to Security Council Resolution 1306 (2000), Paragraph 19, in Relation to Sierra Leone* (New York: 20 December 2000); higher estimates of share of conflict diamonds from Christine Gordon, "Rebels' Best Friend," *BBC Focus On Africa*, October–December 1999, cited in Smillie, Gberie, and Hazleton, op. cit. note 14, and from Fatal Transactions Campaign, op. cit. note 2.

24. Mamara quote in Barbara Crossette, "Singling Out Sierra Leone, U.N. Council Sets Gem Ban," *New York Times*, 6 July 2000.

25. Smillie, Gberie, and Hazleton, op. cit. note 14, pp. 8, 14; David Keen, "Going to War: How Rational Is It?" <www.onwar.org/warandmoney/index.html>; Reno, op. cit. note 13, p. 48; International Rescue Committee from Arms Trade Resource Center, "March Update," distributed by e-mail, 7 March 2000; UNDP, op. cit. note 1, Table 1.

26. Human Rights Watch, "Sierra Leone: Priorities for the International Community," June 2000, <www.globalpolicy.org/security/issues/diamond/hrw2.htm>; Smillie, Gberie, and Hazleton, op. cit. note 14, pp. 8, 14–15. Table 7–2 compiled from the following: Human Rights Watch, op. cit. this note; Smillie, Gberie, and Hazleton, op. cit. note 14; Africa Confidential, "Special Reports. Chronology of Sierra Leone: How Diamonds Fuelled the Conflict," <www.africa-confidential.com/special.htm>, viewed 9 September 2001; U.N. Security Council, "Tenth Report of the Secretary-General on the United Nations Mission in Sierra Leone," 25 June 2001; "Sierra Leone: Security Council Approves War Tribunal," *UN Wire*, 25 July 2001; "Sierra Leone: Rebels, Militias Subvert Gem Mining Ban, UN Says," *UN Wire*, 30 July 2001; Douglas Farah, "Rebels in Sierra Leone Mine Diamonds in Defiance of U.N.," *Wash-*

ington Post, 19 August 2001; "Sierra Leone: Disarmament May Be Hurt by Lack of Funds, U.N. Says," *UN Wire*, 23 August 2001.

27. Character of government forces from Keen, op. cit. note 25, and from Kaldor, op. cit. note 6, p. 94.

28. Keen, op. cit. note 25; Keen, op. cit. note 6, pp. 35–36; William Reno, "War and the Failure of Peacekeeping in Sierra Leone," in SIPRI, op. cit. note 3, p. 151.

29. United Nations, op. cit. note 23; Rachel Stohl, "U.N. Imposes Diamond Ban on Sierra Leone," *Weekly Defense Monitor*, 14 July 2000.

30. United Nations, op. cit. note 23; Smillie, Gberie, and Hazleton, op. cit. note 14, pp. 11, 47.

31. Global Witness, *Taylor-Made—The Pivotal Role of Liberia's Forests in Regional Conflict* (London: 2001); Global Witness, "The Role of Liberia's Logging Industry on National and Regional Insecurity," Briefing to the UN Security Council, January 2001, <www.oneworld.org/globalwitness/press/gwliberia.htm>; Global Witness, "Liberian Timber Profits Finance Regional Conflict," 7 May 2001, <www.globalpolicy.org/security/issues/liberia/2001/gwtimber.htm>; Greenpeace Spain, *Logs of War: The Relationship Between the Timber Sector, Arms Trafficking and the Destruction of the Forests in Liberia* (Madrid: 2001).

32. Greenpeace Spain, op. cit. note 31; rising importance of timber revenues from Global Witness, *Taylor-Made*, op. cit. note 31.

33. United Nations, op. cit. note 23; Small Arms Survey, op. cit. note 16, pp. 171–72.

34. Deaths and displacements from Taylor B. Seybolt, "Major Armed Conflicts," in SIPRI, op. cit. note 3, p. 26; foreign troops from "Peace Here Means War Elsewhere," *The Economist*, 23 June 2001, p. 44; Colette Braeckman, "Congo: A War Without Victors," *Le Monde Diplomatique*, April 2001.

35. United Nations, op. cit. note 20, pp. 41–42. Box 7–2 based on the following: ibid., pp. 10–12; "Miners' Rush for Coltan Threatens Rare Gorilla," *Environment News Service*, 13 April 2001; Harden, op. cit. note 20; UNESCO, "The World Heritage List," <www.unesco.org/whc/heritage.htm#debut>, and "World Heritage List in Danger," <www.unesco.org/whc/danglist.htm>, both viewed 11 August 2001; "One Minute to Midnight for Great Apes," *Ecologist*, July/August 2001, p. 15.

36. United Nations, op. cit. note 20, pp. 11, 14; Harden, op. cit. note 20, pp. 37–38; Musifiky Mwanalasi, "The View from Below," in Berdal and Malone, op. cit. note 6, p. 142.

37. United Nations, op. cit. note 20, pp. 3, 7, 14–19, 29–31; Mwanalasi, op. cit. note 36, pp. 140, 145; Vick, op. cit. note 20.

38. Chinese deal from United Nations, op. cit. note 20, pp. 29–36; other concessions from Vick, op. cit. note 20; Ridgepoint from Reno, op. cit. note 13, pp. 57–58; timber from Harden, op. cit. note 2.

39. United Nations, op. cit. note 20, pp. 37–39; "Sabena/Swissair Declares Embargo on Transport of Coltan," *Africa News*, 21 June 2001.

40. United Nations, op. cit. note 20, pp. 3, 41; Norimitsu Onishi, "Political Fever Wanes in Congo, but Patient Is Still Sick," *New York Times*, 11 April 2001; "Militia Clashes Threaten Congo Peace Process," *Reuters*, 18 July 2001.

41. Global Witness, *A Rough Trade: The Role of Companies and Governments in the Angolan Conflict* (London: 1998). MPLA is the acronym for the Popular Movement for the Liberation of Angola; UNITA stands for National Union for the Total Independence of Angola.

42. Virginia Gamba and Richard Cornwell, "Arms, Elites, and Resources in the Angolan Civil War," in Berdal and Malone, op. cit. note 6, pp. 165–67; diamond production from Smillie, Gberie, and Hazleton, op. cit. note 14; oil production from BP Amoco, *1999 BP Amoco*

Statistical Review of World Energy (London: Group Media & Publications, June 1999); allegations of MPLA-UNITA collusion reported by Gamba and Cornwell, op. cit. this note, by Mwanalasi, op. cit. note 36, and by Global Witness, op. cit. note 41.

43. Trends in UNITA diamond income from Global Witness, op. cit. note 41, from Harden, op. cit. note 2, and from Ross, op. cit. note 3; use of diamond income from United Nations, *Final Report of the UN Panel of Experts on Violations of Security Council Sanctions Against Unita* (New York: 10 March 2000).

44. United Nations, op. cit. note 43; Global Witness, op. cit. note 41.

45. De Beers decision to stop buying Angolan diamonds from United Nations, op. cit. note 43; Global Witness, op. cit. note 41; Duffield, op. cit. note 6, p. 84.

46. Lax controls and smuggling routes from United Nations, op. cit. note 43, and from Global Witness, op. cit. note 41; polishing in Israel and Ukraine from Gamba and Cornwell, op. cit. note 42, p. 166.

47. United Nations, op. cit. note 43.

48. Doward, op. cit. note 3; Global Witness, op. cit. note 1, pp. 4, 6–7.

49. Global Witness, op. cit. note 1, pp. 5–7, 11–12.

50. Ibid., pp. 7, 13–16.

51. "Aceh: Ecological War Zone," *Down to Earth*, November 2000; "Exxon's Aceh Plant Shutdowns to Affect Oil and Gas Delivery in Asia," press release, *Far Eastern Economic Review Online*, 21 March 2001; "Exxon 'Helped Torture in Indonesia,'" op. cit. note 17; "Activists Set Sights on ExxonMobil for 'Complicity of Silence,'" <www.corpwatch.org>, 7 June 2001; John McBeth, "Too Hot to Handle," *Far Eastern Economic Review Online*, 29 March 2001; Ross, op. cit. note 3, pp. 24–25; International Crisis Group, *Aceh: Can Autonomy Stem the Conflict?* ICG Asia Report No. 18 (Brussels: 27 June 2001), p. 5.

52. Human Rights Watch, "Indonesia: Why Aceh is Exploding," press backgrounder (New York: August 1999); "Aceh: Ecological War Zone," op. cit. note 51.

53. Human Rights Watch, op. cit. note 52; International Crisis Group, op. cit. note 51, p. 3; Ross, op. cit. note 3, pp. 23–26; Dini Djalal, "Silencing the Voices of Aceh," *Far Eastern Economic Review Online*, 5 July 2001; death toll from Seybolt, op. cit. note 34, p. 38.

54. McBeth, op. cit. note 51; "Exxon's Aceh Plant Shutdowns," op. cit. note 51; Wayne Arnold, "ExxonMobil, in Fear, Exits Indonesian Gas Fields," *New York Times*, 24 March 2001; "Violence Spirals in Troubled Aceh," *BBC News Online*, 18 March 2001; "Exxon Back on Stream in Indonesia," *BBC News Online*, 19 July 2001; new counterinsurgency operation from Sidney Jones, "For Indonesia, A Sea of Troubles" (op-ed), *New York Times*, 27 July 2001.

55. Robert Jereski, "Activist and Press Backgrounder on ExxonMobil Activities in North Aceh," International Forum for Aceh, 27 May 2001, viewed at East Timor Action Network, <www.etan.org>, 24 July 2001; "Mobil Oil and Human Rights Abuse in Aceh," *Down to Earth*, November 1998; "Exxon 'Helped Torture in Indonesia,'" op. cit. note 17.

56. Ross, op. cit. note 3, p. 27; "No Flags for Papua," *The Economist*, 12 October 2000; Human Rights Watch, *Violence and Political Impasse in Papua* (New York: July 2001), p. 2.

57. "Risky Business: The Grasberg Gold Mine," Project Underground Reports, <www.moles.org/index.htm>, viewed 9 July 2001; "The Strains on Indonesia," *The Economist*, 3 December 2000; "Provocation," *The Economist*, 30 November 2000; Abrash, op. cit. note 17, pp. 38–39; Michael Shari, "Freeport-McMoRan—A Pit of Trouble," *Business Week*, 31 July 2000.

58. Ross, op. cit. note 3, p. 28; Javanese immigration and disproportionate division of benefits from Human Rights Watch, op. cit. note 56, p. 19; *Rape and Other Human Rights Abuses by the Indonesian Military in Irian Jaya (West Papua), Indonesia* (Washington, DC: Robert F. Kennedy Memorial Center for Human Rights, May 1999); Abrash, op. cit. note 17, p. 40.

59. Human Rights Watch, op. cit. note 56, pp. 2–3, 10–11; Jim Lobe, "Indonesia's Hard Line Strengthens Secessionists in West Papua," *Foreign Policy in Focus*, 1 July 2001.

60. "Talking About a Devolution," *The Economist*, 4 January 2000; "Megawati Sorry for Rights Abuses," *BBC News Online*, 16 August 2001; autonomy bill from Jim Lobe, "Indonesia: Aceh Arrests Could Portend Increased Polarization, Violence," *Foreign Policy in Focus*, 1 July 2001.

61. Jeff Atkinson, "Defending the Victims of Mining," *Inside Indonesia*, <www.insideindonesia.org/edit47/mining.htm>, viewed 26 July 2001; Kathryn Robinson, "Revisiting Inco," *Inside Indonesia*, January–March 2001; Roger Moody, "Dirty Landlord," *Inside Indonesia*, January–March 2001. Box 7–3 based on the following: Klare, op. cit. note 5, pp. 203–07; Dan Murphy, "Behind Ethnic War, Indonesia's Old Migration Policy," *Christian Science Monitor*, 1 March 2001; Robin Broad, "The Political Economy of Natural Resources: Case Studies of the Indonesian and Philippine Forest Sectors," *The Journal of Developing Areas*, April 1995, pp. 322–26.

62. Michael Renner, *Fighting for Survival* (New York: W.W. Norton & Company, 1996), pp. 55–56; social disruptions from Ross, op. cit. note 3, p. 29.

63. Renner, op. cit. note 62; Klare, op. cit. note 5, pp. 196–98.

64. Klare, op. cit. note 5, pp. 196–98; June 2001 agreement from "Papua New Guinea: Security Council Members Back Peace Plan," *UN Wire*, 15 August 2001.

65. Renner, op. cit. note 62, pp. 57–58; Human Rights Watch, *The Price of Oil: Corporate Responsibility and Human Rights Violations in Nigeria's Oil Producing Communities* (New York: January 1999); Global Exchange and Essential Action, *Oil for Nothing: Multinational Corporations, Environmental Destruction, Death and Impunity in the Niger Delta* (San Francisco, CA, and Washington, DC: January 2000); Marina Ottaway, "Reluctant Missionaries," *Foreign Policy*, July/August 2001, p. 48.

66. Human Rights Watch, op. cit. note 65; Renner, op. cit. note 62, pp. 57–58.

67. Human Rights Watch, op. cit. note 65; Global Exchange and Essential Action, op. cit. note 65; Shell from Reno, op. cit. note 13, p. 52.

68. Human Rights Watch, op. cit. note 65; Human Rights Commission from Global Exchange and Essential Action, op. cit. note 65, from Chris Simpson, "Shell Overtures to Ogonis," *BBC News Online*, 25 July 2001, and from Barnaby Phillips, "No End to Saro-Wiwa's Struggle," *BBC News Online*, 15 January 2001.

69. Table 7–3 compiled from United Nations, Security Council Documents Full Search, <www.un.org/Docs/sc>, from United Nations, op. cit. note 23, and from United Nations, op. cit. note 20, pp. 41–45; France and China from Global Witness, *Taylor-Made*, op. cit. note 31, p. 6.

70. Human Rights Watch, "Neglected Arms Embargo on Sierra Leone Rebels," Briefing Paper, 15 May 2000, as posted on Global Policy Forum, <www.globalpolicy.org/security/issues/sierra/00-05sl4.htm>; United Nations, op. cit. note 23.

71. Alan Cowell, "New 'Labels' for Diamonds Sold by Sierra Leone," *New York Times*, 28 October 2000; diamond fingerprinting technology from Smillie, Gberie, and Hazleton, op. cit. note 14, pp. 63–64.

72. "Angola: Diamonds Worth $1 Million

Smuggled Daily, UN Says," *UN Wire*, 16 October 2001; Norimitsu Onishi, "Africa Diamond Hub Defies Smuggling Rules," *New York Times*, 2 January 2001; Andrew Parker, "Checks 'May Not Halt All Illicit Diamond Exports,'" *Financial Times*, 25 April 2001; Global Witness, op. cit. note 41; United Nations, op. cit. note 23; idem, op. cit. note 43; Judy Dempsey and Andrew Parker, "Belgium, UK in Drive to Halt War Gems," *Financial Times*, 26 June 2001; U.S. efforts from Ken Silverstein, "Diamonds of Death," *The Nation*, 23 April 2001, pp. 19–20, from Rachel Stohl, "Diamonds Are Forever," *Weekly Defense Monitor*, 13 October 2000, from Anna Franklin and Rachel Stohl, "Attempts Made to Control Conflict Diamonds," *Weekly Defense Monitor*, 23 August 2001, from Campaign to Eliminate Conflict Diamonds, "Questions and Answers About Conflict Diamonds and the 'Clean Diamonds Act,'" <www.phrusa.org/campaigns/sierra_leone/diam_q&a.html>, viewed 10 September 2001, and from Congressman Tony P. Hall, "Hall Joins Leaders of Key Committee in Introducing Compromise on Conflict Diamonds," press release (Washington, DC: 2 August 2001).

73. Silverstein, op. cit. note 72, p. 20; "Diamonds: EU to Discuss Conflict Gems; US Senators Propose Ban," *UN Wire*, 26 June 2001; worries of nongovernmental organizations from Franklin and Stohl, op. cit. note 72, and from Campaign to Eliminate Conflict Diamonds, "Governments and Industry: Stop Blood Diamonds Now!" 21 August 2001, <www.phrusa.org/campaigns/sierra_leone/jewel_release082101.html>.

74. Forest Stewardship Council from Gary Gereffi, Ronie Garcia-Johnson, and Erika Sasser, "The NGO-Industrial Complex," *Foreign Policy*, July/August 2001, pp. 60–61; United Nations, op. cit. note 20, pp. 41–45.

75. In the mid-1990s, for example, human rights and environmental organizations launched campaigns aimed at Shell (for its role in Nigeria) and at Amoco, Texaco, ARCO, and Petro-Canada (for their roles in Myanmar); Ottaway, op. cit. note 65, pp. 47–48.

76. Harden, op. cit. note 2; Nicole Gaouette, "Israel's Diamond Dealers Tremble," *Christian Science Monitor*, 1 June 2001; Smillie, Gberie, and Hazleton, op. cit. note 14, p. 9; electronics companies' reaction from Harden, op. cit. note 20, p. 38, and from Essick, op. cit. note 20.

77. Hilary French, "Socially Responsible Investing Surges," in Worldwatch Institute, *Vital Signs 2001* (New York: W.W. Norton & Company, 2001), pp. 114–15.

78. Small Arms Survey, op. cit. note 16, pp. 251–83; the Declaration of a Moratorium on the Importation, Exportation and Manufacture of Small Arms and Light Weapons in West Africa was adopted by the members of Economic Community of West African States, see United Nations, op. cit. note 23.

79. Rachel Stohl, "United States Weakens Outcome of UN Small Arms and Light Weapons Conference," *Arms Control Today*, September 2001.

80. Michael Renner, "U.N. Peacekeeping: An Uncertain Future," *Foreign Policy in Focus*, September 2000; Michael Renner, "Peacekeeping Expenditures Rebound," in Worldwatch Institute, op. cit. note 77, pp. 84–85.

Chapter 8. Reshaping Global Governance

1. Alessandra Stanley and David E. Sanger, "Genoa Summit Meeting: The Overview; Italian Protester is Killed by Police at Genoa Meeting," *New York Times*, 21 July 2001; Serge Schmemann, "Hijacked Jets Destroy Twin Towers and Hit Pentagon," *New York Times*, 12 September 2001; 4,500–5,000 deaths from "Dead and Missing," *New York Times*, 4 November 2001; "For Now, A Global Movement is Stymied," *Boston Globe*, 30 September 2001.

2. World Bank Group, "World Bank Group and IMF Will Not Hold Annual Meetings," press release (Washington, DC: 17 September 2001); Mobilization for Global Justice, "Mobilization for Global Justice Cancels its Call for

Street Demonstrations Against World Bank/IMF at End of September," press release (Washington, DC: 16 September 2001).

3. Inequality trends from U.N. Development Programme (UNDP), *Human Development Report 2001* (New York: Oxford University Press, 2001); poverty figures from World Bank, *World Development Report 2000/2001* (New York: Oxford University Press, September 2000), p. vi; Toepfer quote from "UN Environment Chief Urges World to Fight Root Causes of Civil Unrest That Can Lead to Terrorism," press release (Nairobi: U.N. Environment Programme (UNEP), 21 September 2001).

4. Paul Blustein, "Cause, Effect and the Wealth of Nations," *Washington Post*, 4 November 2001; Christian E. Weller, Robert E. Scott, and Adam S. Hersh, *The Unremarkable Record of Liberalized Trade*, Briefing Paper (Washington, DC: Economic Policy Institute, October 2001); Martin Khor, "Globalisation and the Crisis of Sustainable Development," paper presented to World Summit on Sustainable Development International Eminent Persons Meeting on Inter-linkages, United Nations University Centre, Tokyo, 3–4 September 2001.

5. Michael Grubb et al., *The Earth Summit Agreements: A Guide and Assessment* (London: Royal Institute of International Affairs, 1993); Maurice Strong, *Where on Earth Are We Going?* (New York: TEXERE LLC, 2000), pp. 189–239.

6. Length of Uruguay Round from Jeffrey S. Thomas and Michael A. Meyer, *The New Rules of Global Trade* (Scarborough, ON, Canada: Carswell Thomson Professional Publishing, 1997), p. 25; United Nations, *Agenda 21: the United Nations Programme of Action from Rio* (New York: U.N. Department of Public Information, undated).

7. Don Kirk, "Worldwide Outlook for Tourism Poor," *New York Times*, 27 September 2001; "Stocks Fall as New Data Reignites Fears on Economy," *Reuters*, 25 October 2001.

8. Current population from United Nations, *World Population Prospects: The 2000 Revision* (New York: 2001).

9. "WTO Chief Proposes World Environment Organization," *Environmental News Service*, 15 March 1999; Renato Ruggiero, Director-General, World Trade Organization (WTO), "Opening Remarks to the High Level Symposium on Trade and the Environment," 15 March 1999; "Lessons from Seattle" (editorial), *Washington Post*, 1 December 1999.

10. UNEP, "International Environmental Governance: Multilateral Environmental Agreements (MEAs)," paper prepared for the Open-Ended Intergovernmental Group of Ministers or their Representatives on International Environmental Governance, Bonn, Germany, 17 July 2001, pp. 3–7.

11. Ibid., p. 17; U.S. Department of State, Bureau of Oceans and International Environmental and Scientific Affairs, "Calendar of Events," at <www.state.gov/g/oes/cal>, viewed 18 August 2001.

12. UNEP, "International Environmental Governance, Report of the Executive Director," paper prepared for the Open-Ended Intergovernmental Group of Ministers or Their Representatives on International Environmental Governance, New York, 18 April 2001, pp. 17–19; Hilary F. French, "Learning from the Ozone Experience," in Lester R. Brown et al., *State of the World 1997* (New York: W.W. Norton & Company, 1998), pp. 158–59; "Sinking Kyoto Protocol with Sinks," *Bulletin of the World Rainforest Movement,* July 2001; Ashley T. Mattoon, "Bogging Down in the Sinks," *World Watch*, November/December 1998, pp. 28–36.

13. "Convention" is a legal term that describes what is known in more common parlance as a treaty. Table 8–1 based on the following: Convention on Biological Diversity (CBD), at <www.biodiv.org>; Monique Chiasson, National Reports Unit of CBD, e-mails to Jessica Dodson, Worldwatch Institute, 13–21 August 2001;

CBD News, January/March 2001; Center for International Earth Science Information Network , "The Convention on Biological Diversity," CIESIN Thematic Guides, at <www.ciesin.org/TG/PI/TREATY/bio.html>, viewed 24 July 2001; Global Environment Facility (GEF), *GEF Contributions to Agenda 21: The First Decade* (Washington, DC: June 2000); U.N. Framework Convention on Climate Change (UN FCCC), at <www.unfccc.de>; Jon Hanks et al., *Earth Negotiation Bulletin*, 30 July 2001; United Nations Treaty Collection, at <untreaty.un.org>; Convention to Combat Desertification, at <www.unccd.int>; *Down to Earth: Newsletter of the Convention to Combat Desertification*, December 2000, pp. 1–3; "Desertification: Parties to U.N. Convention to Open Meeting in Geneva," *UN Wire*, 1 October 2001; Convention on Straddling Fish Stocks, at <www.un.org/Depts/los/convention_agreements/convention_overview_fish_stocks.htm>; World Wildlife Fund, "Top Fishing Nations Drag Feet on UN Fish Stocks Agreement," press release (Washington, DC: 27 November 1997); "Straddling Stocks Agreement Important to Large Migratory Fish," *Dispatches* (newsletter of TRAFFIC), February 2000; Prior Informed Consent Convention, at <www.pic.int>; Persistent Organic Pollutants, at <www.chem.unep.ch/pops/default.html>; WWF's Global Toxic Chemicals Initiative, "Summary of Key Elements in the Global POPs Treaty," 14 December 2000, at <www.worldwildlife.org/toxics/progareas/pop/treaty_summary.pdf>, viewed on 27 July 2001; UNEP, "Stockholm Convention on POPs," *UNEP Chemicals*, June 2001.

14. Andrew C. Revkin, "178 Nations Reach a Climate Accord; U.S. Only Looks On," *New York Times*, 24 July 2001; "Climate Deal Reached in Bonn," *Environmental News Service*, 23 July 2001; "CLIMATE CHANGE: Countries Accused of Attempting To Renegotiate Kyoto," *UN Wire*, 31 October 2001.

15. Kyoto Protocol, at <www.unfccc.int/resource/docs/convkp/kpeng.pdf>, viewed 29 October (Annex I includes industrial nations and 38 economies in transition); 60–80 percent from J. T. Houghton et al., eds., *Climate Change 2001: The Scientific Basis*, Contribution of Working Group I to the Third Assessment Report of the Intergovernmental Panel on Climate Change (Cambridge, U.K.: Cambridge University Press, 2001), pp. 75–76.

16. "EU Makes its Move to Ratify Kyoto Protocol," *Environment News Service*, 23 October 2001; French, op. cit. note 12.

17. Biosafety Protocol at <www.biodiv.org/biosafety/protocol.asp>; updated information on the status of ratifications is available at <www.biodiv.org/biosafety/signinglist.asp>.

18. Critiques of the convention from author's conversations with authorities on the biodiversity convention, from Joy Hyvarinen, *The Convention on Biological Diversity: Future Issues* (Bedfordshire, U.K.: The Royal Society for the Protection of Birds, July 2001), and from idem, *Strengthening the Convention on Biological Diversity* (Bedfordshire, U.K.: The Royal Society for the Protection of Birds, August 2001).

19. UNEP, op. cit. note 10, pp. 33–34; WTO, Committee on Trade and Environment, "Compliance and Dispute Settlement Provisions in the WTO and in Multilateral Environmental Agreements," Note by the WTO and UNEP Secretariats, 6 June 2001.

20. Eric Neumayer, *Greening Trade and Investment* (London: Earthscan, 2001), pp. 158–84.

21. UNEP and International Institute for Sustainable Development (IISD), *Environment and Trade: A Handbook* (Winnipeg, MN, Canada: IISD, 2000), pp. 53–59.

22. Aaron Cosbey and Stas Burgiel, "The Cartagena Protocol on Biosafety: An Analysis of Results," Briefing Note (Winnipeg, MN, Canada: IISD, 2000).

23. Law of the Sea dispute resolution procedures from United Nations, Division for Ocean Affairs and Law of the Sea, "Settlement of Disputes," at <www.un.org/Depts/los/los_disp.htm>, viewed 29 October 2001, and from

Abram Chayes and Antonia Handler Chayes, *The New Sovereignty* (Cambridge, MA: Harvard University Press, 1995), pp. 217–18.

24. Lee A. Kimball, "The Debate Over a World/Global Environment Organization (W/GEO): A First Step Toward Improved International Institutional Arrangements for Environment and Development?" unpublished paper, 5 March 2001, p. 8; Calestous Juma, "Stunting Green Progress," *Financial Times*, 5 July 2000.

25. GEF, op. cit. note 13, pp. 6–9; idem, *Joint Summary of the Chairs, GEF Council Meeting, May 9–11, 2001* (Washington, DC: 15 May 2001), p. 2.

26. GEF, op. cit. note 13, pp. 3–5; idem, "GEF Projects—Allocations and Disbursements," 3 October 2001, paper submitted to Meeting on the Third Replenishment of the GEF Trust Fund, 11–12 October 2001.

27. UNEP's budget for two years (2000–01) was $196.7 million, per UNEP Governing Council, Nairobi, 5–9 February 2001, "Global Ministerial Environment Forum, Programme, the Environment Fund and Administrative and other Budgetary Matters, Report of the Executive Director," 2 October 2000, p. 10. This figure includes the resources from the U.N. regular budget, from the Environment Fund, and from Trust Funds, along with counterpart contributions; $173.4 billion of this total was devoted to program resources and the remainder to the support budget. Secretariat budgets from UNEP, op. cit. note 10, pp. 41–43, and from Hilary French and Lisa Mastny, "Controlling International Environmental Crime," in Lester R. Brown et al., *State of the World 2001* (New York: W.W. Norton & Company, 2001), p. 171. U.S. military budget amounted to $300,767 million in Fiscal Year (FY) 2000 and to $311,271 million in FY2001, according to <www.whitehouse.gov/omb/budget/fy2002/budget.html>, viewed on 12 October 2001. The Environmental Protection Agency received just over $7.8 billion in fiscal years 2000 and 2001 and is expected to receive a similar amount in 2002, according to "FY2002 Annual Performance Plan and Congressional Justification Appropriation (EPA's Proposed Budget)," at <www.epa.gov/ocfo/budget/2002/2000cj.htm>, viewed on 12 October 2001. World military expenditures of $784 billion in 2000 from Stockholm International Peace Research Institute, "World and Regional Military Expenditure Estimates, 1991–2000," at <projects.sipri.se/milex/mex_wnr_table.html>, viewed 11 October 2001.

28. IISD, "Summary of the Expert Consultations on International Environmental Governance 28–29 May 2001," *Sustainable Developments*; idem, "Summary of the Second Meeting of the Open-Ended Intergovernmental Group of Ministers or their Representatives on International Environmental Governance: 17 July 2001," *Earth Negotiations Bulletin*, 18 July 2001; idem, "Summary of the Third Open-Ended Intergovernmental Group of Ministers or their Representatives on International Environmental Governance: 9–10 September 2001," *Earth Negotiations Bulletin*, 12 September 2001.

29. UNEP, Report of the Chair, Open-Ended Intergovernmental Group of Ministers or their Representatives on International Environmental Governance, Second Meeting, Bonn, Germany, 17 July 2001; IISD, "Summary of the Third Open-Ended Intergovernmental Group," op. cit. note 28.

30. See especially Chapters 2 and 33 of United Nations, op. cit. note 6.

31. IISD, "Summary of the UNECE Regional Ministerial Meeting for the World Summit on Sustainable Development: 24–25 September 2001," *Earth Negotiations Bulletin*, 28 September 2001, pp. 9–10; Denmark, "World Summit on Sustainable Development—A Global Deal," first circulated at the UNECE Regional Ministerial Meeting for the World Summit on Sustainable Development, Geneva, 24–25 September 2001.

32. United Nations, op. cit. note 6, Chapter

33; official development assistance in 1992 of $60.42 billion (in current dollars) from Organisation for Economic Co-operation and Development (OECD), Development Assistance Committee, *Development Co-operation 1993* (Paris: 1994), pp. 168–69.

33. Figure 8–1 from Development Assistance Committee, *Development Assistance Committee Online* (DAC/o), OECD database, Table 1, updated 25 April 2001; aid as share of gross domestic product in 2000 and 2000 data in Table 8–2 from OECD, "Development Assistance Committee Announces ODA Figures for 2000," news release (Paris: 23 April 2001); 1992 data in Table 8–2 from OECD, op. cit. note 32.

34. Strong, op. cit. note 5, pp. 383–84.

35. For history of environmental critiques, see Bruce Rich, *Mortgaging the Earth* (Boston: Beacon Press, 1994); recent trends in sectoral breakdown of World Bank lending from World Bank, "10 Things You Never Knew About the World Bank," at <www.worldbank.org/ten things/intro.htm>, viewed 12 October 2001; environmental impacts of private-sector lending from Urgewald, Friends of the Earth, and Campagna per la Reforma della Banca Mondiale, "Risky Business: How the World Bank's Insurance Arm Fails the Poor and Harms the Environment," July 2001, at <www.foe.org>, and from Hilary French, *Vanishing Borders* (New York: W.W. Norton & Company, 2000), pp. 134–36.

36. Robert Weissman, "Why We Protest: The IMF and World Bank Hurt Poor Countries and Undermine Democracy" (op ed), *Washington Post*, 10 September 2001; Friends of the Earth International, "Environmental Consequences of the IMF's Lending Policies," at <www.foe.org/international/imf/page1.html>, viewed 9 October 2001; "IMF Tells Starving Nicaraguans to Tighten Their Belts, Cuts Off Debt Relief," Social Justice Committee Action Alert, 12 October 2001, at <www.s-j-c.net/nicaraguaOct 2001.htm>, viewed 1 November 2001; Chistopher Barr, *Banking on Sustainability: Structural Adjustment and Forestry Reform in Post-Suharto Indonesia* (Washington, DC: World Wildlife Fund and Center for International Forestry Research, 2001).

37. Figure 8–2 and shares of total based on World Bank, *Global Development Finance 2000*, electronic database (Washington, DC: 2000), with updates from idem, *Global Development Finance 2001*, electronic database (Washington, DC: 2001); debt-service payments as share of government expenditures from David Malin Roodman, *Still Waiting for the Jubilee: Pragmatic Solutions for the Third World Debt Crisis*, Worldwatch Paper 155 (Washington, DC: Worldwatch Institute, April 2001), p. 24.

38. Figure 8–3 based on numbers provided in World Bank, *Global Development Finance 2001*, op. cit. note 37, p. 36; decline in 2001 from Institute of International Finance, Inc., *Capital Flows to Emerging Market Economies* (Washington, DC: 20 September 2001).

39. China and India examples from French, op. cit. note 35, p. 104; "Foreign Direct Investment: A Lead Driver for Sustainable Development?" *Towards Earth Summit 2002*, Economic Briefing Series No. 1 (London: UNED Forum, undated), p. 4.

40. Capital flow trends based on World Bank, op. cit. note 37, p. 36; foreign-exchange transactions from "Sustainable Finance: Seeking Global Financial Security," Sustainable Finance Briefing Paper, Towards Earth Summit 2002, at <www.earthsummit2002.org>; poverty related to Asian financial crisis from James D. Wolfensohn, President, World Bank Group, "The Other Crisis," Annual Meetings Address, Washington, DC, 6 October 1998, at <www.worldbank.org/html/extdr/am98/jdw-sp/am98-en.htm>, viewed 8 October 1998.

41. United Nations, op. cit. note 6, Chapter 2; on the differential impact of WTO rules on industrial and developing countries, see Transcript of Press Briefing by Nobel Laureate and former World Bank Chief Economist Joseph

Stiglitz, World Bank, Washington, DC, 11 October 2001, at <www.worldbank.org/html/extdr/transcripts/ts101101.htm>, viewed 12 October 2001; role of developing countries in Seattle breakdown from Martin Khor, Third World Network, "Seattle Debacle: Revolt of the Developing Nations," posted to MAI-NOT listserve, 10 December 1999.

42. Need for "development round" and possible income gains from Ernesto Zedillo et al., "Recommendations of the High-level Panel on Financing for Development," commissioned by the Secretary-General of the United Nations, New York, 22 June 2001, pp. 8–10; developing-country concerns on implementation from "New Ministerial Text to Hand Ministers a Challenge In Doha" and "New Implementation Draft—Another Round for Gains or Grounds to Refrain," *BRIDGES Weekly Trade News Digest* (International Centre for Trade and Sustainable Development), 30 October 2001; nongovernmental organization (NGO) skepticism from "Our World is Not for Sale. WTO: Shrink or Sink," endorsed by 360 organizations as of 24 October 2001, at <www.canadians.org/campaigns/campaigns-trade-notforsale-2.html>, viewed 31 October 2001.

43. For information on the Financing for Development Summit, see <www.un.org/esa/ffd>.

44. James Tobin, "A Tax on International Currency Transactions," in UNDP, *Human Development Report 1994* (New York: Oxford University Press, 1994), p. 70; revenue estimates from a Tobin tax from Zedillo et al., op. cit. note 42, p. 20; Michael Renner, "U.N. Funds Stay on Roller Coaster," in Worldwatch Institute, *Vital Signs 2001* (New York: W.W. Norton & Company, 2001), p. 60.

45. "Letter to the UN High Level Panel," signed by 64 NGOs in 26 countries, circulated via e-mail by Robin Round, Halifax Initiative, Vancouver, BC, Canada; Halifax Initiative, "Taxing Currency Transactions for Development," United Nations Financing for Development Submission, January 2001, at <www.halifaxinitiative.org/hi.php/Tobin/112>, viewed 31 October 2001; Association for the Taxation of Transactions to Aid Citizens, at <www.attac.org>.

46. Barbara Unmüßig, "New World Conferences: New Prospects for Global Environment and Development Financing?" at <www.weedbonn.org/unreform/unconf2002.htm>, viewed 1 November 2001; Zedillo et al., op. cit. note 42, pp. 21–22.

47. Number of NGOs at Global Forum from Robert Weissman, "Citizen Summit," *Multinational Monitor*, July–August 1992, p. 29; number accredited from Strong, op. cit. note 5, p. 231; alternative treaties from Adam Rogers, *The Earth Summit: A Planetary Reckoning* (Los Angeles: Global View Press, 1993), pp. 253–87.

48. On different eras of NGO activism in the international arena, see Steve Charnovitz, "Two Centuries of Participation: NGOs and International Governance," *Michigan Journal of International Law*, winter 1997, pp. 183–286.

49. Union of International Organizations, *Yearbook of International Organizations* (Munich: K. G. Sauer Verlag, 1999/2000 and 2000/2001), Appendix 3, Table 1.

50. Climate Action Network, at <www.climatenetwork.org>, viewed on 29 October 2001; Third World Network, at <www.twnside.org.sg>, viewed 29 October 2001.

51. Number of transnational corporations in 1970 from Joshua Karliner, *The Corporate Planet* (San Francisco: Sierra Club Books, 1997), p. 5; numbers in 2000 from U.N. Conference on Trade and Development, *World Investment Report 2001* (New York: United Nations, 2001), pp. xiii, 10.

52. History of efforts to negotiate U.N. Code of Conduct from Virginia Haufler, *A Public Role for the Private Sector: Industry Self-Regulation in a Global Economy* (Washington, DC: Carnegie Endowment for International Peace, 2001), pp. 16–17. Table 8–3 based on the following:

OECD, "OECD Guidelines for Multinational Enterprises," at <www.oecd.org/publications/ Library/webook/00-2001-97-1/index.htm>, viewed 14 August 2001; CERES Principles, at <www.ceres.org>; Natural step, at <www. naturalstep.org>; International Chamber of Commerce, "Business Charter for Sustainable Development," at <www.iccwbo.org/home/ environment/charter.asp>, viewed 16 August 2001; Eco-Management and Audit Scheme, at <www.emas.org.uk>; ISO, "ISO 14000—Meet the Whole Family!" at <www.iso.ch/iso/en/ iso9000-14000/pdf/iso14000.pdf>, viewed 17 August 2001; idem, "ISO Survey of ISO 9000 and ISO 14000 Certificates 10th Cycle: up to and including 31 December 2000," at <www.iso.ch/iso/en/iso9000-14000/pdf/ survey10thcycle.pdf>, viewed 16 August 2001; J. Timmons Roberts, "Emerging Global Environmental Standards," *Journal of Developing Societies*, March 1998, pp. 144–63; "Going Green with Less Red Tape," *Business Week*, 23 September 1996, pp. 75–76; Global Reporting Initiative, "Sustainability Reporting Guidelines on Economic, Environmental, and Social Performance," June 2000, at <www.global reporting.org>, viewed 17 August 2001; Edward Goodell, ed., "Standards of Corporate Social Responsibility," Social Venture Network, 1999, at <www.svn.org>, viewed 16 August 2001; UN Global Compact, at <www.unglobal compact. org>; "Principled Partnership with World Business," *Financial Times*, 6 September 2000; Jennifer Nash and John Ehrenfeld, "Code Green," *Environment*, January/February 1996, pp. 16–20, 36–45.

53. U.N. Department of Public Information, "The Global Compact: Shared Values for the Global Market," brochure (New York: December 1999); Planet Ark Environment News, "More than 300 Firms Sign Up for UN Global Compact," 27 July 2001, at <www.planetark. org/dailynewsstory.cfm?newsid=11788>, viewed 1 August 2001; "26 July: High Level Meeting of Business, Labour, and Civil Society, Views from Participants," at <www.un. org/partners/business/gcevent/second_page. htm>, viewed 31 October 2001; Kenny Bruno and Joshua Karliner, "Tangled up in Blue: Corporate Partnerships at the United Nations," at <www.corpwatch.org/trac/globalization/un/ tangled.html>, viewed 29 October 2001.

54. Susan Ariel Aaronson, "Oh Behave! Voluntary Codes Can Make Corporations Model Citizens," *The International Economy*, March/April 2001.

55. Andrew Revkin, "Some Energy Executives Urge U.S. Shift on Global Warming," *New York Times*, 1 August 2001; Cat Lazaroff, "Lieberman, McCain Call for Greenhouse Gas Caps," *Environmental News Service*, 3 August 2001.

56. David Ignatius, "Think Globally, Build Networks," *Washington Post*, 28 January 2001; Wolfgang H. Reinicke and Francis Deng, *Critical Choices: The United Nations, Networks, and the Future of Global Governance* (Ottawa, ON, Canada: International Development Research Centre, 2000).

57. World Commission on Dams, *Dams and Development: A New Framework for Decision-Making* (London: Earthscan, 2000), pp. 27–28.

58. Ibid.; Reinicke and Deng, op. cit. note 56, pp. 37–40.

59. *Report of the Third World Commission on Dams Forum Meeting at the Spier Village, 25–27 February 2001*, Cape Town, South Africa, pp. 40, 45, 52, 55; Navroz K. Dubash et al., *A Watershed in Global Governance?: An Independent Assessment of the World Commission on Dams*, Executive Summary (Washington, DC: World Resources Institute, Lokayan, and Lawyer's Environmental Action Team, 2001).

60. "NGOs Protest Against World Bank Position on World Dams Report," e-mail from Peter Bosshard, Berne Declaration, 20 March 2001; World Bank, "World Bank Group Appoints Dr. Emil Salim to Head Extractive Industries Review Consultative Process," press release (Washington, DC: 23 July 2001).

61. "What the 'Extractive Industries Review' is," at <www.eireview.org>, viewed 25 October

2001; Daphne Wysham, "NGO Letter to James Wolfensohn, President, World Bank Group," Sustainable Energy & Economy Network et al., Washington, DC, 8 October 2001.

62. Michael Hardt and Antonio Negri, "What the Protesters in Genoa Want," *New York Times*, 20 July 2001; Joseph S. Nye, Jr., "Globalization's Democratic Deficit," *Foreign Affairs*, July/August 2001.

63. French, op. cit. note 35, pp. 111–23.

64. Ibid.; Lori Wallach and Michelle Sforza, *Whose Trade Organization* (Washington, DC: Public Citizen, 1999), pp. 195–203.

65. Steve Charnovitz, "Opening the WTO to Nongovernmental Interests," *Fordham International Law Journal*, November–December 2000, pp. 173–216.

66. World Bank, "World Bank Revises Disclosure Policy," press release (Washington, DC: 7 September 2001); Tahir Mirza, "Openness: WB's Move Found Insufficient," *DAWN* (Internet edition), 10 September 2001, at <www.dawn.com/2001/09/10/top14.htm>, viewed 11 September 2001; Robert Naiman, Center for Economic Policy and Research, Washington, DC, "Why We Must Open the Meetings of the IMF and the World Bank Boards: the Case of User Fees on Primary Health in Tanzania," circulated to the Stop-IMF list-serve, 1 June 2001; Joseph Stiglitz, "The Insider: What I Learned at the World Economic Crisis," *New Republic*, 17 April 2000, p. 60.

67. "Reference Document on the Participation of Civil Society in United Nations Conferences and Special Sessions of the General Assembly During the 1990s," Office of the President of the Millennium Assembly, 55th session of the United Nations General Assembly, August 2001; Motoko Mekata, "Building Partnerships Toward a Common Goal: Experiences of the International Campaign to Ban Landmines," in Ann M. Florini, ed., *The Third Force: The Rise of Transnational Civil Society* (Washington, DC: Carnegie Endowment for International Peace and Japan Center for International Exchange, 2000), pp. 143–76; Coalition for an International Criminal Court, at <www.igc.org/icc>, viewed 31 October 2001.

68. Number at annual forums in 1993 and 2000 from Minu Hemmati et al., *Multi-Stakeholder Processes for Governance and Sustainability—Beyond Deadlock and Conflict* (London: Earthscan, forthcoming), p. 26 (in version available online at <www.earthsummit2002.org/msp/book.htm>).

69. Ibid., pp. 26–29; United Nations, "Major Groups, Report of the Secretary-General," Commission on Sustainable Develoment acting as the preparatory committee for the World Summit on Sustainable Development, 30 April–2 May 2001.

70. National Councils on Sustainable Development from Earth Council, *NCSD Report, 1999–2000* (San José, Costa Rica: undated), p. 1; Judy Walker, Director of Local Agenda 21 Campaign, International Council for Local Environmental Initiatives, Toronto, discussion with Molly Sheehan, Worldwatch Institute, 29 August 2001.

71. U.N. General Assembly, "Rio Declaration on Environment and Development," Annex I of the Report of the United Nations Conference on Environment and Development (Rio de Janeiro, 3–14 June 1992), at <www.un.org/documents/ga/conf151/aconf15126-1annex1.htm>, viewed 23 October 2000; Elena Petkova with Peter Veit, "Environmental Accountability Beyond the Nation-State: The Implications of the Aarhus Convention," *Environmental Governance Note* (Washington, DC: World Resources Institute, April 2000); Latin America from UNEP, *Global Environment Outlook 2000* (London: Earthscan, 1999), pp. 289–91; Africa from Peter Veit, World Resources Institute, Washington, DC, discussion with Jessica Dodson, Worldwatch Institute, 1 November 2001; Paula J. Dobriansky, Under Secretary of State for Global Affairs and Head of the United States Delegation, "Governance as a Foundation for Sustainable Development,"

Remarks to the UN Economic Commission for Europe Regional Ministerial Meeting for the World Summit on Sustainable Development, Geneva, 24 September 2001. Box 8–1 based on the following: Paul Taylor, Director of Good Urban Governance Campaign, U.N. Centre for Human Settlements (Habitat), discussion with Molly Sheehan, Worldwatch Institute, 27 August 2001; Samuel Paul, "Report Cards on Urban Services in Bangalore," in Jim Antoniou, ed., *Implementing the Habitat Agenda: In Search of Urban Sustainability* (London: Development Planning Unit, University College London, 2001), pp. 178–79.

72. Thom Shanker, "White House Says the U.S. Is Not a Loner, Just Choosy," *New York Times*, 30 July 2001; Alan Sipress, "U.S. Draws Abortion Line at U.N.," *Washington Post*, 28 August 2001.

73. Former President Bush quote from Patrick E. Tyler and Jane Perlez, "World Leaders List Conditions on Cooperation," *New York Times*, 19 September 2001.

Index

INDEX

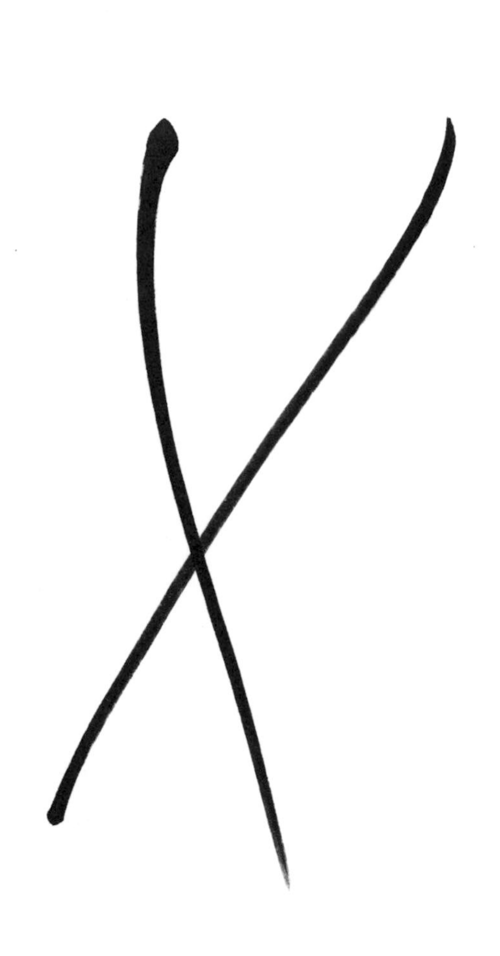